Lecture Notes in Physics

The Lecture Notes in Physics

The series Lecture Notes in Physics (LNP), founded in 1969, reports new developments in physics research and teaching – quickly and informally, but with a high quality and the explicit aim to summarize and communicate current knowledge in an accessible way. Books published in this series are conceived as bridging material between advanced graduate textbooks and the forefront of research to serve the following purposes:

- to be a compact and modern up-to-date source of reference on a well-defined topic;
- to serve as an accessible introduction to the field to postgraduate students and nonspecialist researchers from related areas;
- to be a source of advanced teaching material for specialized seminars, courses and schools.

Both monographs and multi-author volumes will be considered for publication. Edited volumes should, however, consist of a very limited number of contributions only. Proceedings will not be considered for LNP.

Volumes published in LNP are disseminated both in print and in electronic formats, the electronic archive is available at springerlink.com. The series content is indexed, abstracted and referenced by many abstracting and information services, bibliographic networks, subscription agencies, library networks, and consortia.

Proposals should be sent to a member of the Editorial Board, or directly to the managing editor at Springer:

Dr. Christian Caron
Springer Heidelberg
Physics Editorial Department I
Tiergartenstrasse 17
69121 Heidelberg/Germany
christian.caron@springer-sbm.com

Martin Kröger

Models for Polymeric and Anisotropic Liquids

Springer

Author

Martin Kröger
Polymer Physics
ETH Zürich
Wolfgang-Pauli-Str. 10
8093 Zürich
Switzerland
E-mail: mk@mat.ethz.ch
Website: www.complexfluids.ethz.ch

Martin Kröger, *Models for Polymeric and Anisotropic Liquids*,
Lect. Notes Phys. 675 (Springer, Berlin Heidelberg 2005), DOI 10.1007/b105182

ISBN 978-3-642-06564-4 e-ISBN 978-3-540-31519-3
ISSN 0075-8450

Springer is a part of Springer Science+Business Media
springeronline.com

Softcover reprint of the hardcover 1st edition 2005

Simplicity is the ultimate sophistication
Leonardo da Vinci (1452–1519)

Preface

Part I of this monograph is concerned with the theoretical, analytical as well as numerical prediction of field-induced dynamics and structure for simple models describing soft matter. It presents selected results and demonstrates ranges of applications for the methods described in Part II. Special emphasis is placed on the finitely extendable nonlinear elastic (FENE) chain models for polymeric liquids, their dynamical and rheological behavior and the description of their inherently anisotropic material properties by means of deterministic and stochastic approaches. A number of representative examples are given on how simple (but high-dimensional) models can be implemented in order to enable the analysis of the microscopic origins of the dynamical behavior of polymeric materials. These examples are shown to provide us with a number of routes for developing and establishing low-dimensional models devoted to the prediction of a reduced number of significant material properties. Concerning the types of complex fluids, we cover the range from flexible polymers in melts and solutions, wormlike micelles, actin filaments, rigid and semiflexible molecules in flow-induced anisotropic, and also liquid crystalline phases. Fokker–Planck equations and molecular and brownian dynamics computer simulation methods are involved to formulate and analyze the model fluids.

Part II allows the reader to redo simulations and motivates for further investigation of polymeric and anisotropic fluids. It contains computational recipes for devising simulation methods and codes, including Monte Carlo, molecular and brownian dynamics (written in Mathwork's Matlab, thus allowing for simple visualization and animation). A special chapter on isotropic and irreducible tensors allows for comfortable conversion between stochastic differential equations, tensorial balances, and equations for coefficients, including the testing of closure approximations. We explicitly derive coupled equations for alignment tensors for arbitrary tensor fields suitable for *n*th order approximations strictly valid close to equilibrium, and also highly anisotropic states.

Switzerland
March, 2005

Martin Kröger
ETH Zürich

Contents

Part I Illustrations & Applications

Part I

Illustrations & Applications

1

Simple Models for Polymeric and Anisotropic Liquids

We hope that the complexity of the world is neither in contrast with the simplicity of the basic laws of physics [1] nor with the simple physical models to be reviewed or proposed in the following. However, physical phenomena occurring in complex materials cannot be encapsulated within a single numerical paradigm. In fact, they should be described within hierarchical, multi-level numerical models in which each sub-model is responsible for different spatio-temporal behavior and passes out the averaged parameters to the model, which is next in the hierarchy (Fig. 1.1). Polymeric liquids far from equilibrium belong to the class of anisotropic liquids.[1] This monograph is devoted to the understanding of the anisotropic properties of polymeric and complex fluids such as viscoelastic and orientational behavior of polymeric liquids, the rheological properties of ferrofluids and liquid crystals subjected to external fields, based on the architecture of their molecular constituents. The topic is of considerable concern in basic research for which models should be as simple as possible, but not simpler. Certainly, it is also of technological relevance. Statistical physics and nonequilibrium thermodynamics are challenged by the desired structure-property relationships. Experiments such as static and dynamic light and neutron scattering, particle tracking, flow birefringence etc. together with rheological measurements have been essential to adjust or test basic theoretical concepts, such as a 'linear stress-optic rule' which connects orientation and stress, or the effect of molecular weight, solvent conditions, and external field parameters on shape, diffusion, degradation, and alignment of molecules.

During the last decade the anlaysis of simple physical particle models for complex fluids has developed from the molecular computation of basic systems (atoms, rigid molecules) to the simulation of macromolecular 'complex' system with a large number of internal degrees of freedom exposed to external forces. This monograph should be in certain aspects complementary to others. The foundations of molecular

[1] Greek: an (non) iso (equal) trop (to turn): Anisotropic materials exhibit properties with different values when measured in different directions. Material properties are rotation-invariant, usually either due to boundary conditions, anisotropic applied external fields, or the presence of nonspherical constituents.

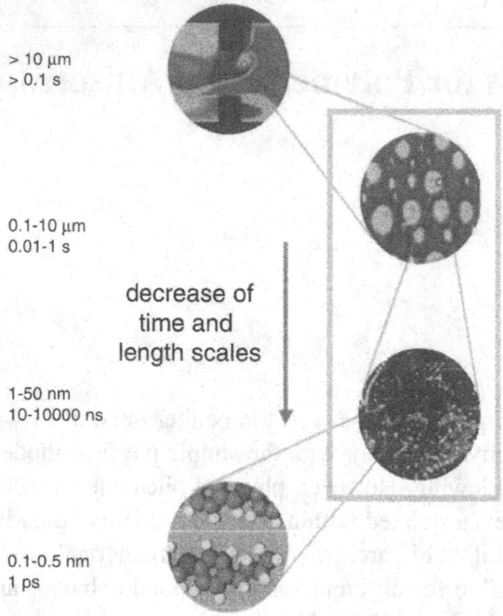

Fig. 1.1. Time and length scales of a typical polymer problem. In this review we are concerned with micro- and mesoscopic models (framed) which aim to describe physical behavior beyond equilibrium, beyond chemical details (*bottom*), and may be implemented into the macro-computation of complex flows (*top*)

and brownian dynamics methods for simple microscopic models for macromolecular systems have been extensively revisited [2]. Multiscale simulation in polymer science with special emphasis on coarse-grained models (incl. a soft-ellipsoid model) has been recently reviewed by Kremer and Muller–Plathe [3]. In the light of modern reviews on physical micro- and mesoscopic models to be mentioned below our focus is placed onto aspects which have been less extensively considered. Upon these are, in part 1 of this monograph, orientation and entanglement effects, the implications of stretchability, flexibilty, order parameters, scission and recombination on material properties of anisotropic, dilute and concentrated polymeric bulk fluids in the presence of macroscopic flow and electromagnetic fields. Part II is an attempt to collect the minimum amount of information to implement and develop analytic theory and computational tools.

In part 1 this monograph is first of all concerned with the applicability and suitability of bead-spring multi chain models which incorporate finite extensibility of segments (so called FENE models, cf. Page 203), molecular architecture and flexibility, and capture topological interactions. Second, it aims to give an overview about the range of applications of simple mesoscopic theories, in particular primitive path models and elongated particle models, where topological aspects are either approximatly treated or disregarded. In view of a rapidly growing amount of research and number of publications on these topics, we try to present a balanced selection of simple, representative examples, connect them with related research, and thereby get in touch with a large – still not exhaustive – number of classical and modern approaches. In order to keep the monograph short, we do not summarize basic knowledge available from standard text books. We therefore do not provide an introduction to the the theory of stochastic differential equations, the statistical physics of simple, molecular, and macromolecular liquids, linear response theory, rheology, or experimental methods. We are going to cite the relevant original literature where implementation details can be found. However, Part II of this monograph provides the reader with the basic ingredients needed to devise a simulation scheme and to derive equations of change for moments for a given model. In particular, it contains sample codes for various applications.

The existence of universality classes is significant for the theoretical description of polymeric complex fluids. Any attempt made at modeling polymer properties might expect that a proper description must incorporate the chemical structure of the polymer into the model, since this determines its microscopic behavior. Thus a detailed consideration of bonds, sidegroups, etc. may be envisaged. However, the universal behavior that is revealed by experiments suggests that macroscopic properties of the polymer are determined by a few large scale properties of the polymer molecule. Structural details may be ignored even for microscopic (beyond-atomistic) models since at length scales in the order of nanometers, different polymer molecules become equivalent to each other, and behave in the same manner. This universal behavior justifies the introduction of crude mechanical models, such as bead-spring chain models, to represent real polymer molecules (Fig. 1.2).

The FENE chain model and its variations can be considered as a maximum coarse-grained, still brute force simulation model to the physical properties of polymeric fluids. These models didn't fail to describe rheooptical material properties quite satisfactory when solved without approximation, but are often numerically expensive while conceptually simple. FENE chains constitute the appropriate level of description in order to test polymer kinetic theory [4, 5], and assumptions made to simplify their analysis. This monograph discusses several realizations in detail, and hopes to stimulate for advanced treatments, therefore disregards many others (FENE chain models for star polymers, co-polymers, polymer blends, brushes, polyelectrolytes, in order to mention a few).

The dynamics of a single, fluorescing, DNA macromolecule held at one end by 'optical tweezers' and subjected to a uniform flow was successfully compared with simulations [6] of a FENE chain that accounted for the molecule's entropic elasticity, brownian motion, and hydrodynamic drag. Using self-diffusion data and analytical expressions to obtain this drag in the limits of the undeformed coil and of the fully stretched thread, these results once more confirmed the success of the FENE chain model in predicting the rheological properties of simple polymeric systems. Excellent agreement between the theoretical predictions based on the FENE models and data from experimentation indicated that the model also seemed able [7] to interpret the underlying physical mechanisms for the dynamics of polymer solutions [8–10], melts [11–13], copolymer melts [14, 15], brushes [16] not only in the qiescent state, but also subjected to flow fields [7, 8, 17–26]. During the last decade, the FENE chain

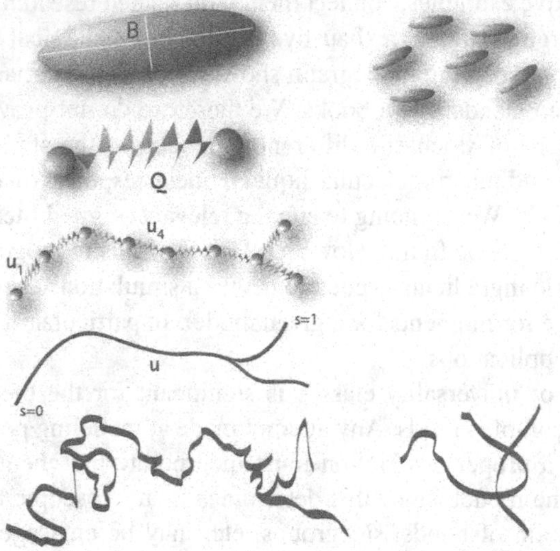

Fig. 1.2. Simple microscopic models for complex fluids with increasing level of abstraction and decreasing degrees of freedom (lhs, *bottom* to *top*), and their sketched range of application: (**a**) Atomstically detailed polymer which accounts for anisotropic intermolecular interactions incl. entanglements, (**b**) coarse grained model via a mapping (Sect. 8.10.1) to a 'primitive path', (**c**) further approximated by a multibead (nonlinear FENE) chain, (**d**) further coarse-grained to a (FENE) dumbbell which accounts for entropic elasticity and orientation but not for entanglement effects, and (**e**) ellipsoids of revolution – incl. rigid rods, dissipative particles, with spherical or mean-field interaction. Models must meet the requirement of being thermodynamically admissible

model has been extended to incorporate the effect of scission, recombination (FENE-C) and branching of chains in order to investigate the formation and development of complex micellar systems and networks [14, 27–30], cf. Fig. 1.3. The model has been further extended (FENE-B) to incorporate semiflexibility of chains [31–34], and studied in confined geometries. To give an overview about the range of applicability of the sufficiently detailed and simple microscopic models, we restrict ourself to the formulation and analysis of models for particulate fluids and validate them against experimental data.

The nomenclature given at Page 203 is recommended in order make the search for results obtained for extensions of the original FENE dumbbell more comfortable. Actually, the most complete summary of the various 'analytic' FENE models may be found in [35]. Configuration tensor models such as the FENE-P and more general quasi-linear models (Johnson-Segalman, Gordon-Schowalter, Phan-Thien/Tanner etc.) have been also developed in a fully nonisothermal setting [36–38]. NEMD together with a dissipative particle dynamics (DPD) thermostat had been successfully applied to study the shear-induced alignment transition of diblock copolymer melts, surfactants and liquid crystals in a large-scale system [14], based on an effective simplified continuum model for FENE dumbbells [39] biased towards phase separation. Simplified versions of FENE chain models neglect flexibility or finite

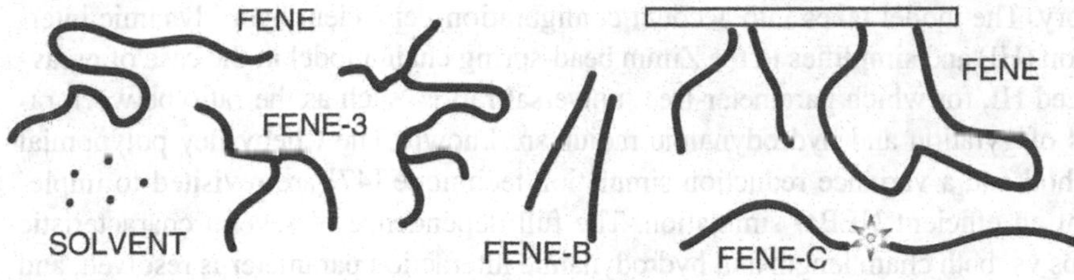

Fig. 1.3. Simple FENE models for a range of macromolecular fluids to be treated in a unified fashion: with/without solvent (simple fluid) for linear/star/branched, flexible/semiflexible, bulk/confined/tethered, non-/breakable macromolecules, cf. Table 14.1. Charged, tethered polymers have been excluded from the review since an excellent review is available in this series [40]

extensibilty and have been widely used. Rigid elongated particles further neglect stretchability. Models dealing with these objects will be reviewed in certain detail as long as the simplified description turns out to be appropriate (unentangled: dilute polymers, rigid molecules: liquid crystals). Some space will be reserved for the discussion on the connection between the different levels of description, projection operators, coarse-graining procedures, and the theory of nonequilibrium thermodynamics which sets a framework for simple physical models.

1.1 Section-by-Section Summary

Part I

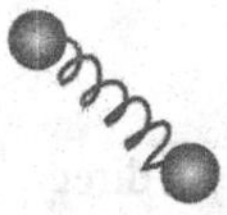

Chapter 2: In the quiescent state, polymers in dilute solution should have negligible interactions with each other on purely geometrical grounds, in contrast to semi-dilute or concentrated solutions and melts. The flow behavior of polymer solutions is, however, more complex than that of the familiar Newtonian fluids. Within these solutions shear thinning and the Weissenberg effect [4] are typical phenomena of technological importance. These effects are found to be strongly correlated with flow-induced conformational changes of the dissolved polymer chains and they can be dramatic in dilute solutions. Orientation and deformation of chain molecules can, and has been measured in flow birefringence light scattering and neutron scattering experiments (for methods and references see [41]), and via computer simulation [42–45]. For a review on molecular orientation effects in viscoelasticity we refer to [46]. For this introductory section we will be concerned with approximate solutions for FENE dumbbells (with $N = 2$ beads) in the infinitely dilute and semi-dilute regimes.

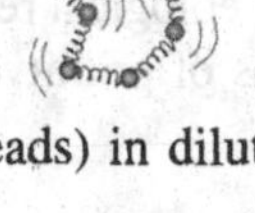

Chapter 3 is next on the hierarchy and treats multibead chains ($N > 2$ beads) in dilute solutions. We start from a stochastic approach to polymer kinetic theory. The model takes into account configuration-dependent hydrodynamic interaction (HI) and simplifies to the Zimm bead-spring chain model in the case of preaveraged HI, for which parameter-free 'universal ratios' such as the ratio between radius of gyration and hydrodynamic radius are known. The Chebyshev polynomial method and a variance reduction simulation technique [47] are revisited to implement an efficient NEBD simulation. The full dependence of several characteristic ratios vs. both chain length and hydrodynamic interaction parameter is resolved, and compared with analytical and experimental results. Polymer solutions under good solvent conditions have been also studied extensively via NEMD by taking into account explicit solvent particles, e.g., in [42–45, 48]. In that case, hydrodynamic interactions and excluded volume are incorporated through momentum transfer and a WCA potential between beads, respectively.

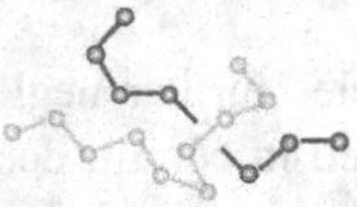

Chapter 4 demonstrates insights obtained by NEMD into the microscopic origin of the nonlinear viscoelastic properties of (dense) polymer melts by using a FENE chain model. Stress-strain relationships for polymer melts are the main requirement for the conventional flow simulation of polymer processing, useful in modelling industrial applications including injection moulding, film blowing, and extrusion. The reliability and accuracy of such simulations depends crucially on the constitutive equations. Although closed-form phenomenological models have been widely used in research and commercial codes, their degree of success is limited because of a lack of physical ingredient on the molecular level. For the purpose of realistic modelling, and further development of semiempirical constitutive equations, full FENE chain models are shown to be uniquely suited.

Chapter 5 extends the FENE chain system in several directions. We offer explicit examples of recently established models: wormlike micellar systems modelled by a FENE-C potential, model liquid crystals composed of semiflexible FENE chains, as well as a model for semiflexible (FENE-B, actin) filaments and networks. Results for the models are obtained by NEMD or NEBD, though we will also discuss analytic descriptions that are able to guide the interpretation of important aspects of the results.

Chapter 6 offers illustrative examples on how to formulate and handle kinetic model equations for primitive paths (coarse-grained atomistic chains) by approximate analytical or 'exact' numerical treatments. The role of topological interactions is particularly important, and has given rise to a successful theoretical framework: the 'tube model'. Progress over the last 30 years had been reviewed in the light of specially-synthesized model materials, an increasing palette of experimental techniques, simulation and both linear and nonlinear rheological response in [5]. Here we review a selected number of improved versions of primitve path models which allow to discuss the effect of approximations on the linear and nonlinear rheological behavior of polymer melts. Brute force FENE chain simulation results summarized in the preceeding chapters are used to test the assumptions made in the formulation of these kinetic models.

Chapter 7 deals with elongated particle models. There are many early approaches in the literature to the modelling of fluids with simple microstructure. For example, equations for suspensions of rigid particles have been calculated by averaging the detailed motion of the individual particles in a Newtonian fluid. In particular, the solution for the motion of a single ellipsoid of revolution in a steady shear [49] in terms of a Fokker–Planck equation has been used to determine the governing equations for the slow flow of a dilute suspension of non-interacting particles. In more concentrated systems, various approximations to the particle motions have been used. Hinch and Leal [50] have named this approach, based upon a detailed analysis of the microstructure, 'structural'. Alternatively, 'phenomenological' continuum theories for anisotropic fluids have been postulated. These theories tend to be quite general, being based upon a small number of assumptions about invariance. Perhaps the most successful and well-known example is the Ericksen-Leslie (EL) director theory for uniaxial nematic liquid crystals. Additionally, numerous models have been developed and discussed in terms of symmetric second and higher order tensorial measures of the alignment. Given these diverse methods of derivation and apparently diverse domains of application, one may ask if, and how, such diverse approaches may be interrelated. The answer and several examples (incl. concentrated suspensions of rod-like polymers, liquid crystals, ferrofluids) are given in this section.

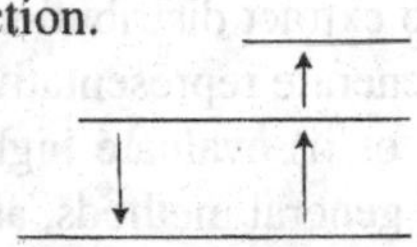

Chapter 8 is an attempt to review several strategies and open questions concerning the thermodynamically admissible description of complex nonequilibrium fluids on different levels (conc. length and time scales or structural details) of description. We will touch the theory of projection operators which act on the space coordinates of atoms such that the resulting quantities serve as slow variables needed to proceed with a separation of time scales in the corresponding Langevin equations. Attempts being made to characterize the system with (a few) structural quantities, known to be within reach of analytical theoretical descriptions and/or accessible through experimentation will be reviewed. A similar formal structure, namely a symplectic structure, for thermodynamics and classical mechanics was noted early by Peterson [51] in his work about the analogy between thermodynamics and mechanics. He notes that the equations of state, by which he means identical relations among the thermodynamic variables characterizing a system, are actually first-order partial differential equations for a function that defines the thermodynamics of the system. Like the Hamilton–Jacobi equation, such equations can be solved along trajectories given by Hamilton's equations, the trajectories being quasi-static processes, obeying the given equation of state. This gave rise to the notion of thermodynamic functions as infinitesimal generators of quasi-static processes, with a natural Poisson bracket formulation. In this case the formulation of thermodynamic transformations is invariant under canonical coordinate transformations, just as with classical mechanics. These illuminating ideas have been further developed [52, 53] and generalized Poisson structures are now recognized in many branches of physics (and mathematics). We are therefore also concerned with the formulation of so called 'thermodynamically admissible' simple models for complex fluids, where admissibility is assumed whenever the complete set of state variables characterizing the systems possess the 'General Equation for the Non-Equilibrium Reversible-Irreversible Coupling' (GENERIC) structure [38,54]. This structure (a special representation of a less predictive 'Dirac' structure which also contains the Matrix model by Jongschaap [55] as a special case, connections between thermodynamic formalism are revisited in [56]) requires a Poisson bracket for the reversible part of the dynamics. Specifically, the time-structure invariance of the Poisson bracket as manifested through the Jacobi identity has been used to derive constraint relationships on closure approximations [57]. Explicit coarsening procedures from connected or disconnected atomistic chains (or FENE chains, Chap. 4) to primitive paths (Chap. 6, Fig. 1.2) are given in Sects. 8.10.1–8.10.2.

Part II

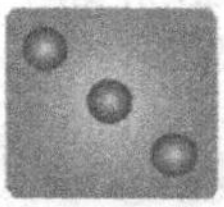

Chapter 9. Monte Carlo methods use random numbers, or 'random' sequences, to sample from a known shape of a distribution, or to extract distributions by other means. and, in the context of this monograph, to i) generate representative equilibrated samples prior being subjected to external fields, or ii) evaluate high-dimensional integrals. Recipes for both topics, and some more general methods, are summarized in this chapter. Advanced Monte Carlo 'moves' for polymers, required to optimize the speed of algorithms for a particular problem at hand, are outside the scope of this brief introduction.

$\Delta^{(2)}_{\mu\nu,\lambda\gamma}$

Chapter 10 summarizes definitions and properties of cartesian, anisotropic, irreducible and isotropic tensors and related tensor operators. Tensors rather than scalars allow to describe the anisotropic behavior of structural fluids subjected to external fields. The formulas presented in this chapter help to evaluate tensor operators (differentiation, integration) without performing a differentiation or an integral, to rewrite arbitrary tensors of arbitrary rank made of unit vectors in terms of the corresponding dyadics, and vice versa. This sets us in position to write down (coupled) moment equations starting from a given differential equation for (orientational) distribution functions, to derive approximate sets of coupled equations for moments of the distribution functions.

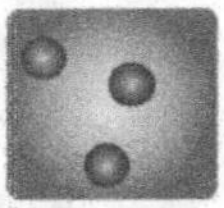

Chapter 11 introduces Fokker–Planck, Smoluchowski, and stochastic differential equations, their interrelation and methods to solve them numerically. We focus on the dynamics, in particular, the orientational dynamics of structured fluids subjected to orienting fields. The dynamics and anisotropy is properly modeled by using orientational distribution functions, their equation of change, and the corresponding balance equations for moments (here, alignment tensors) of the distribution function. We restrict ourself to discuss the case of one-particle (single-link) orientational distribution functions and explictely derive coupled set of moment equations which cover all cases discussed in Part I.

Chapter 12 offers basic recipes and sample applications which allow the reader to immediately start his/her own simulation project on topics we dealt with in the foregoing chapters. The chapter provides simulation codes and underlying equations. We concentrate on the necessary, and skip anything more sophisticated. Codes have been used in classrooms, they are obviously open for modifications and extensions. Codes are short, run without changes, demonstrate the main principle in a modular fashion, and are thus in particular open regarding efficiency issues and extensions. Algorithms are presented in the Matlab^{TM}language, which is mostly directly portable to programming languages like fortran, c, or Mathematica^{TM}.

2

Dumbbell Model for Dilute and Semi-Dilute Solutions

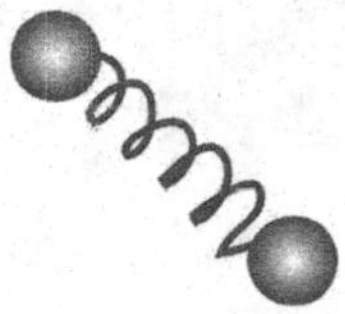

Dumbbell models are very crude representations of polymer molecules. Too crude to be of much interest to a polymer chemist, since it in no way accounts for the details of the molecular architecture. It certainly does not have enough internal degrees of freedom to describe the very rapid motions that contribute, for example, to the complex viscosity at high frequencies. On the other hand, the elastic dumbbell is orientable and stretchable, and these two properties are essential for the qualitative description of steady-state rheological properties and those involving slow changes with time. For dumbbell models one can go through the entire program of endeavor – from molecular model to fluid dynamics – for illustrative purposes, in order to point the way towards the task that has ultimately to be performed for more realistic models. According to [4], dumbbell models must, to some extend then, be regarded as mechanical playthings, somewhat disconnected from the real world of polymers. When used intelligently, however, they can be useful pedagocically and very helpful in developing a qualitative understading of rheological phenomens.

Before we turn to FENE chain models with increasing complexity and predicitve power for entangled polymeric systems, we should summarize some of the efforts undertaken to analyze various approximations to the original FENE dumbbell model for infinitely dilute solutions. This model can be rigorously solved by brownian dynamics (BD) and had been used in the pioneering micro-macro simulations [58].

A FENE dumbbell consists of two beads (mass points) connected with a nonlinear spring. Its internal configuration is described by a connector vector $\boldsymbol{Q}$. The FENE spring force law is given by [4,59,60]

$$\boldsymbol{F}^{\mathrm{FENE}} = -\frac{H\boldsymbol{Q}}{1-Q^2/Q_0^2}, \tag{2.1}$$

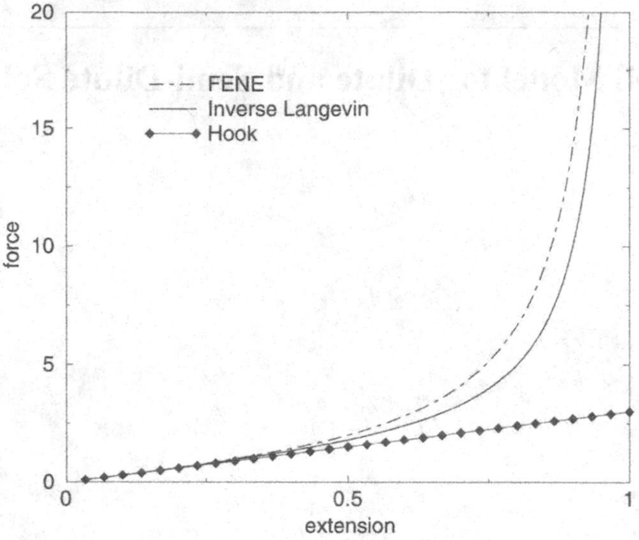

Fig. 2.1. For the freely-jointed chain, the force-extension (***F***–***Q***) relationship is an inverse Langevin function (Langevin function $L(F) = \coth(F) - 1/F$), which is obvious due to its immediate analogy with the case of magnetic moment (dimensionless 'extension' $\boldsymbol{R} = \langle \boldsymbol{u} \rangle$) subjected to an external magnetic field (dimensionless 'force' $\boldsymbol{F} = \boldsymbol{h}$), which is worked out in Sect. 7.5. There, in equilibrium, one has for the 'extension' $\boldsymbol{Q} \propto \langle \boldsymbol{u} \rangle_{\mathrm{eq}} = L(h)\boldsymbol{h}/h$, i.e., for the scalar $Q(F) = L(F)$, therefore $F(Q) \propto L^{-1}(Q)$. For the FENE and Hookean forces, we have $Q^{\mathrm{FENE}}(F) = Q_0(\sqrt{4F^2 + (HQ_0)^2} - HQ_0)/(2F)$ and $Q^{\mathrm{Hook}}(F) = F/H$, respectively. Since $\lim_{F\to\infty} Q(F) = 1$, $Q^{\mathrm{FENE}}(F) = Q_0$, and for small F, $Q(F) = F/3$ and $Q^{\mathrm{FENE}} = Q^{\mathrm{Hook}} = F/H$, in the graph we plot $Q(F)$ for all three cases with $Q_0 = 1, H = 3$, and switch the F and Q axis for the presentation, just to avoid computing the inverse Langevin

with H and Q_0 denoting the (harmonic) spring coefficient and the upper limit for the dumbbell extension. The singularity of the force at $Q^2 = Q_0^2$ is the mathematical implementation of the dumbbell's finite extensibility, and aims to approximate the inverse Langevin function, as further explained and demonstrated in Fig. 2.1. The FENE spring is a valid approximation to a chain of freely rotating elements (the Kramers chain) as long as the number of elements is large, and it gives a reasonable approximation for the entropy of chains of finite length. An inifintely dilute FENE polymer solution is modeled by a suspension of FENE dumbbells in a continuous, Newtonian solvent, where the dumbbell beads are centers of a hydrodynamic drag force, exerted by the surrounding solvent. Assuming Stokes law the drag force is considered being proportional to the relative velocity between solvent and bead, with a constant ζ, the friction coefficient. Point of departure for the statistical analysis is the diffusion equation for the configurational distribution function $\psi(\boldsymbol{Q},t)$

$$\frac{\partial \psi}{\partial t} = \frac{2k_BT}{\zeta}\Delta\psi + \frac{2}{\zeta}\nabla\cdot\{\boldsymbol{F}\psi\} - \nabla\cdot\{(\boldsymbol{\kappa}\cdot\boldsymbol{Q})\psi\} . \tag{2.2}$$

Here, T is the absolute temperature, k_B denotes Boltzmann's constant, and $\boldsymbol{F} = \boldsymbol{F}^{\mathrm{FENE}}$ denotes the determinstic force. The Laplacian and nabla operators refer to derivatives in configuration space. Time dependent expectation values with respect to ψ will subsequently be denoted by angular brackets $\langle \ldots \rangle$, and the FENE parameter $b \equiv HQ_0^2/k_BT$, the relaxation time $\tau \equiv \zeta/4H$ and a dimensionless shear parameter $\Gamma \equiv \tau\gamma$ will be often used. We will be (throughout this monograph) concerned with homogeneous flow whose transposed velocity gradient is denoted as $\boldsymbol{\kappa} \equiv (\boldsymbol{\nabla v})^\dagger$, i.e., $\boldsymbol{v} = \boldsymbol{\kappa} \cdot \boldsymbol{r}$. This enables us to carry out the calculations in the frame of a special coordinate system, the one fixed by the center of mass of the dumbbell, the directions of the axes are specified by the flow geometry. Notice, that (2.2) can be solved analytically only for potential flows [4]. A more detailed motivation for (2.2) is given in Chap. 12 of this monograph.

The FENE dumbbell model has been originally used to describe non-newtonian rheological effects in monodisperse and idealized infinitely dilute polymer solutions with [61–63] or without hydrodynamic interaction [59,60], and to interpret scattering patterns [63–65]. Analytic theories – except those we are going to illustrate in more detail in the next section – have been restricted to infinitely dilute solutions based on a one-particle-description, in which interactions with surrounding molecules have not been considered. The FENE dumbbell with the pre-averaging Peterlin approximation (FENE-P) has been used extensively to describe the rheological behavior of dilute [4] polymer solutions. The model is, however, severely limited, since it cannot describe the broad distribution of relaxation times that real polymer molecules possess. Detailed comparisons of various FENE dumbbell models for dilute solutions conc. its rheological behavior in shear, elongational [66, 67] and also turbulent flows [68] are available. It was shown that while in the linear viscoelastic limit and in elongational flow the behaviour is close, in shear and turbulent flows serious deviations appear. Fairly understood (in terms of a FENE-P model, cf. [67]) is the effect of drag reduction upon adding small amounts of polymers to highly viscous liquid, which are transported through (long) pipelines.

The FENE-P chain, which is conceptually located between FENE-dumbbell models and full FENE chain models, however, has not been as widely used because of the large number of coupled equations that must be solved simultaneously in order to calculate the stress tensor. In [69] the FENE-PM chain, as a 'good' and efficient approximation to the FENE-P chain had been introduced. The reduced number of equations greatly expedites calculations for longer chains. It had been demonstrated [70, 71] by means of standard and stochastic numerical techniques that the pre-averaging Peterlin approximation used to derive the FENE-P macroscopic constitutive equation has also a significant impact on the statistical and rheological properties of the full FENE chain model.

2.1 FENE-PMF Dumbbell in Finitely Diluted Solution

Results of light scattering experiments on dilute polymer solutions in various concentrations below the (equilibrium) overlap concentration have revealed a strong concentration dependence of the polymer conformation in shear flow [72]. In order to present yet another candidate for describing the observed phenomena in an approximate fashion, for illustrative purposes, in order to introduce the Peterlin approximation and basis tensors for later use, and before turning to the recommended full FENE models in the next sections, let us treat the FENE dumbbell model supplemented by a mean field term which describes the concentration dependence in the frame of a one-particle description. The basic idea [44] is to consider interactions between different molecules in an averaged approximation. The notation 'FENE-PMF' follows the recommendations on Page 203.

2.2 Introducing a Mean Field Potential

The mean field term models the effect of concentration induced anisotropy caused by inter- as well as intramolecular interactions in the polymer solution. An expression for the mean field potential can be adapted from theories for concentrated solutions of rodlike polymers [73] and liquid crystals [74, 75] or obtained by carrying out a finite multipole expansion of the intermolecular pair potentials, in which the unknown multipole moments are taken to be phenomenological coefficients [76]. The series has to be written down to an order, which, after averaging with the configuration distribution function, leads to a non-constant and anisotropic expression involving the tensor of gyration, i.e. up to the quadrupole-quadrupole-interaction. The corresponding mean field force reads

$$\boldsymbol{F}^{\mathrm{MF}} = \frac{k_B T}{Q_0^2} f\left(\frac{c}{c^*}\right) \langle \boldsymbol{Q}_{[2]} \rangle^* \cdot \boldsymbol{Q} \,. \tag{2.3}$$

The symbol $\boldsymbol{Q}_{[2]} = \overline{\boldsymbol{Q}\boldsymbol{Q}}$ denotes the irreducible (symmetric traceless) part of the dyadics, $\boldsymbol{Q}_{[2]} = \boldsymbol{Q}_{(2)} - 1/3$, and $\boldsymbol{Q}_{(2)} \equiv \boldsymbol{Q}\boldsymbol{Q}$. This notation, using square and round brackets, cf. Page 199, will be used throughout this monograph. In (2.3), c is the concentration (mass density) of the polymers in solution, c^* is a reference concentration. The scalar function f represents a phenomenological coefficient. If it is assumed to be zero for infinitely dilute solutions data of [72] suggest $f = (c/c^*)^{1/3}$ with a characteristic concentration c^*. This means f is proportional to the reciprocal average distance between the molecules. The ansatz differs from the ones used in [73–75] in the respect that a connector vector $\boldsymbol{Q}$ with variable length enters the expression for the potential instead of a unit vector specifying the direction of a rod.

2.3 Relaxation Equation for the Tensor of Gyration

By multiplying (2.2) for homogeneous flows with $\boldsymbol{Q}_{(2)}$ and subsequent integration by parts, with $\boldsymbol{F} = \boldsymbol{F}^{\mathrm{FENE}} + \boldsymbol{F}^{\mathrm{MF}}$, we obtain

$$\frac{d}{dt}\langle \boldsymbol{Q}_{(2)}\rangle = \frac{4k_BT}{\zeta}\mathbf{1} + \frac{4}{\zeta}\{\langle \boldsymbol{F}^{\mathrm{FENE}}\boldsymbol{Q}\rangle + \langle \boldsymbol{F}^{\mathrm{MF}}\boldsymbol{Q}\rangle\}$$
$$+\boldsymbol{\kappa}\cdot\langle \boldsymbol{Q}_{(2)}\rangle + \langle \boldsymbol{Q}_{(2)}\rangle\cdot\boldsymbol{\kappa}^{\dagger}. \tag{2.4}$$

The second moment will be expressed in a dimensionless form

$$\boldsymbol{g} \equiv \langle \boldsymbol{Q}_{(2)}\rangle^* \equiv \frac{\langle \boldsymbol{Q}_{(2)}\rangle}{Q_0^2}. \tag{2.5}$$

For a stationary shear flow (plane Couette geometry) with shear rate $\dot\gamma$ the second-rank gradient tensor $\boldsymbol{\kappa}$ is given by $\kappa_{\mu\nu} = \dot\gamma\delta_{\mu 1}\delta_{2\nu}$ if we denote with $\boldsymbol{e}^{(1)}$ the flow direction, $\boldsymbol{e}^{(2)}$ the gradient direction, and $\boldsymbol{e}^{(3)} = \boldsymbol{e}^{(1)}\times\boldsymbol{e}^{(2)}$ the vorticity direction. For this geometry the orientation angle χ and the mean square dumbbell elongation $\langle Q^2\rangle$ are related to the tensor $\boldsymbol{g}$ by [77] $\tan 2\chi = (2g_{12})/(g_{11}-g_{22})$, and $\langle Q^2\rangle/Q_0^2 = g_{\lambda\lambda} = \mathrm{Tr}\,\boldsymbol{g}$, while the tensor of gyration $\frac{1}{4}\langle \boldsymbol{Q}_{(2)}\rangle$ equals [1] $\frac{1}{4}Q_0^2\boldsymbol{g}$. In dilute solutions the tensor of gyration is assumed to be isotropic under equilibrium conditions. By construction the mean field potential vanishes under equilibrium conditions, since it is linear in the irreducible part of the gyration tensor.

Next, we wish to obtain a closed approximate set of equations for a stationary solution of the relaxation equation (2.4). Inserting (2.1) and (2.3) and the explicit expression for $\boldsymbol{\kappa}$ into (2.4) yields

$$\frac{1}{b}\delta_{\mu\nu} = \left\langle \frac{Q_\mu Q_\nu}{Q_0^2 - Q^2}\right\rangle - \frac{1}{b}f(\frac{c}{c^*})\left\langle \overline{Q_\mu Q_\lambda}\right\rangle^* \langle Q_\lambda Q_\nu\rangle^*$$
$$-\Gamma\left\{\delta_{\mu 1}\langle Q_2 Q_\nu\rangle^* + \delta_{\nu 1}\langle Q_2 Q_\mu\rangle^*\right\}. \tag{2.6}$$

We choose a standard decoupling approximation, referred to as Peterlin approximation [4, 78, 79], modified such that it is exact in equilibrium. Thus, a term equal to zero is added and subsequently approximated by carrying out the involved averaging under equilibrium conditions. This can be done, because the equilibrium distribution function ψ_{eq} for the given problem is known [4, 59]. Coupled moment equations may be alternatively derived by making use of a Taylor series expansion for the expectation value associated with the FENE force term, cf. [4, 60, 80, 81]. One obtains

$$\left\langle \frac{Q_\mu Q_\nu}{Q_0^2 - Q^2}\right\rangle \approx \frac{\langle Q_\mu Q_\nu\rangle^*}{1-\langle Q^2\rangle^*} - \left\{\frac{\langle Q_\mu Q_\nu\rangle^*_{\mathrm{eq}}}{1-\langle Q^2\rangle^*_{\mathrm{eq}}} - \left\langle \frac{Q_\mu Q_\nu}{Q_0^2 - Q^2}\right\rangle_{\mathrm{eq}}\right\}$$
$$= \frac{\langle Q_\mu Q_\nu\rangle^*}{1-\langle Q^2\rangle^*} - \left\{\frac{1}{b+2} - \frac{1}{b}\right\}\delta_{\mu\nu}. \tag{2.7}$$

Use had been made of the isotropic moments (after Taylor expansion) which become $\forall_n \langle Q^{2n}\rangle^*_{\mathrm{eq}} \approx \prod_{k=1}^n (2k+1)/(b+2k+3)$. Insertion of the (2.7) into (2.6) yields the desired closed set of nonlinear equations

[1] Definition gyration tensor for multibead chains: $\boldsymbol{R}_g \equiv (1/N)\sum_{i=1}^N \boldsymbol{r}_i\boldsymbol{r}_i$ where $\boldsymbol{r}_i$ is the bead position vector with respect to the center of mass of the chain. Radius of gyration $R_g \equiv \mathrm{Tr}\boldsymbol{R}_g$.

$$\frac{\boldsymbol{g}}{1-\mathrm{Tr}\,\boldsymbol{g}} - \frac{1}{b} f\left(\frac{n}{n^*}\right) \overline{\boldsymbol{g}} \cdot \boldsymbol{g} - \tau(\boldsymbol{\kappa}\cdot\boldsymbol{g} + \boldsymbol{g}\cdot\boldsymbol{\kappa}^\dagger) = \frac{1}{b+2} . \tag{2.8}$$

Explicit equations for the components $g_{\mu\nu}$ can be derived most conveniently in a symmetry-adapted form.

2.4 Symmetry Adapted Basis

The symmetric second-rank tensor of gyration has six independent components. In the plane Couette geometry two more components vanish for symmetry reasons, because invariance under the transformation $\boldsymbol{e}^{(3)} \rightarrow -\boldsymbol{e}^{(3)}$ is required. An exception will be discussed in Sect. 7.6. The corresponding four independent components of the second moment are g_{11}, g_{12}, g_{22}, and g_{33}. We transform (2.8) to a version which separates the irreducible and trace-dependent parts of the tensor of gyration, since these are especially emphasized in the terms associated with the FENE and mean field forces. The irreducible part of the tensor is decomposed with respect to a set of pseudospherical cartesian basis tensors. This will result in a simple expression for the orientation angle and in a more tractable expansion for small shear parameters. The resulting equations are easily decoupled in this case. A set of orthonormal basis tensors $\boldsymbol{T}^{(k)}$ with $k = 0, 1, 2, \mathrm{Tr}$ is chosen according to [82, 83] whose elements are given by

$$\boldsymbol{T}^{(0)} = (3/2)^{1/2}\, \overline{\boldsymbol{e}^{(3)}\boldsymbol{e}^{(3)}} , \quad \boldsymbol{T}^{(1)} = 2^{-1/2}(\boldsymbol{e}^{(1)}\boldsymbol{e}^{(1)} - \boldsymbol{e}^{(2)}\boldsymbol{e}^{(2)}) ,$$
$$\boldsymbol{T}^{(2)} = 2^{1/2}\, \overline{\boldsymbol{e}^{(1)}\boldsymbol{e}^{(2)}} , \quad \boldsymbol{T}^{(tr)} = 3^{-1/2}(\boldsymbol{e}^{(1)}\boldsymbol{e}^{(1)} + \boldsymbol{e}^{(2)}\boldsymbol{e}^{(2)} + \boldsymbol{e}^{(3)}\boldsymbol{e}^{(3)}) \tag{2.9}$$

with the orthonormality relation

$$\forall_{k,l} T^{(k)}_{\mu\nu} T^{(l)}_{\mu\nu} = \delta_{kl} . \tag{2.10}$$

Note, that $\boldsymbol{T}^{(0)}$, $\boldsymbol{T}^{(1)}$, and $\boldsymbol{T}^{(2)}$ are symmetric traceless, while $\boldsymbol{T}^{(tr)}$ is associated with the trace of a tensor. Two more 'symmetry braking' basis tensors

$$\boldsymbol{T}^{(3)} = 2^{1/2}\, \overline{\boldsymbol{e}^{(1)}\boldsymbol{e}^{(3)}} , \quad \boldsymbol{T}^{(4)} = 2^{1/2}\, \overline{\boldsymbol{e}^{(2)}\boldsymbol{e}^{(3)}} \tag{2.11}$$

will be used in connection with 'rheochaotic states' in Sect. 7.6. The tensor $g_{\mu\nu}$ can be decomposed according to

$$g_{\mu\nu} = \sum_k g_k T^{(k)}_{\mu\nu} , \quad g_k = T^{(k)}_{\mu\nu} g_{\mu\nu} . \tag{2.12}$$

The orientation angle χ and the (mean square) dumbbell elongation $\langle Q^2\rangle^*$ now take the form $\tan 2\chi = g_2/g_1$, $\langle Q^2\rangle^* = \sqrt{3}\, g_{\mathrm{Tr}}$. Using the decomposition and the orthonormality relation a set of coupled non-linear equations for the pseudospherical and trace-dependent components of $\boldsymbol{g}$ is derived from (2.8):

$$
\begin{aligned}
g^{(0)} &= -\Gamma\frac{\sqrt{3}}{3}g_2 - J\{(g_1^2+g_2^2-g_0^2)+\sqrt{2}g_0g_{\mathrm{Tr}}\}\,,\\
g^{(1)} &= \Gamma g_2 - J(2g_1g_0-\sqrt{2}g_1g_{\mathrm{Tr}})\,,\\
g^{(2)} &= \Gamma\left(\frac{g_{\mathrm{Tr}}\sqrt{6}}{3}+\frac{g_0\sqrt{3}}{3}+g_1\right) - J(2g_2g_0-\sqrt{2}g_2g_{\mathrm{Tr}})\,,\\
g^{(\mathrm{Tr})} &= \Gamma\frac{\sqrt{6}}{3}g_2 + J(g_0^2+g_1^2+g_2^2)+\frac{\sqrt{3}}{b+2}\,,
\end{aligned}
\tag{2.13}
$$

with the abbreviations

$$
g^{(i)} \equiv \frac{g_i}{1-\sqrt{3}g_{\mathrm{Tr}}}, \qquad J \equiv b^{-1} f(c/c^*)/\sqrt{6} \tag{2.14}
$$

Note that (2.7, 2.8, 2.13) correct some misprints in [80]. We cannot give an analytical solution of the system without carrying out further approximations, which would result in a significant change of the model. For small dimensionless shear rates Γ, however, exact aanalytical expressions for the orientation angle and the dumbbell elongation are $\tan 2\chi = (1-\phi)/(\tilde{b}\Gamma)$ and

$$
\langle Q^2\rangle^* = \frac{3}{b+5}\left\{1+\frac{2}{3}\left(1-(1-\frac{1}{\sqrt{2}})\phi\right)(1-\phi)^{-2}\tilde{b}^3\Gamma^2\right\} \tag{2.15}
$$

with $\phi = \phi(c) \equiv f(c/c^*)(b+2)/(b(b+5)^2)$ and $\tilde{b} \equiv (b+2)/(b+5)$. These expressions show that for a given shear rate the orientation angle decreases and the radius of gyration increases with rising (still small) concentration. Of course, they reduce to the ones known for FENE dumbbels at zero concentration ($c = \phi = 0$). For Hookean dumbbell the relations for χ and $\langle Q^2\rangle^*$ are obtained for $b \to \infty$, $\tilde{b} = 1$.

For larger shear rates the system of coupled non-linear equations (2.13) has to be solved numerically. Solutions are restricted to a limited range of f (or ϕ). To illustrate the influence of the mean field term, results are presented for a fixed value of $b = 1$ for the FENE parameter (the significance of b in the original theory has been well analyzed in [59,60]). For comparison, we will show plots for the dumbbell elongation and the orientation angle for various b and different concentration parameters.

In Fig. 2.2 the radius of gyration in units of the equilibrium radius is given for different concentrations vs dimensionless shear rate Γ. For given rate, the radius of gyration increases with rising concentration. The relative increase is larger for smaller shear rates, because with rising shear, the deformation is limited by Q_0. Figure 2.3 shows the related plot for the orientation angle. For all concentrations the curve differs from the simple law $\tan 2\chi \propto \dot{\gamma}^{-1} \propto \Gamma^{-1}$, which results from linear theories or from perturbation results of low order. A dashed curve referring to the simple law is given for comparison.

The quantity $g \equiv \sqrt{g_1^2+g_2^2}$ shown in Fig. 2.4 is a measure for the degree of alignment into the shear plane. As expected, we find an increasing anisotropy with rising concentration. The influence of the FENE parameter b is presented in Figs. 2.5, 2.6. The shear rate is given in units of a characteristic time constant $\lambda = \tau b/3$ for

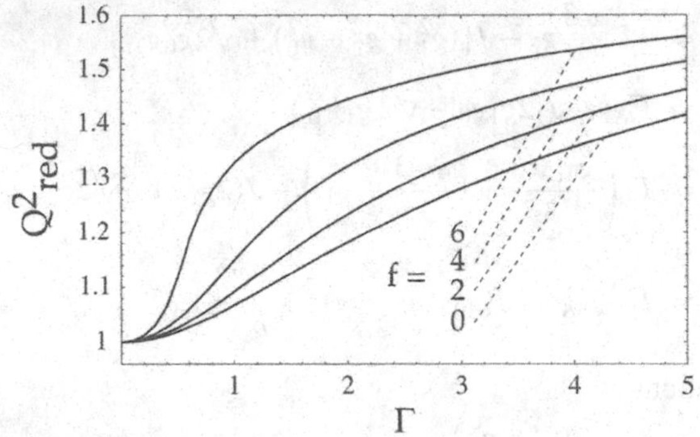

Fig. 2.2. Radius of gyration in units of its equilibrium value versus shear parameter Γ for concentration parameters of $f = 0, 2, 4$, and 6, and a FENE parameter $b = 1$ [80]

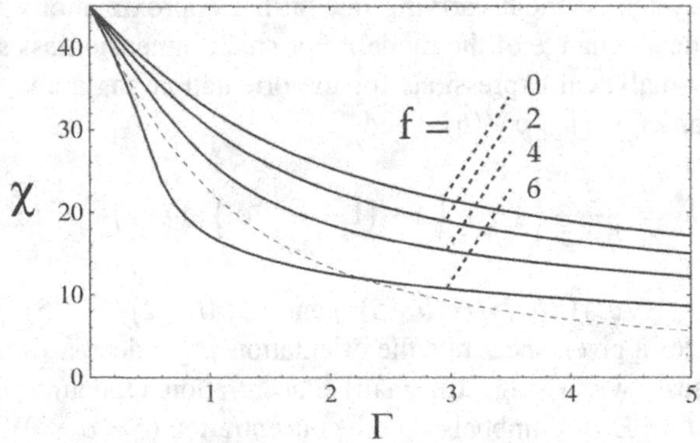

Fig. 2.3. Orientation angle versus shear parameter Γ, see Fig. 2.2 for the choice of parameters. *Dashed curve* according to a linear bead spring theory resulting in $\tan 2\chi = \Gamma^{-1}$ [80]

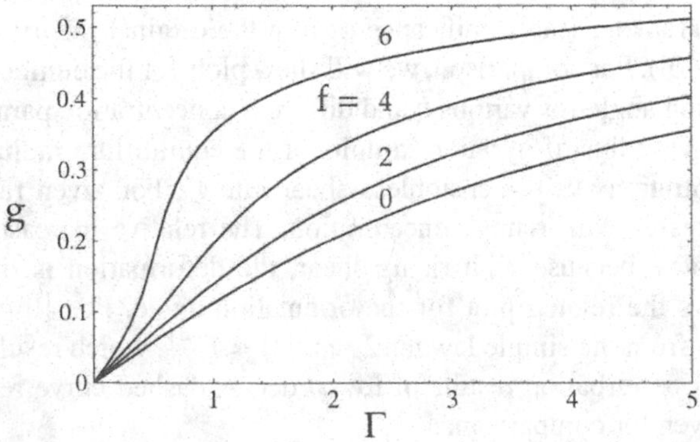

Fig. 2.4. Quantity $g = (g_1^2 + g_2^2)^{1/2}$ related to pseudospherical components of the tensor of gyration versus shear parameter Γ [80]

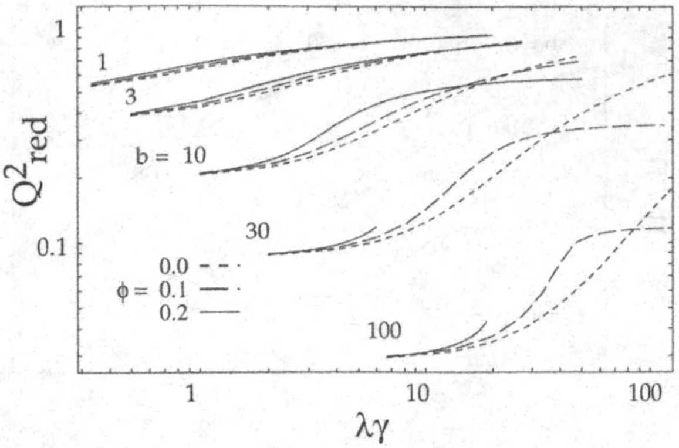

Fig. 2.5. FENE dumbbell elongation versus shear parameter $\lambda\dot{\gamma} = b\Gamma\dot{\gamma}/3$ for various b and different concentration parameters ϕ [80]

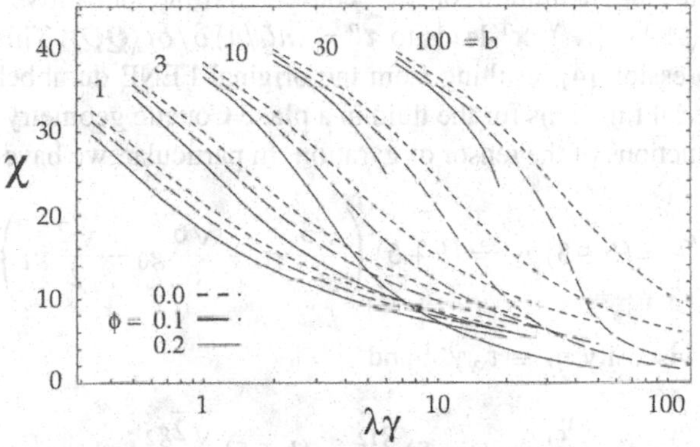

Fig. 2.6. Orientation angle versus shear parameter $\lambda\dot{\gamma}$ for various b and different concentration parameters ϕ [80]

FENE dumbbells in this case to achieve comparability with results from the original theory [59, 60]. The mean field influence is controlled by variation of ϕ which characterizes the mean field magnitude independently of b in the case of small shear rates. In the range of higher shear rates the dumbbell elongation falls with rising concentration parameter (Fig. 2.5). Especially for higher b, the elongation is now limited by the mean field, not by the finite extensibility.

2.5 Stress Tensor and Material Functions

The polymer contribution to the stress tensor $\boldsymbol{\tau}^p$ for the FENE dumbbell takes the form of an extended Kramers expression [4], cf. Sect. 8.7,

$$\boldsymbol{\tau}^p = n\left\langle(\boldsymbol{F}^{\mathrm{FENE}} + \boldsymbol{F}^{\mathrm{MF}})\boldsymbol{Q}\right\rangle + nk_BT\,\mathbf{1}\,. \qquad (2.16)$$

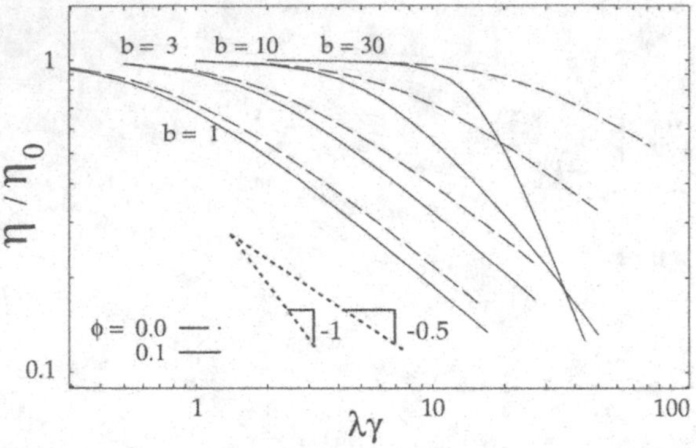

Fig. 2.7. Reduced viscosity versus shear parameter $\lambda\dot{\gamma}$ for various b and different concentration parameters ϕ [80]

Using (2.4) and the definition of the convected time derivative $\delta/\delta t(\ldots) \equiv d/dt(\ldots) - \boldsymbol{\kappa}\cdot(\ldots) - (\ldots)\cdot\boldsymbol{\kappa}^\dagger$ leads to $\boldsymbol{\tau}^p = (n\zeta/4)\,\delta/\delta t\,\langle \boldsymbol{QQ}\rangle$. This is similar to a Giesekus expression [4] resulting from the original FENE dumbbell theory. The shear flow material functions for the fluid in a plane Couette geometry [4] are therefore given as functions of the tensor of gyration. In particular, we have

$$\frac{\eta_p}{\eta_{p,0}} = (b+5)\,g_{22} = (b+5)\left(\frac{\sqrt{3}}{3}\,g_{\mathrm{Tr}} - \frac{\sqrt{6}}{6}\,g_0 - \frac{\sqrt{2}}{2}\,g_1\right) \tag{2.17}$$

for the reduced viscosity $\eta_p \equiv \tau_{xy}\dot{\gamma}^{-1}$ and

$$\frac{\Psi_1}{\Psi_{1,0}} = (b+5)\,\frac{g_{12}}{\Gamma} = (b+5)\,\frac{\sqrt{2}g_2}{2\Gamma} \tag{2.18}$$

for the reduced first viscometric function $\Psi_1 \equiv (\tau_{xx} - \tau_{yy})\dot{\gamma}^{-2}$. The 2nd viscometric function $\Psi_2 \equiv (\tau_{yy} - \tau_{zz})\dot{\gamma}^{-2}$ is equal to zero in the present case. Figure 2.7 shows the reduced viscosity versus shear parameter $\lambda\dot{\gamma}$ for various b and two different concentration parameters ϕ. There is a stronger shear thinning effect for $\phi \neq 0$. These results compare well with data from light scattering experiments [72, 80, 81] such that there is no need to present detailed comparisons (which can be also found in [84, 85]).

With increasing concentration (close to and above the overlap concentration) correlations between different molecules become stronger and the one-particle description has to be abandoned [86, 87]. Scattering experiments have been performed on semi-dilute polymer solutions at rest and in laminar shear flow at different temperatures by SANS [88] and by (small angle) light scattering (SALS) [89–91] as well as by dynamic light scattering [92].

2.6 Reduced Description of Kinetic Models

Numerical implementation of kinetic models in direct numerical flow calculations is in general computationally expensive. This is especially true for chain models to be discussed in later sections. However, kinetic models of polymer dynamics may serve as a starting point for the derivation of constitutive equations. Derivations are not straightforward but require approximations to the underlying kinetic model. The need for so–called closure approximations occurs also in other branches of statistical physics and several suggestions for such approximations have been proposed in the literature (see e.g. [93] and references therein). The frameworks 'reduced description' and 'invariant manifolds' have been developed to efficiently obtain an approximate solution for Fokker–Planck equations for FENE dumbbells and liquid crystals [94] and of the types to be discussed later in this monograph. In [95] the authors give a compact non-technical presentation of two basic principles for reducing the description of nonequilibrium systems based on the quasiequilibrium approximation. These two principles are: Construction of invariant manifolds for the dissipative microscopic dynamics, and coarse-graining for the entropy-conserving microscopic dynamics. It had been demonstrated in general and illustrated how canonical distribution functions are obtained from the maximum entropy principle, how macroscopic and constitutive equations are derived therefrom and how these constitutive equations can be implemented numerically [94, 96]. A measure for the accuracy of the quasiequilibrium approximation had been proposed that can be evaluated while integrating the constitutive equations. Within the framework of reduced description, equations of change for the 'dual' variables appearing in an ansatz for the distibution function play a major role. The method has been further applied to ferrofluids in [97]. Constructive methods of invariant manifolds for kinetic problems are going to be reviewed elsewhere [98]. A closely related approach using projectors will be shortly discussed in Sect. 8.7.

3

Chain Model for Dilute Solutions

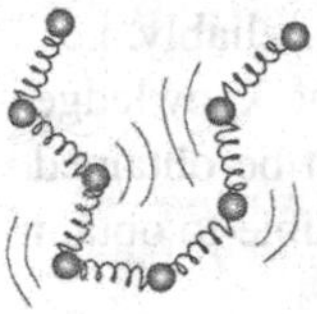

Various experimental observations reveal an important aspect of the behavior of polymer solutions which is not captured by FENE dumbbell models. When the experimental data for high molecular weight systems is plotted in terms of appropriately normalized coordinates, the most noticeable feature is the exhibition of universal behavior. By this it is meant that curves for different values of a parameter, such as the molecular weight, the temperature, or even for different types of monomers can be superposed onto a single curve. For example, when the reduced intrinsic viscosity is plotted as a function of the reduced shear rate, the curves for polystyrene in different types of good solvents at various temperatures collapse onto a single curve [4]. There is, however, an important point that must be noted. While polymers dissolved in both theta solvents and good solvents show universal behavior, the universal behavior is different in the two cases. An example of this is the observed scaling behavior of various quantities with molecular weight. The scaling is universal within the context of a particular type of solvent. The term universality class is used to describe the set of systems that exhibit common universal behavior [99]. Thus theta and good solvents belong to different universality classes.

3.1 Hydrodynamic Interaction

As pointed out in 1948 [100], the perturbation of the solvent flow field induced by suspended spherical particles ('beads') leads to an additional interaction between beads, the so called HI. Incorporation of this effect into the classical Rouse model for dilute polymer solutions makes the resulting model equations – containing a HI matrix – nonlinear. Predictions for some material properties were found to become much more realistic when HI is accounted for [4, 13, 58, 64, 65, 101, 102]. In the usual discussion of HI, one linearizes the Navier-Stokes equation (NSE) and assumes that the propagation of solvent flow perturbations is infinitely fast. If the beads are point particles one obtains for the perturbation of the flow at position $\boldsymbol{r}$: $\Delta \boldsymbol{v}(\boldsymbol{r}) = \boldsymbol{\Omega}(\boldsymbol{r}-\boldsymbol{r}')\cdot\boldsymbol{F}(\boldsymbol{r}')$, where $\boldsymbol{F}(\boldsymbol{r}')$ is the force exerted by a bead at point $\boldsymbol{r}'$ on the solvent, and $\boldsymbol{\Omega}(\boldsymbol{r})$ is the Green's function of the time-dependent linearized NSE, known as Oseen-Burgers tensor (one has to require $\boldsymbol{\Omega}(\boldsymbol{0}) = \boldsymbol{0}$ in order to avoid hydrodynamic self-interactions).

There appear to be two routes by which the universal predictions of models with HI have been obtained so far, namely, by extrapolating finite chain length results to the limit of infinite chain length where the model predictions become parameter free, and by using renormalization group theory methods. In the former method, there are two essential requirements. The first is that rheological data for finite chains must be generated for large enough values of N so as to be able to extrapolate reliably, i.e., with small enough error, to the limit $N \to \infty$. The second is that some knowledge of the leading order corrections to the infinite chain length limit must be obtained in order to carry out the extrapolation in an efficient manner. It is possible to obtain universal ratios in the zero shear rate limit in all the cases [58].

The diffusion equation, sometimes referred to as Fokker–Planck equation, for the configurational distribution function $f(t,\boldsymbol{r}_1,\boldsymbol{r}_2,..,\boldsymbol{r}_N)$ for a chain with N beads reads [58, 103] subject to homogeneous flows ($\boldsymbol{\kappa}$ was defined in Chap. 2)

$$\frac{\partial f}{\partial t} = -\sum_{i=1}^{N} \frac{\partial}{\partial \boldsymbol{r}_i}\cdot\left(\boldsymbol{\kappa}\cdot\boldsymbol{r}_i + \frac{1}{\zeta}\sum_j \boldsymbol{H}_{ij}\cdot\boldsymbol{F}_j\right) f + \frac{k_B T}{\zeta}\sum_{i,j}\frac{\partial}{\partial \boldsymbol{r}_i}\cdot\boldsymbol{H}_{ij}\cdot\frac{\partial}{\partial \boldsymbol{r}_j} f \tag{3.1}$$

with the HI matrix $\boldsymbol{H}_{ij} \equiv \boldsymbol{H}(\boldsymbol{r}_{ij}) = \delta_{ij}\boldsymbol{1} + \zeta\,\boldsymbol{\Omega}(\boldsymbol{r}_{ij})$. In the Itô approach, the stochastic differential (Langevin) equations of motions for bead positions equivalent to the Fokker–Planck equation (3.1) are

$$d\boldsymbol{r}_i = \left(\boldsymbol{\kappa}\cdot\boldsymbol{r}_i + \frac{1}{\zeta}\sum_j^N \boldsymbol{H}_{ij}\cdot\boldsymbol{F}_j\right) dt + \sqrt{\frac{2k_B T}{\zeta}}\, d\boldsymbol{S}_i\,, \tag{3.2}$$

where $d\boldsymbol{S}_i \equiv \sum_j \boldsymbol{B}_{ij}\cdot d\boldsymbol{W}_j(t)$; $\boldsymbol{W}$ denotes a Wiener process (Gaussian white noise vector); $\boldsymbol{B}$ is related to the HI matrix through the fluctuation-dissipation theorem $\boldsymbol{H}_{ij} = \sum_k^N \boldsymbol{B}_{ik}\cdot\boldsymbol{B}_{jk}^T$ and $\boldsymbol{F}_j$ denotes the sum of (other than HI, i.e. spring) forces on bead j. Equation (3.2) is the starting point for a NEBD computer simulation, the only tool available for treating chains with HI rigorously. There are two possibilities for restoring a positive-semidefinite diffusion term when the assumption of point particles fails (one implicitly introduces a bead radius through Stokes monomer ıFriction!coefficient!Stokes friction coefficient ζ): one can prevent the beads from overlapping, or one can modify the Oseen-Burgers HI tensor. In the following application we will use $\boldsymbol{\Omega}$ according to the regularization proposed by Rotne, Prager and Yamakawa [104]. The Langevin equation (3.2) can't be solved in closed form. In order to obtain a tractable form, in 1956 Zimm replaced the random variables $\boldsymbol{\Omega}\boldsymbol{r}_{ij}$ by their equilibrium (isotropic) averages, i.e., $\boldsymbol{H}_{ij} \to H_{ij}\boldsymbol{1}$ with the $N\times N$ matrix $H_{ij} = \delta_{ij} + h^*(1-\delta_{ij})(2/|i-j|)^{1/2}$ and a HI parameter [105]

$$h^* \equiv \frac{\zeta}{6\pi\eta_s}\sqrt{\frac{H}{\pi k_B T}}\,, \tag{3.3}$$

where H denotes the harmonic bead-spring coefficient. The parameter h^* can be expressed as $h^* = a_b/(\pi k_B T/H)^{1/2}$ which is roughly the bead radius a_b over the root-mean-square distance between two beads connected by a spring at equilibrium, hence $0 < h^* < 1/2$. For analytical and experimental estimates of h^* see [4,103,106]. For the Zimm model $h^* = 1/4$ minimizes the effect of chain length and the very short and long chain limits can be elaborated analytically.

3.2 Long Chain Limit, Cholesky Decomposition

For several reasons, the long chain limit is important. It is independent of the details of the mechanical model, and hence is a general consequence of the presence of HI and equilibrium averaged HI for the Zimm model [58]. respectively. For long chains it should be observed that h^* occurs only in the combination ζ/h^* in all material properties. Therefore, the parameter h^* has no observable effect on the material properties of long chains. Power law dependences of various material properties on molecular weight $M \propto N$ with universal exponents are expected (see Sect. 8.2.2.1 of [107]) and, from the prefactors, one can form universal ratios [58]. The universal exponents and prefactors are ideally suited for a parameter-free test of the model by means of experimental data for high molecular weight polymer solutions. We obtained estimates by extrapolation from extensive and efficient simulation.

3.3 NEBD Simulation Details

A coarse-grained molecular model represents the polymer molecules: the FENE bead-spring chain model, i.e., N identical beads joined by $N-1$ (anharmonic) springs. The solvent is modeled as an incompressible, isothermal Newtonian homogeneous fluid characterized by its viscosity η_s. The solution is considered to be infinitely diluted, and the problem is limited to the behavior of one single molecule. In combination with the variance reduction scheme, chain lengths comparable to real conditions (e.g., $N = 300$, cf. Chap. 4) are now coming within reach of simulations.

The decomposition of the diffusion matrix $\boldsymbol{H}$ to obtain a representation for $\boldsymbol{B}$ (e.g., Cholesky decomposition) for long chains is expensive and scales with N^3. A highly efficient method [108] is based on an approximation of the square root function in Chebyshev (tensor) polynomials $\boldsymbol{T}_k$ of the first kind, following the notation in [109],

$$\boldsymbol{B} = \sqrt{\boldsymbol{H}} \approx \sum_{k=1}^{L} c_k \boldsymbol{T}_{k-1}(\boldsymbol{H}) - \frac{\mathbf{1}}{2}c_1\,, \tag{3.4}$$

where the recursive formula

$$T_{k+1}(H) = 2H \cdot T_k(H) - T_{k-1}(H)\,, \tag{3.5}$$

together with $T_0(H) = 1$ and $T_1(H) = H$ define these polynomials. For a fixed L, (3.4) is a polynomial in H which approximates B in the interval [–1,1] (concerning the eigenvalues of H), where all the zeros of T_k are located. The sum can be truncated in a very graceful way, one that does yield the 'most accurate' approximation of degree L (in a sense which can be made precise). The convergence of the Chebyshev polynomial approximation requires that the eigenvalues of the matrix H are within the interval [–1,1]. Actually, this is not the case, and one introduces shift coefficients, h_a and h_b in order to apply the recursion formula to the 'shifted' matrix $H' \equiv h_a H + h_b \mathbf{1}$ whose eigenvalues should be within the desired range. This requirement is fulfilled for $h_a = 2/(\Lambda_M - \Lambda_0)$, $2h_b = -h_a(\Lambda_M + \Lambda_0)$, where Λ_0 and Λ_M denote the minimum and maximum eigenvalues of the original HI matrix H, respectively [102]. The coefficients of the series are readily obtained by standard methods [109,110]: $c_j = L^{-1}\sum_{k=1}^{L} \alpha_{kj}^{L}\,(b_+ + b_- \cos[\pi(k-1/2)/L])^{1/2}$, with coefficients $b_+ \equiv (h_a + h_b)/2$, $b_- \equiv (h_b - h_a)/2$, and the abbreviation $\alpha_{kj}^{L} \equiv 2\cos[\pi(j-1)(k-1/2)/L]$. Instead of calculating the square root matrix first, thus implying several time consuming matrix by matrix products for the evaluation of the polynomials of the series, and afterwards its product with the random W vector, the desired vector is obtained directly as a result of a series of different vectors V, recursively calculated only through less expensive matrix (H) by vector (V) products, i.e., one replaces dS_i in (3.2) by

$$\begin{aligned} dS_i &\equiv B_{ij} \cdot dW_j \\ &= \left(\sum_k^L c_k T_{k-1}(H') - \mathbf{1}\tilde{c}_1 \right) \cdot dW_j(t) \\ &= \sum_k^L c_k\, dV_{k-1}^i - \tilde{c}_1 dW_j\,, \end{aligned} \tag{3.6}$$

with $\tilde{c}_1 = c_1/2$. The recursion formula for $dV_k^i \equiv T_k(H') \cdot dW_i$ is immediately obtained from (3.5). Its evaluation requires an effort $\propto N^2$ for every $k = 1,2,..,L$. The overall computational demand of the method we use scales with $N^2 L \propto N^{9/4}$ per time step as shown in [102]. The eigenvalue range applied in the implementation of this idea is specific for the problem under study. In general, one has to ensure that the degree of violation of the fluctuation-dissipation theorem (with respect to an elegible matrix norm) is small enough to obtain exact moments of the distribution function with a desired accuracy, e.g., along the lines indicated in [111], in order to prevent a direct calculation of eigenvalues. There is an increasing interest in using iterative schemes to decompose the HI matrix, e.g., [65, 108, 111–122].

In addition to this decomposition method a variance reduction simulation technique has been implemented in [102] to reduce the statistical error bars (see also [58], p. 177). For this purpose two simulations are run in parallel, one at equilibrium, and another undergoing steady shear flow but using the same sequence of random numbers. After a certain time interval the desired magnitudes are sampled, and the chain simulated under steady shear flow is (periodically) reset to the state of the chain in equilibrium. Simulations for this model have been further performed, e.g., for the case of step shear deformation in [123]. The Cholesky decomposition has been recently applied within an accelerated Stokesian dynamics algorithm for brownian suspensions [124] and for simulations of supercooled DNA [125]. For a sample brownian dynamics code, as well as an algorithm for the Chebyshev decomposition see Chap. 12.

Table 3.1. Analytical, experimental and numerical results for the zero shear rate limit. E.g., Fixman estimated $U_{\mathrm{RD}} = 1.42$ [126] but couldn't estimate $U_{\eta\mathrm{R}}$ due to the slow convergence of rheological properties η (and also $\Psi_{1,2}$). The asterisk marks results obtained *taking into account* excluded volume. The estimates of de la Torre et al. and Bernal et al. [115, 127, 128] were obtained by extrapolation from their results for $h^* = 1/4$ [102]

	U_{RD}	$U_{\eta\mathrm{R}}$	$U_{\Psi\eta}$	$U_{\Psi\Psi}$	$U_{\eta\lambda}$	$U_{\Psi S}$
			Theory			
Rouse [58]	$\propto N^{-1/2}$	$\propto N^{+1/2}$	0.8	0	1.645	$\propto N$
Zimm [58]	1.47934	1.66425	0.413865	0	2.39	20.1128
Consist. averag. [103]		1.66425	0.413865	0.010628		
Gaussian approx. [129]	–	1.213(3)	0.560(3)	–0.0226(5)	1.835(1)	14.46(1)
Twofold normal Zimm [129]	–	1.210(2)	0.5615(3)	–0.0232(1)	1.835(1)	14.42(1)
Renormalization [106]	–	1.377(1)	0.6096(1)	–0.0130(1)	–	20.29(1)
Oono* [130]	1.56(1)	–	–		–	–
Öttinger* [131]	–	–	0.6288(1)	–	–	10.46(1)
			Experiment			
Schmidt [132, 133]	1.27(6)	–	–	–	–	–
Miyaki [134]	–	1.49(6)	–	–	–	–
Bossart [63]	–	–	0.64(9)	–	–	–
Bossart* [63]	–	–	0.535(40)	–	–	–
			Simulation			
Fixman [126] (NEBD)	1.42(8)	–	–		–	–
de la Torre [127] (NEBD)	1.28(11)	1.47(15)	–	–	2.0	–
Rubio [135] (MC)	–	>1.36(5)	–	–	–	–
Garcia Bernal* [128] (NEBD)	1.48(15)	1.11(10)		–	–	–
Aust* (NEMD) [45]	1.41(6)	–	–	–	–	–
Kröger (NEBD) [102]	1.33(4)	1.55(6)	0.45(7)	0.05(4)	–	19(2)

3.4 Universal Ratios

The most interesting theoretical predictions for experimentally accessible quantities are those which are independent of any physical parameters. In the limit of infinitely long chains the Zimm model predicts a diffusion coefficient $\lim_{N\to\infty} D_h = ch^* k_B T/(\zeta\sqrt{N})$, radius of gyration $\lim_{N\to\infty} R_g = (N k_B T/2H)^{1/2}$, and spectrum of relaxation times

$$\lim_{N\to\infty} \lambda_j^{\text{Zimm}} = c_j (N/j)^{2/3} \zeta/(4h^* H\pi^2) \tag{3.7}$$

with $c_1 = 1.22$ and $c_j = 2\pi j/(2\pi j - 1)$ for $j > 1$ [136].

Having established these relationships for the Zimm model one can construct and define a number of universal ratios for experimentally accessible quantities. The universal quantity

$$U_{\text{RD}} \equiv \frac{R_g}{R_h} \equiv \frac{6\pi\eta_s D_h R_g}{k_B T}, \tag{3.8}$$

is the ratio between radius of gyration and hydrodynamic radius, the latter quantity can be actually measured experimentally in a dynamic experiment, e.g., by observing the relaxation time of the dynamic scattering function $S(q,t)$ for small momentum transfers $qR_g \ll 1$. The universal ratio

$$U_{\eta\text{R}} \equiv \lim_{c\to 0} \frac{\eta_p}{c\eta_s(4\pi R_g^3/3)} \stackrel{\text{Zimm}}{\rightarrow} \frac{9}{2} \frac{U_{\eta\lambda} D_h \lambda_1^{\text{Zimm}}}{U_{\text{RD}} R_g^2} \tag{3.9}$$

is a measure for the specific polymer contribution η_p to the reduced shear viscosity,

$$U_{\Psi\eta} \equiv \lim_{c\to 0} \frac{ck_{\text{B}} T \Psi_1}{\eta_p^2}, \tag{3.10}$$

gives the ratio between first viscometric function and squared polymer contribution to the shear viscosity,

$$U_{\Psi\Psi} \equiv \frac{\Psi_2}{\Psi_1}, \tag{3.11}$$

is the ratio between the second and first viscometric function,

$$U_{\eta\lambda} \equiv \lim_{c\to 0} \frac{\eta_p}{ck_{\text{B}} T \lambda_1} \stackrel{\text{Zimm}}{=} \frac{\lambda_\eta}{\lambda_1} = \frac{\pi^{5/2}}{4[\Gamma(3/4)]^2 c_1}, \tag{3.12}$$

reflects the proportionality between η_p and the longest relaxation time, and

$$U_{\Psi S} \equiv \frac{k_{\text{B}} T \Psi_1}{c\eta_s^2 R_g^6} = U_{\Psi\eta} U_{\eta\text{R}}^2 (4\pi/3)^2, \tag{3.13}$$

(also introduced in [58]) is just a combination of two of the above universal ratios. Results for the Zimm model are also given in the above defiining equations. From these ratios one can, for example, eliminate the unspecified proportionality coefficients in the 'blob' theory of polymer statistics [137, 138].

Universal ratios are collected in Table 3.3. It contains results for diverse theoretical approaches such as obtained by the Zimm model, the Gaussian approximation, a consistent averaging procedure, and renormalization group calculations, together with experimental and numerical findings. The estimates for the exact long-chain limit are extrapolated from NEBD data, where the polymer contribution to the stress tensor and radius of gyration needed to analyze universal ratios are calculated directly from bead trajectories. In particular, the monomer diffusion coefficient

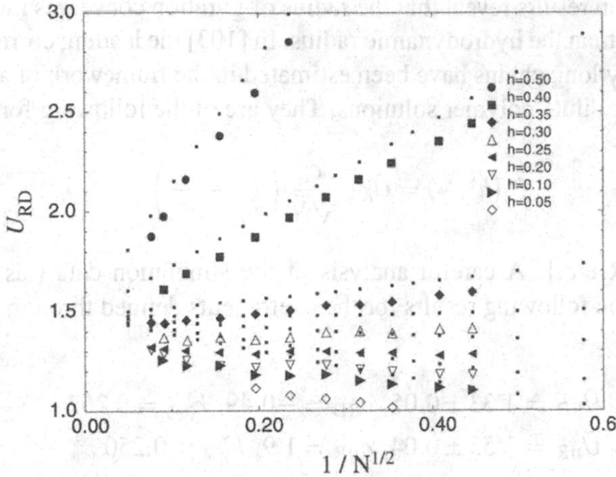

Fig. 3.1. The ratio U_{RD} between gyration and hydrodynamic radii vs the inverse square root of chain length for different values of the HI interaction parameter h^*. As a reference, results for the Zimm model are also shown (*small dots*). By extrapolation to $N \to \infty$ the universal ratio is obtained (see Table 3.3). Apparently, U_{RD} depends linearly on $1/\sqrt{N}$ [102]

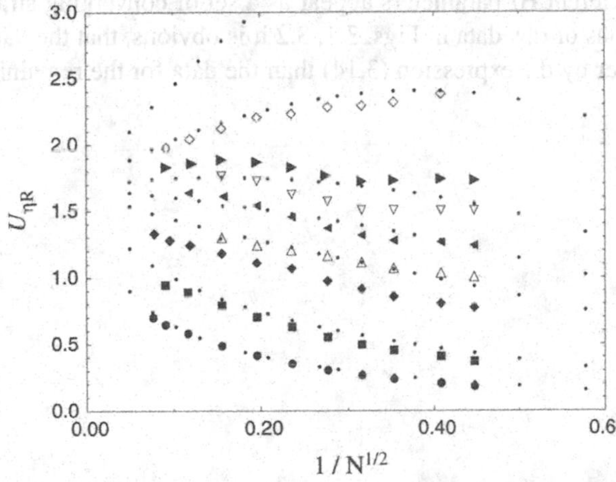

Fig. 3.2. The ratio $U_{\Psi\eta}$, cf. Fig. 3.1 [102]

D and radius of gyration R_g are sampled from bead trajectories $\{\boldsymbol{r}_i(t)\}$ according to $D = \lim_{t\to\infty}(\sum_{i=1}^{N}[\boldsymbol{r}_i(t) - \boldsymbol{r}_i(0)]^2)/(6Nt)$ and $R_g^2 = \sum_i[\boldsymbol{r}_i - \boldsymbol{r}_c]^2/N$, respectively, where $\boldsymbol{r}_c$ denotes the center of mass of the molecule. The simulation reveals that the power law regime for monomer diffusion D will be obtained earlier than the one for the more 'global' R_g. By analogy to classical results for the diffusion of a sphere embedded in a Newtonian liquid the hydrodynamic radius (of the corresponding sphere) is defined by $R_h = k_BT/(6\pi\eta_s D)$. An independent discussion about relaxation times for this system, needed to determine $U_{\eta\lambda}$ can be found in [127]. As for the Zimm model, simulation results reveal that the radius of gyration converges more fast to its long chain limit than the hydrodynamic radius. In [103] the leading corrections to the limit of infinitely long chains have been estimated in the framework of a generalized Zimm model for dilute polymer solutions. They are of the following form:

$$U_i(h^*,N) = \tilde{U}_i + \frac{c_i}{\sqrt{N}}\left(\frac{1}{h_i^*} - \frac{1}{h^*}\right), \tag{3.14}$$

for $i \in \{\mathrm{RD}, \eta\mathrm{R}, etc.\}$. A careful analysis of the simulation data (last row of Table 3.3) yields the following results for the coefficients defined through (3.14):

$$\begin{aligned}
\tilde{U}_{\mathrm{RD}} &= 1.33 \pm 0.05,\ c_{\mathrm{RD}} = -0.49,\ h^*_{\mathrm{RD}} = 0.267\,,\\
\tilde{U}_{\eta\mathrm{R}} &= 1.55 \pm 0.04,\ c_{\eta\mathrm{R}} = 1.9,\ h^*_{\eta\mathrm{R}} = 0.250\,,\\
\tilde{U}_{\Psi\eta} &= 0.29 \pm 0.1,\ c_{\Psi\eta} = -0.20,\ h^*_{\Psi\eta} = 0.261\,,\\
\tilde{U}_{\Psi\Psi} &= 0.05 \pm 0.1,\ c_{\Psi\Psi} = 0.05,\ h^*_{\Psi\Psi} = 0.247\,.
\end{aligned} \tag{3.15}$$

As expected from [103] the values h_i^* for which the leading order corrections are absent do not coincide for the various functions U_i. Since the functions (3.14) for a given i and different HI parameters appear as a set of converging straight lines in the representations of raw data in Figs. 3.1, 3.2 it is obvious, that the data for U_{RD} is represented better by the expression (3.14) than the data for the remaining universal ratios.

4

Chain Model for Concentrated Solutions and Melts

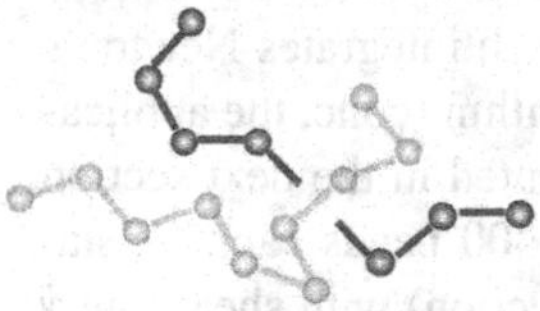

A dense collection of repulsive FENE chains serves as a suitable microscopic model for both entangled and unentangled polymer melts. We will consider once more linear and monodisperse chains although FENE models are immediately applicable to polydisperse polymers with aribtrary architectures. Besides its success for the study of polymer melts at equilibrium [13, 139–141], the nonlinear viscoelastic and structural properties of FENE chain models such as viscosities and scattering patterns are in accordance with experimental results for shear- and elongational flows [17, 18, 142–146]. Due to the computational demands caused by the strong increase of relaxation time with molecular weight (M) only recently it has been observed, that the basic model also exhibits the experimentally observed rheological crossover, certainly related to the ability of polymers to form knots (topological constraints) between macromolecules which is further discussed in [5, 147–154]. The crossover manifests itself in a change of power law for the zero shear viscosity at a certain M.

For FENE melts, FENE forces of the type (2.1) act between all adjacent beads (next neighbors) within chains, and the repulsive part of the radially symmetric Lennard–Jones (LJ) potential (often called WCA potential, introduced by Weeks, Chandler and Anderson [155]) is added between ALL pairs of beads – within cutoff distance – in order to model excluded volume,

$$\boldsymbol{F}^{\mathrm{WCA}}(\boldsymbol{r}) = \varepsilon\, \boldsymbol{F}^{\mathrm{WCA}*}(\boldsymbol{r}/\sigma)$$
$$\boldsymbol{F}^{\mathrm{WCA}*}(\boldsymbol{r}) = -4\nabla_{\boldsymbol{r}}\left(\frac{1}{r^{12}} - \frac{1}{r^{6}} + \frac{1}{4}\right) = -24\left(\frac{r^{6}-2}{r^{12}}\right)\frac{\boldsymbol{r}}{r^{2}}, \quad r \leq 2^{1/6}, \quad (4.1)$$

and $\boldsymbol{F}^{\mathrm{WCA}*}(r \geq 2^{1/6}) = 0$. where r denotes the distance between two interacting beads. The Lennard–Jones and WCA potentials can be read off from (4.1). Here and in the following all dimensionless quantities which are reduced to the usual

Lennard–Jones units of [156–158] are denoted by an asterisk *only if* otherwise ambiguities could arise. We refer to [159] for the discussion of an alternative short range repulsive potential.[1]

4.1 NEMD Simulation Method

The total radially symmetric force $\boldsymbol{F}$ between pairs of beads for the FENE multichain system is $\boldsymbol{F} = \boldsymbol{F}^{\mathrm{WCA}} + \boldsymbol{F}^{\mathrm{FENE}}$ and $\boldsymbol{F} = \boldsymbol{F}^{\mathrm{WCA}}$ for adjacent and non-adjacent beads, respectively. As in [17, 139, 140] for melts the FENE spring coefficients $H = 30$ and $Q_0 = 1.5$ (at temperature $T = 1$, Lennard–Jones units) chosen strong enough to make bond crossings energetically infeasible and small enough to choose a reasonable integration time step during the NEMD simulation, which integrates Newton's equation of motion for this system via a velocity Verlet algorithm (conc. the application reviewed in this section). The simulated systems presented in the next section consist of 3×10^5 beads arranged in chains with $N = 4 - 400$ beads each. A stationary, planar Couette flow in x-direction (gradient in y-direction) with shear rate $\dot{\gamma}$ will be imposed [17]. Neighbor lists, Lees-Edwards boundary conditions [156], and layered link cells [161] are used to optimize the computer routines, In contrast to the standard procedure for equilibrium simulations we update the list of pair dependencies on an upper limit for the increase of the relative separation of these pairs, not on the absolute motion of individual particles. Temperature is kept constant by rescaling the magnitude of the peculiar particle velocities which corresponds to the Gaussian constraint of constant kinetic energy [158] for small integration time steps. Alternative constraint mechanisms (configurational, Nose-Hoover thermostats, SLLOD, etc.) have been extensively discussed elsewhere, and are still under discussion. Since simulation runs are CPU time consuming it should be mentioned that the generation of well quasiequilibrated dense samples for simulations is of particluar relevance. Several codes have been developed which attempt to reach pre-equilibration (at given density) using Monte Carlo, tree-based, fuzzy logic, neural network strategies, to mention a few. The NEMD simulation method is – in principle – independent of the choice for a particular FENE model. For a sample code see Chap. 12, where also temperature control, integration scheme etc. will be introduced. Generating a pre-equilibrated initial configuration is a difficult task for dense polymeric systems. Efficient sample generators are available in the literature, e.g., [32], where chains are 'blown' up dynamically, or one can try to statically 'walk ' into a local potential energy minimum using the method of conjugated gradients, for which codes are available, e.g., in [109].

---

[1] Short range repulsive (SR) potential [159,160]: $U^{\mathrm{SR}*} \equiv (9-8r)^3$, $r < r_{\mathrm{csh}} = 9/8$ (reduced LJ units). The parameters are chosen such that at $r = 1$, the values of the potential functions SR and (4.1) and of their first derivatives are equal. Short range attractive (SA) potential: $U^{\mathrm{SA}*} = (512/27)\,(1-r)\,(3-2r)^3$, $r < 3/2$.

4.2 Stress Tensor

The stress tensor $\boldsymbol{\sigma}$ (equals the negative friction pressure tensor), a sum of kinetic and potential parts, is calculated from its tensorial virial expression

$$\boldsymbol{\sigma} = -\frac{1}{V}\left\langle \sum_{i=1}^{N_b} \boldsymbol{c}^{(i)}\boldsymbol{c}^{(i)} + \frac{1}{2}\sum_{i=1}^{N_b}\sum_{j=1}^{N_b} \boldsymbol{r}^{(ij)}\boldsymbol{F}(\boldsymbol{r}^{(ij)}) \right\rangle, \tag{4.2}$$

where V is the volume of the simulation cell, N_b is the total number of beads, $\boldsymbol{r}^{(i)}$ and $\boldsymbol{c}^{(i)}$ are the spatial coordinate and the peculiar velocity of bead i within a polymer chain, respectively, $\boldsymbol{r}^{(ij)} \equiv \boldsymbol{r}^{(i)} - \boldsymbol{r}^{(j)}$, and $\boldsymbol{F}$ is the pair force. The stress tensor is accessible as time average from the calculated bead trajectories. For dense fluids, the main contribution to the rheological properties stems from the potential part of the stress tensor, except for the case of highly aligned samples. Material function such as viscosities and shear moduli are defined in terms of the stress tensor and flow parameters [4]. The official nomenclature is periodically published by the Journal of Rheology.

4.3 Lennard–Jones (LJ) Units

For any measurable quantity A with dimension $[A]$,

$$[A] = \mathrm{kg}^{\alpha}\mathrm{m}^{\beta}\mathrm{s}^{\gamma} \tag{4.3}$$

one has

$$A = A_{\mathrm{dimless}} \times A_{\mathrm{ref}}\,, \tag{4.4}$$

with

$$A_{\mathrm{ref}} = m^{\alpha+\gamma/2} r_0^{\beta+\gamma} \varepsilon^{-\gamma/2}\,, \tag{4.5}$$

where σ, ε provide the length and energy scales via the Lennard–Jones potential and the monomeric mass m via Newton's equations of motion. Specifically, the reference quantities for density, temperature, time and viscosity are $n_{\mathrm{ref}} = \sigma^{-3}$, $T_{\mathrm{ref}} = \varepsilon/k_B$, $t_{\mathrm{ref}} = \sigma\sqrt{m\varepsilon}$, $\eta_{\mathrm{ref}} = \sigma^{-2}\sqrt{m\varepsilon}$. We therefore have to deal exclusively with $\varepsilon = \sigma = 1$ in (4.1). See Sect. 4.9 for a comment on how to intepret dimensionless simulation numbers.

4.4 Flow Curve and Dynamical Crossover for Polymer Melts

For the FENE chain melt, rheological properties were extracted for various shear rates over eight decades from $\dot{\gamma} = 10^{-8}$ to $\dot{\gamma} = 1$ for $N = 4 - 400$ [142, 162]. For the short chains ($N < 20$) a weak shear dilatancy is detected. With increasing shear rate the trace of the pressure tensor decreases due to the intramolecular bond stretching. The non-newtonian viscosity $\eta \equiv \sigma_{xy}/\dot{\gamma}$ is shown for different chain lengths and rates

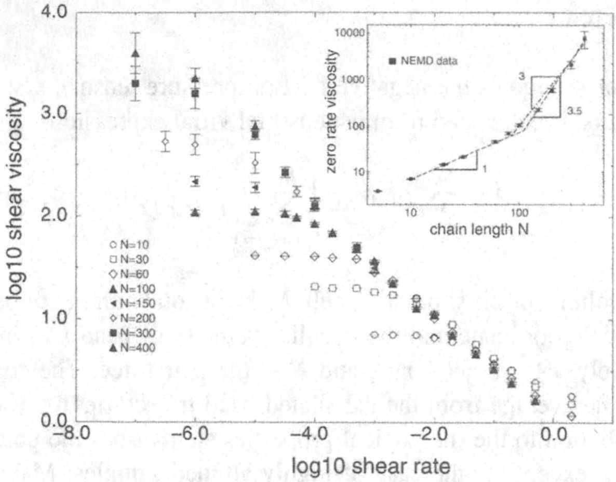

Fig. 4.1. Non-Newtonian shear viscosity η of the FENE model vs shear rate $\dot{\gamma}$ (LJ units) for different chain lengths N. Inset: Zero rate shear viscosity η_0 vs chain length. Adapted from [162]

in Fig. 4.1. The FENE chain melt is shear thinning, and approaches a power law curve $\eta \propto \dot{\gamma}^{-\alpha}$ independent of M with the exponent $\alpha = 0.5 \pm 0.2$. From the non-newtonian viscosity η in Fig. 4.1 the zero rate viscosity η_0 [4] can be estimated. This quantity clearly exhibits a crossover from a Rouse-type regime $\eta_0 \propto N^1$ to $\eta_0 \propto N^{\nu \geq 3}$ (inset of Fig. 4.1) It is well represented by the expression $\eta_0 = 0.7N(1 + Z^{\nu-1})$ with a number of 'rheologically relevant' entanglements per chain $Z \equiv N/N_c$ and exponent $\nu = 3.3 \pm 0.2$. The zero rate first viscometric function $\Psi_1 \propto (\sigma_{yy} - \sigma_{xx})/\dot{\gamma}^2$ [4] is found to exhibit a crossover at the same critical chain length.

Elliptical contours in the structure factor of single chains and their rotation against flow gradient direction have been analyzed and plotted against wave number in order to visualize the (different) degree of orientation on different length scales inside a polymer during shear flow, see also Fig. 4.2 for a schematic drawing.

4.5 Characteristic Lengths and Times

For the characteristic relaxation times τ_N defined from the onset of shear thinning at shear rate $\dot{\gamma} = \dot{\gamma}_N \equiv 1/\tau_N$ we obtain from the NEMD simulations: $\tau_N \propto N^{\approx 2}$ for short chains, in accordance with the Rouse model predictions. Based on careful measurements of monomer diffusion coefficients and further properties for the FENE chain melt obtained from MD simulations [139, 140] with up to $N = 400$ beads per chain a 'dynamical' crossover has been observed. A characteristic length was found which marks the crossover between 'Rouse' to 'reptation' diffusion regimes, for which the diffusion coefficients ideally scale as $D \propto 1/N$ and $D \propto 1/N^2$, respectively. The plateau modulus G_N^0, from which the entanglement M_e can be rigorously

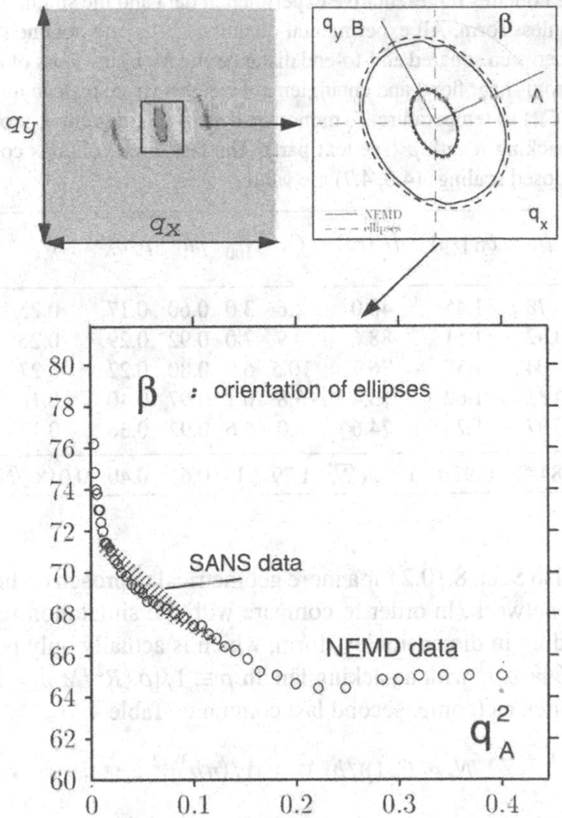

Fig. 4.2. Differences between local and global order of polymeric FENE chains under shear flow conditions are revealed via the NEMD structure factor of single chains. (*Top left*) Structure factor extracted by NEMD, projected to shear plane. (*Top right*) Contour fit allows to extract the half axes (half wave numbers) of ellipses and the rotation angle β. (*Bottom*) Rotation angle vs wave number. Experimental results by [163] serve as a reference

deduced [164] has been reported for the FENE chain melt in [147] for chains up to $N = 10^4$ from the shear stress plateau during relaxation after step strain. The reported value for N_e is about a factor 2.3 larger than the one reported for the dynamical crossover in [139], and thus rather close to the critical weight $N_c = 100 \pm 10$ obtained via NEMD in [162].

The commonly experimentally accessible quantities characterizing a polymer melt at certain temperature are its monomer density ρ, average M, monomer mass m, squared end-to-end distance per monomer $b^2 \equiv \langle R^2/N \rangle$, the critical and entanglement weights, $M_c = mN_c$ and $M_e = mN_e$, respectively, and the Kuhn length b_K. These quantities are related to the bond length $b_0 = b^2/b_K$, the characteristic ratio $C_\infty = b_K/b_0$, and the so called tube diameter $d_T = b\sqrt{N_e}$. It has been suggested recently [165] that both N_e and N_c can be calculated from ρ, b^2 and a fixed length

Table 4.1. The table contains representative experimental data and the simulation data (FENE model) in dimensionless form. All experimental quantities listed are obtained from literature data for i) the ratio between squared end-to-end distance and M, ii) the mass of a repeating unit m, iii) the critical (from shear flow) and entanglement weights (from plateau modulus), and iv) bond length b_0 (or C_∞) at temperature T, monomer density ρ (in g/cm^3), monomer number density $n = \rho/m$, packing length p (see text part). The last three columns contain universal numbers, if the proposed scalings (4.6, 4.7) are valid

Polymer	T	ρ	b_0 [ra]	d_T [ra]	C_∞	$\frac{N_c}{100}$	$pn^{\frac{1}{3}}$	p/b_K	$\frac{N_e^{1/2}}{N_c C_\infty}$	$\frac{N_e b_K^2}{C_\infty p^2}$	$\frac{N_c b_K}{C_\infty^{\frac{3}{2}} p}$
PE	443 K	0.78	1.45	40.0	7.6	3.0	0.60	0.17	0.25	453	84
PS	490 K	0.92	1.51	88.6	9.9	7.0	0.92	0.29	0.26	454	81
PαMS	459 K	1.04	1.57	76.7	10.5	6.9	0.80	0.22	0.27	451	85
PIB	490 K	0.82	1.62	73.4	5.8	6.1	0.97	0.40	0.18	384	109
PDMS	298 K	0.97	1.70	74.6	6.0	6.6	0.92	0.36	0.17	417	119
FENE	ε/k_B	$0.84\frac{m}{\sigma^3}$	0.97σ	$1.3\sigma\sqrt{N_e}$	1.79	1	0.66	0.40	$0.018\sqrt{N_e}$	$3.4\,N_e$	103

$p \approx 10^{-9}$m. See also Sect. 8.10.2 for a more geometrical approach on how to analyze the entanglement network. In order to compare with the simulation results one has to rewrite this finding in dimensionless form, which is actually only possible for N_e and then states: $N_e \propto \rho p^3$ with a packing length $p \equiv 1/[\rho \langle R^2/M \rangle] = 1/(nb^2)$. This definition ia rewritten as (comp. second last column of Table 4.5)

$$N_e \propto C_\infty (p/b_K)^2 = [1/(nb^3)]^2 \,, \tag{4.6}$$

or $nd_T b^2 = c_e$ with a proportionality coefficient $c_e = 21 \pm 2$, where n denotes monomer number density. A corresponding relationship for M_c was also proposed [162] (comp. last column of Table 4.5)

$$N_c \propto C_\infty^{3/2} (p/b_K) = 1/(nb_0^2 b) \,, \tag{4.7}$$

in agreement with the simulation data, and a proportionality coefficient of about $c_e^2/5$ such that $C_\infty \sqrt{N_e} \approx 4N_c$. Thus, one is led to the prediction $N_e n b_0^3 > N_c$ for very flexible chains with $C_\infty < 1.9$. Predictions are summarized in Fig. 4.3. The possibility for the existence of materials with $N_c < N_e$ has been proposed earlier by Fetters et al. [165]. The statement (4.7) has the advantage upon the one in [165] that it exclusively contains dimensionless quantities, and thus allows for a verification by computer simulation. Equations (4.6, 4.7) imply, that N_c is inversely proportional to the number of monomers in the volume bb_0^2, whereas $\sqrt{N_e}$ is inversely proportional to the number of monomers in the volume b^3. Under equilibrium conditions the simulated FENE chains exhibit an average bond length $b_0 = 0.97$, $b = 1.34 b_0$, hence $C_\infty = b^2/b_0^2 = 1.79$ and $p/b_K = 0.404$. Relationship (4.6) predicts a simulation value $N_e \approx 120$ which is slightly above the one reported for N_c, a factor of 3–4 above the one reported for a dynamical crossover in [139, 140], and just by a factor of 1.5 above the one reported from direct measurements of the relaxation modulus [147].

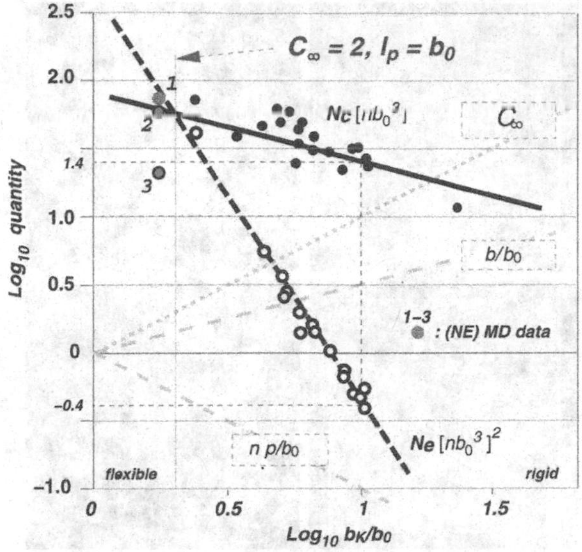

Fig. 4.3. Scaling behavior of crossover and entanglement molecular weights according to (4.6), 4.7. The figure contains the predicted behavior (*lines*) as well as experimental (*full symbols*) and simulation results (*open symbols*, symbol 1 for N_c [162], 2 for N_e [147], 3 for N_e [140])

The reported findings underline the relevance of the FENE model in predicting static, dynamic and flow behaviors of real polymers for arbitrary weights. Beside the investigation of rheological behaviors of FENE melts the simulation of bead trajectories allows to analyze, for example, the degree of flow-induced orientation of chain segments, the validity of the so called 'stress-optic rule', the degree of entanglement [167] anisotropic tube renewal, and therefore renders possible the test of coarse-grained descriptions in later sections.

4.6 Linear Stress-Optic Rule (SOR) and Failures

Shear flow together with elongational flows are essential for the understanding of the flow properties of fluids in complex flows [6, 18, 144, 168–170]. We wish to further demonstrate the impact of the FENE chain melt model for the investigation of the microscopic origins of experimentally observable transport and optical phenomena. One of the aspects of practical relevance (in particular for rheooptics) concerns the validity of the linear stress-optic rule (SOR), cf. (4.10), a proportionality between stress and alignment (better, birefringence) tensors, which is fulfilled for polymer melts under 'usual' conditions. Along with the spirit of this monograph, we focus on studies in the nontrivial regime, where the proportionality is known to be at least partially lost, i.e., at temperatures close to the glass transition temperature T_g or at high elongation rates. To this end we discuss results obtained during constant rate

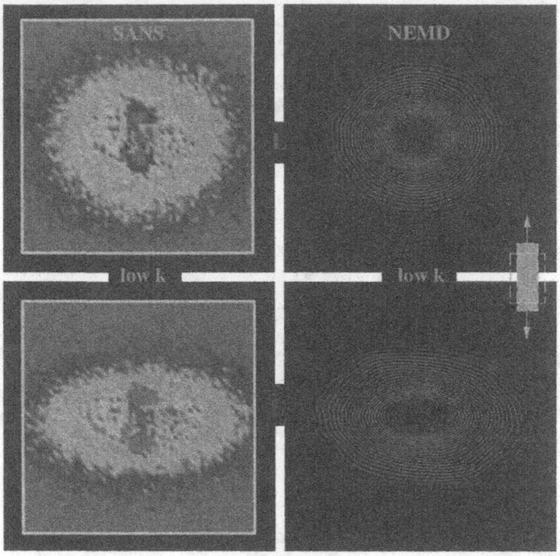

Fig. 4.4. The single chain structure factor for stretched samples with *equal* values of flow birefringence for samples fulfilling (*bottom*) or not (*top*) the SOR. The figure compares data from SANS experiments (*left*) [166] and NEMD simulation (*right*). Due to the fact, that orientational relaxation is fast on a local scale, the overall extension of the polymer has to be much larger for samples fulfilling the SOR, i.e., at high temperatures or low rates, in order to exhibit the same local alignment. Adapted from [233]

uniaxial elongational flow followed by relaxation after reaching a constant stretching ratio [144]. Experimentally measured rate dependencies of the stress-optical behavior of amorphous polymers undergoing elongational flow at temperatures close above T_g are reported in Fig. 4.5. For the lowest rates only small deviations from the 'equilibrium curve' have been detected, where the SOR is valid. For the higher elongation rates the curves exhibit a stress overshoot, and a stress offset σ_{off} for which approximate values vs the reduced elongation rate $a_T\dot{\varepsilon}$ are given in Fig. 4.5b. The phenomenological description of the viscoelastic behavior of amorphous polymers in the region where deviations of the SOR appear has been adjusted many times within the last decades, cf. [41, 144] and refs. cited herein.

In the NEMD simulation, a time-dependent uniaxial isochoric homogeneous elongational flow in x-direction with elongation rate $\dot{\varepsilon} = \partial v_x/\partial x$ is imposed via rescaling of the dimension of the central box [18, 171]. Rheological information under uniaxial flow is contained in the 'uniaxial' component of the stress tensor (4.2) or 'tensile stress': $\sigma \equiv \sigma_{xx} - (\sigma_{yy} + \sigma_{zz})/2$. The (2nd rank) alignment tensor, the anisotropic second moment of the orientation distribution function of segments [4],

$$\boldsymbol{a}_{[2]} \equiv \sum_{i=1}^{N-1} \boldsymbol{a}_{[2]}{}^{(i)} \equiv \left\langle \boldsymbol{u}^{(i)}\boldsymbol{u}^{(i)} \right\rangle - \frac{1}{3}\mathbf{1}\,, \tag{4.8}$$

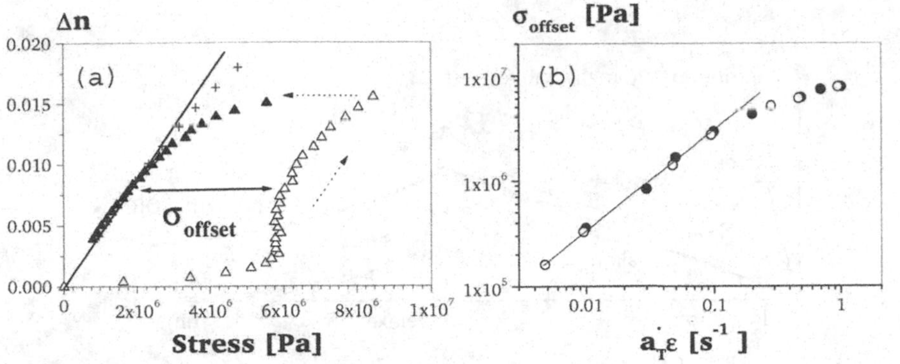

Fig. 4.5. (*Left*) Experimental data taken from for birefringence (Δn) vs tensile stress for a commercial polystyrene subjected to uniaxial elongational flow (open symbols, at $T=$ 102.7°C, rate $\dot{\varepsilon} = 0.2\,\mathrm{s}^{-1}$) and subsequent relaxation (filled symbols). The crosses represent the behavior at high temperatures ('equilibrium curve' [163]). A 'stress-offset' and thus a failure of the stress-optic rule is evident and interpreted through NEMD results for FENE chains in the text part. (*Right*) Corresponding stress offset values vs the reduced elongation rate $\dot{\varepsilon}a_T$. Adapted from [144]

is extracted directly as an ensemble average from the dyadic constructed of the normalized segment vectors between beads (adjacent beads accordingly labeled) $\boldsymbol{u}^{(i)} \equiv \boldsymbol{r}^{(i+1)} - \boldsymbol{r}^{(i)}$ tangential to the chains contour. The alignment tensor is considered being proportional to the refractive index tensor of the fluid [41, 172] whose relevant information for the case of uniaxial elongational flow in x-direction we denote by $\Delta n \equiv a_{\mathrm{xx}} - (a_{\mathrm{yy}} + a_{\mathrm{zz}})/2$. The stress-alignment diagram, obtained by NEMD in [144] compared very well with the experimental data, cf. Fig. 4.5a, and thus motivated to investigate microscopic origin of the observed behavior. In particular, results for diverse (intra/intermolecular, kinetic/potential, attractive/repulsive, non/nearest neighbor) contributions to the stress tensor tensor as revealed in Figs. 5,6 of [144] and also results for shear flow [17] imply that the stress tensor $\boldsymbol{\sigma}$ for the FENE chain melts can be written essentially as the superposition of three terms

$$\boldsymbol{\sigma} = \boldsymbol{\sigma}_{\mathrm{bonded}} + \boldsymbol{\sigma}_{\mathrm{nonbonded}}\,,$$
$$\boldsymbol{\sigma}_{\mathrm{nonbonded}} \approx C^{-1}\boldsymbol{a}_{[2]} + \tilde{\boldsymbol{\sigma}}_{\mathrm{simple}}\,, \tag{4.9}$$

where $\boldsymbol{\sigma}_{\mathrm{bonded}}$ denotes the stress contribution from nearest neighbors within polymer chains (bond pushing/stretching and/or bond orientation), C is the linear stress-optic coefficient for the regime where the SOR is valid, and $\tilde{\boldsymbol{\sigma}}_{\mathrm{simple}}$ is proportional to the stress which is measured for a corresponding simple fluid by removing all bonds (i.e. FENE springs) within the system. A value $C = 0.32$ has been independently confirmed from NEMD simulation on weak shear flow in [17, 144]. See Fig. 4.6 for a schematic drawing. For 'small' flow rates and/or temperatures large compared with the 'bonded' ('intra', non-significant stretch) and 'simple' (proportional to flow rate) contributions become small compared to the SOR contribution such that – according

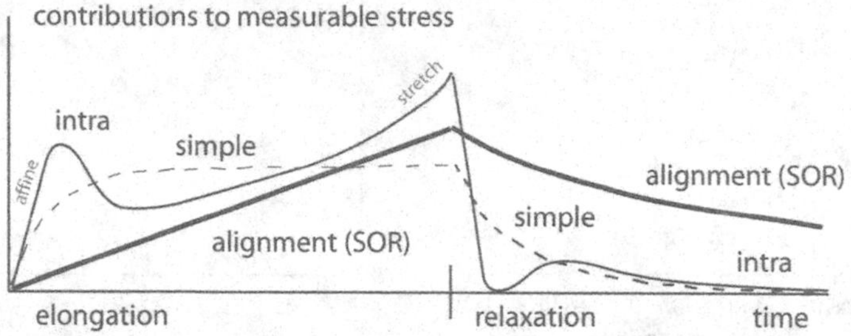

Fig. 4.6. Schematic drawing clarifying the origin of hysteretic behavior (deviations from the stress-optic rule SOR) in the stress-optic diagram for uniaxial elongational flow of FENE polymer melts according to [144]. The measured (total, tensile) stress is the sum of bonded (intra) and nonbonded interactions, where the nonbonded interactions appear to carry a part which is proportional to alignment (i.e. fulfilling the SOR) and another one, which is behaving like the one for a corresponding simple 'newtonian' fluid (FENE bonds removed). The simple and intra stresses become increasingly relevant with decreasing temperature (or increasing rate due to the time temperature superposition principle). The intra stress dominates if bond stretch (due to finite extensibility of chains) comes into play

to (4.9) the validity of the SOR is expected in these regimes. The nonbonded stress hence originates the SOR for the microscopic FENE model. This finding has been further discussed in [143, 144]. In this context one should notice, that the splitting (4.9) is qualitatively different from the one into stresses of predominantly entropy-elastic and energy-elastic origin as discussed in [173, 174].

4.7 Nonlinear Stress-Optic-Rule

The onset of failure of the linear stress-optical rule for polymeric liquids had been often discussed, cf. [175], in terms of a critical stretch, critical stress, or critical dimensionless flow rate. If we wish to determine rheological properties solely from optical data we need a nonlinear generalization of the linear SOR, since stretch, stress and Deborah number are not directly available from the birefringence measurements [176–178]. It seems, that the stress-optical coefficient is well characterized in terms of a single invariant of the refractive index tensor. Thus, if stresses could be uniquely determined by optical means, the nonlinear stress-optical coefficient can be determined from a single stress-optical diagram in uniaxial elongational flow where deviation from the linear SOR appears most 'easily'. The linear SOR connects the anisotropic stress tensor $\boldsymbol{\sigma}_{[2]} = \boldsymbol{\sigma} - \mathrm{Tr}(\boldsymbol{\sigma})/3$ and anisotropic refractive index tensor $\boldsymbol{n}_{[2]}$ [172, 179]

$$\boldsymbol{\sigma}_{[2]} = C^{-1} \boldsymbol{n}_{[2]} , \tag{4.10}$$

independent of frame, with a stress-optical coefficient C which is independent of $\boldsymbol{n}$. Here, the quantity $\boldsymbol{\sigma}_{[2]}$ stands for the contribution to the total stress tensor due to the presence of polymers. Deviations from a linear SOR due to newtonian solvent stresses, which are independent of the orientation of polymers, are not within the scope of this section, they were discussed in Sect. 4.6. The same is true for deviations which occur in the vicinity of the glas transition temperature [144].

It is often convenient to write down the linear SOR in both the laboratory components and principal components (eigenvalues). Let us consider shear flow (with directions: flow x, gradient y, vorticity z) and elongational flows. In these cases both tensors $\boldsymbol{\sigma}_{[2]}$ and $\boldsymbol{n}_{[2]}$ are sufficiently characterized by three components $n_{\mathrm{x},1,2}$ (laboratory frame) and $n_{A,B,C}$ (princpial frame) characterizing their traceless symmetric part. In the following equations n is a placeholder for σ or n. For the relevant combinations of laboratory components $n_{..}$ (of stress or refractive index tensors) we introduce the notation

$$
\begin{aligned}
n_{\mathrm{x}} &\equiv n_{xy}\,, \\
n_1 &\equiv n_{xx} - n_{yy}\,, \\
n_2 &\equiv n_{yy} - n_{zz}\,,
\end{aligned}
\tag{4.11}
$$

such that

$$
\boldsymbol{n}_{[2]} = \frac{1}{3}\begin{pmatrix} 2n_1 + n_2 & 3n_{\mathrm{x}} & 0 \\ 3n_{\mathrm{x}} & n_2 - n_1 & 0 \\ 0 & 0 & -(n_1 + 2n_2) \end{pmatrix}
\tag{4.12}
$$

can be used to rewrite (4.10). The eigenvalues of (4.12) define the princpial components $n_{A,B,C}$: $n_{A/B} = (n_1 \mp 3(4n_{\mathrm{x}}^2 + n_1^2)^{1/2} + 2n_2)/6$, and $n_C = -(n_1 + 2n_2)/3$. The type of flow geometry decides on the most suitable choice of components. For example, in uniaxial elongational flow (homogeneous flow field $\boldsymbol{v}(\boldsymbol{r}) = \boldsymbol{\kappa}\cdot\boldsymbol{r}$ with $\kappa_{yy} = \kappa_{zz} = -\kappa_{xx}/2$) principal and laboratory frames coincide, we directly measure the tensile stress σ_1 (cf. notation in 4.11) and birefringence value n_1. In shear flow, these frames do not coincide. We usually measure directly shear stress σ_{x} and first normal stress differences σ_1 (beside σ_2), the flow alignment angle χ and birefringence value (difference between two principial refractive indices) $\Delta n \equiv n_A - n_B$. Due to the identity $(\Delta n)^2 = (\Delta n)^2(\sin^2(2\chi) + \cos^2(2\chi)) = (n_A - n_B)^2 = (2n_{\mathrm{x}})^2 + n_1^2$, the measured quantities are related through the linear SOR as follows $(\Delta n/C)\sin 2\chi = 2n_{\mathrm{x}}/C = 2\sigma_{\mathrm{x}}$, and $(\Delta n/C)\cos 2\chi = n_1/C = \sigma_1$, such that $\tan 2\chi = 2n_{\mathrm{x}}/n_1$. The equation for the shear component alone allows us to determine most conveniently the stress-optical coefficient by plotting $\Delta n \sin 2\chi$ vs σ_{x}, i.e., by measuring χ, Δn and just the shear stress σ_{x} for several shear rates.

The linear SOR for polymers directly connects birefringence data with rheological properties. Experimental failures of the linear SOR have been reported by several authors for elongational flows, but not yet for shear flows. Such a failure prevents the determination of rheological properties by optical techniques as long as a nonlinear generalization of the linear SOR is not available.

The only possible nonlinear generalization of the linear SOR with scalar coefficients reads ('nonlinear SOR')

$$
\boldsymbol{\sigma}_{[2]} = C_1^{-1}\boldsymbol{n}_{[2]} + C_2^{-1}\,\overline{\boldsymbol{n}_{[2]}\cdot\boldsymbol{n}_{[2]}}\,,
\tag{4.13}
$$

because all (anisotropic) powers of $\boldsymbol{n}_{[2]}$ can be expressed in terms of two lower ones where the prefactors of these terms depend on the invariants of $\boldsymbol{n}_{[2]}$. The related theorem, cf. Sect. 10.7, named by Caley and Hamilton, is rooted in the fact that the eigenvectors of $\boldsymbol{n}^i_{[2]}$ are independent of the power i. The linear SOR is recovered for $C_1 = C$ and $1/C_2 = 0$. The coefficients $C_{1,2} = C_{1,2}(I_2, I_3)$ are functions of the second invariant $I_2 = \mathrm{Tr}\,(\boldsymbol{n}_{[2]} \cdot \boldsymbol{n}_{[2]})/2$ and third invariant $I_3 = \det(\boldsymbol{n}_{[2]})$ of the traceless refractive index tensor $\boldsymbol{n}_{[2]}$ (for which $I_1 = \mathrm{Tr}\,\boldsymbol{n}_{[2]}$ vanishes). The invariants can be also expressed in terms of the components defined in (4.11) as follows

$$
\begin{aligned}
I_2(\boldsymbol{n}_{[2]}) &= \frac{1}{3}(3n_x^2 + n_1^2 + n_1 n_2 + n_2^2)\,, \\
I_3(\boldsymbol{n}_{[2]}) &= \frac{1}{27}(n_1 + 2n_2)(9n_x^2 + 2n_1^2 - n_1 n_2 - n_2^2)\,.
\end{aligned}
\tag{4.14}
$$

The components $c_{x,1,2}$ of the traceless squared refractive index tensor (2nd matrix on the right hand side of (4.13)) become $c_x = n_x(n_1 + 2n_2)/3$, $c_1 = n_1(n_1 + 2n_2)/3$, and $c_2 = n_x^2 - n_2(2n_1 + n_2)/3$. With the help of (4.12) we can immediately write down the nonlinear equation (4.13) in terms of components. In order to evaluate $C_{1,2}(I_2, I_3)$ we need to study a huge number of flow situations, such that $I_{2,3}$ are varied independently. From a practical viewpoint, however, this is an impossible task. For uniaxial elongational flows, for example, both invariants are strictly related to each other. For shear flows, we heavily explore a region far from this 'path'. Next, we make use of the model [175, 180] to obtain an explicit expression for the stress-optic coefficient.

4.8 Stress-Optic Coefficient

This section discusses the stress-optic coefficient based on the thermodynamically admissible reptation model for fast flows of polymer melts [180]. This model includes the effect of chain stretch, double reptation, convective constraint release – ingredients which are known to be important when modeling the rheology of polymer melts, as discussed in [175]. The model uses four degrees of freedom, a unit segment vector on the primitive path ($\boldsymbol{u}$), the segment contour position s, and the stretch variable λ. These quantitites will be defined below. (see also Fig. 4.7)

Stress

The model [175, 180] provides a microscopically (finitely extendable nonlinear chain) inspired expression for the (traceless polymeric) stress tensor of the following form

$$
\boldsymbol{\sigma}_{(2)} = 5G\left(1 + \frac{\lambda_{\max}^2(\lambda^2 - 1)}{\lambda_{\max}^2 - \lambda^2}\right)\boldsymbol{a}_{(2)}
\tag{4.15}
$$

with $G = 3Zn_p k_B T/5$, polymer number density n_p, and, in equilibrium, $\boldsymbol{\sigma}^{\mathrm{eq}}_{(2)} = Zn_p k_B T \mathbf{1} = (5/3)G\mathbf{1}$, and $\boldsymbol{\sigma}^{\mathrm{eq}}_{[2]} = 0$, of course. The underlying model considers a

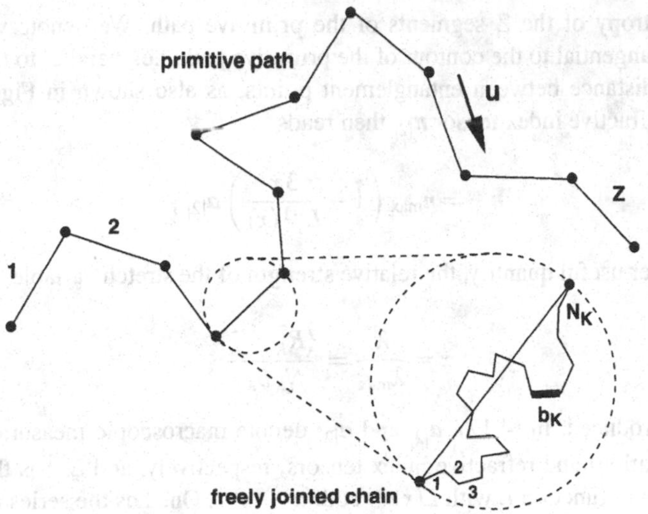

Fig. 4.7. Schematic drawing introducing notation

primitive path of a macromolecule made of Z entanglement segments of equal, but not fixed, length. In order to account for a substructure between adjacent entanglement points along the primitive path one further considers, as depicted in Fig. 4.7, a freely jointed (Kuhn) chain with N_K segments of equal and constant length b_K between each of these entanglement points, as in [178, 179, 181, 182]. The end-to-end vector of a Kuhn chain, which is the segment vector of the primitive path, we denote as $\boldsymbol{R}_K$. An unconstrained Kuhn chain follows Gaussian statistics, i.e., the equilibrium end-to-end distance is $\langle R_K^2 \rangle_{\mathrm{eq}}^{1/2} = \sqrt{N_K}\, b_K$, where b_K is known as Kuhn length and, as well as N_K, determined by the chemistry of the polymer. In equilibrium, but also out of equilibrium, the end-to-end distance determines, according to Fig. 4.7, the mean length l_e of an entanglement segment $l_e \equiv \langle R^2 \rangle^{1/2}$. While we specified already the equilibrium value of l_e, its maximum value $l_e = N_K b_K$ equals the maximum end-to-end distance of the Kuhn chain. The contour length of the primitive path is simply $L = Z l_e$, as inferred from Fig. 4.7. The quantity Z is proportional to the molecular weight and related to the overall number of Kuhn elements N per chain by $Z = N/N_K$. It is then convenient to introduce a dimensionless stretch variable for the primitive path $\lambda \equiv L/L_{\mathrm{eq}} = \langle R_K^2/N_K \rangle^{1/2}/b_K$. It approaches unity in equilibrium, and its maximum value is determined by the number of Kuhn steps,

$$\lambda_{\max} \equiv L_{\max}/L_{\mathrm{eq}} = \sqrt{N_K}\,. \tag{4.16}$$

Birefringence

Kuhn and Grün [181] calculated the optical birefringence for a freely jointed chain with fixed end-to-end distance (l_e in our notation), i.e., eventually out of 'equilibrium'. Their result relates optical anisotropy of the single macromolecule of Fig. 4.7

to the anisotropy of the Z segments of the primitive path. We denote with $\boldsymbol{u}$ the unit vector tangential to the contour of the primitive path, i.e., parallel to the Kuhn's end-to-end distance between entanglement points, as also shown in Fig. 4.7. The deviatoric refractive index tensor $\boldsymbol{n}_{[2]}$ then reads

$$\boldsymbol{n}_{[2]} = n_{\max}\left(1 - \frac{3x}{L^{-1}(x)}\right)\boldsymbol{a}_{[2]} , \tag{4.17}$$

where another useful quantity, the relative strength of the stretch variable,

$$x = \frac{\lambda}{\lambda_{\max}} = \frac{\langle R_K^2\rangle^{1/2}}{N_K b_K} \tag{4.18}$$

has been introduced. In (4.17), $\boldsymbol{a}_{[2]}$ and $\boldsymbol{n}_{[2]}$ denote macroscopic measurable alignment (orientation) and refractive index tensors, respectively, and L^{-1} is the inverse of the Langevin function L with $L(x) = \coth(x) - x^{-1}$. One has the series expansion $L^{-1}(x) = 3x + 9x^3/5 + o[x^5]$ for $x \ll 1$ and $L^{-1}(1) = \infty$. The alignment tensor is obtained as an average over all segments of the primitive path.

In the limit of maximum stretch, where $x = 1$ and where all segments are fully aligned in the same direction, say $\boldsymbol{e}_x$, we have $\boldsymbol{n}_{[2]} = n_{\max}(\boldsymbol{e}_x\boldsymbol{e}_x - \boldsymbol{1}/3)$ since $L^{-1}(1) = \infty$. From (4.15), together with (4.17), we obtain $n_{\max} = (25/3)\lambda_{\max}^2 CG = (25/3)N_K CG$. The coefficient C depends solely on the chemistry of the monomers and is considered as independent on molecular weight, and deformation state of the polymer [179, 181, 183].

Nonlinear Stress-Optical Coefficient

By making use of the expression for the optical birefringence tensor (4.17) the stress tensor (4.15) is readily rewritten in terms of the relative stretch variable $x = \lambda/\lambda_{\max}$ of Sec. 4.8 as

$$\boldsymbol{\sigma}_{[2]} = C_1^{-1}\boldsymbol{n}_{[2]} ,$$
$$C_1^{-1} = C^{-1}\frac{x^2}{(1-x^2)}\frac{3}{5}\left(1 - \frac{3x}{L^{-1}(x)}\right)^{-1} , \tag{4.19}$$

where C is uniquely defined such that the linear SOR with the stress-optical coefficient C holds (i.e. $C_1 = C$ for $\lambda = 1$) and $N_K \gg 1$ has been assumed to arrive at (4.19). The same expression, (4.19), for C_1^{-1} arises from an alternate model summarized in the footnote[2]. Concerning the nonlinear SOR (4.13) we see from (4.19), that these particular models predict $C_2^{-1} = 0$ and $C_1 = C_1(x)$, thus x can be expressed in

[2] Another approach to the orientation and stress in polymer melts has been followed recently [184]. Instead of using stretch and orientation variables as independent quantities, as done in Sect. 2.1 for polymer solutions, the authors start from an equation for the end-to-end distance vector between entanglements $\boldsymbol{R}$. In this section, the average $\langle \boldsymbol{RR}\rangle$ had been decoupled into the separate averages $\boldsymbol{a}_{(2)} = \langle \boldsymbol{RR}\rangle / \langle R^2\rangle$ and $\lambda^2 = \langle R^2\rangle / L_{\mathrm{eq}}^2 = \mathrm{Tr}\,\boldsymbol{g}$ with

terms of the invariants $I_{2,3}$, if a nonlinear generalization of the SOR applies. A good alternate representation of (4.19) to within 1% for all $0 \leq x \leq 1$ is

$$C_1^{-1} \approx C^{-1} \left(\frac{1-2x^2/5}{1-x^2} - c \right) . \tag{4.23}$$

with a tiny correction $c = 3(N_K - 1)^{-1}/5$ which ensures that $C_1 = C$ in equilibrium.

Equation (4.9) offers a crude but useful approximation to the stress in polymer melts. It allows to predict rheological properties for the many chain FENE model, based on a single chain model.

In [144] the degree of stretch and orientation of the polymer chains on different length scales (and 'collective' deformations) have been also measured and analyzed in order to allow for a critical test of alternative pictures which were proposed earlier to describe deviations from the SOR. Upon these models (which have been ruled out) are those which assume stretching of few selected segments, thus leaving the measured anisotropy of chains largely unchanged. Just at a late stage of elongation when segmental stretching leads to a strong increase in σ_{bonded}, local inhomogeneities in bond stretchings/contractions are observed while expression (4.9) remains valid.

Experimentally, flow induced alignment on different length scales is measured via the single chain structure factor S_{sc} (from deuterated samples, definition provided by (5.31)) and flow birefringence or infrared dichroism. While the latter quantities measure the alignment tensor ((4.8), probing the anisotropy of segments), at small wave numbers (Guinier regime), S_{sc} resolves the gyration tensor. cf. Fig. 4.4 for both experimental and FENE chain data for an elongated polymer melt.

$$\boldsymbol{g} \equiv \frac{\langle \boldsymbol{RR} \rangle}{L_{\text{eq}}^2} = \lambda^2 \boldsymbol{a}_{(2)} , \tag{4.20}$$

as in (2.5). The proposed dynamics of $\boldsymbol{g}$ is described by the following equation [184]

$$\frac{d\boldsymbol{g}}{dt} = \boldsymbol{\kappa} \cdot \boldsymbol{g} + \boldsymbol{g} \cdot \boldsymbol{\kappa}^T - \frac{f}{\tau} \overline{\boldsymbol{g}} - \frac{\tilde{g}}{3\tau_R} \mathbf{1} , \tag{4.21}$$

with $f \equiv (\lambda_{\text{max}}^2 - 1)/(\lambda_{\text{max}}^2 - \lambda^2)$, $\tilde{g} \equiv f\lambda^2 - 1$, and

$$\frac{1}{\tau} \equiv \frac{2}{\tau_d} + \left(\frac{1}{\tau_R} - \frac{2}{\tau_d} \right) \frac{\beta g}{1 + \beta g} , \tag{4.22}$$

where $\beta \geq 2$ (we use $\beta = 2$ as in [184]). We can solve (4.21) for $\boldsymbol{g}$, compute λ from its trace (thus also have access to $x = \lambda/\lambda_{\text{max}}$ since λ_{max} is a model parameter) and obtain the stress tensor using (4.17) and (4.19), with $\boldsymbol{a}_{[a]} = \boldsymbol{a}_{(2)} - 1/3$. Accordingly, the equation of change for the trace of $\boldsymbol{g}$, derived from (4.21), does not contain τ, but just the Rouse time τ_R as for the model in the previous section. The stress tensor for this model also coincides with the one used in the current section, (4.15), if $\langle \boldsymbol{uu} \rangle$ is replaced by its analog, $\boldsymbol{g}/\text{Tr}\boldsymbol{g}$. For large N_K, also the equations of change for the stretch variable correspond to each other, where $\tau_R = \tau_s/2$. Hence, the formula for the nonlinear stress-optic coefficient C_1^{-1}, given in (4.19), remains valid. From the analysis of this model, (4.20–4.22), we arrive at the same conclusion – even quantitatively – as for the model outlined in the current section.

4.9 Interpretation of Dimensionless Simulation Numbers

A word of caution concerning the interpretation of dimensionless results is in order. Simulation has to deal with quantities in terms of reference units for mass, length and energy. These have to be obtained by comparing experiment with simulation and provide the basic length (σ) and energy (ε) scale of the Lennard–Jones potential as well as the mass (m) of a bead in solving Newton's equation. Although some freedom exists in how to adjust three dimensionless units, an accepted one is to obtain the reference energy from the measured temperature $\varepsilon_{ref} = Tk_B$, the bead mass from the real N_c divided by the simulated one, and σ^2 from the ratio between measured and simulated end-to-end-distances. Sample data such as reported in Table 4.5 motivates obtaining reference units for any simulated quantity for the study of particular materials. For polyethylene (polystyrene), e.g., we deduce a reference length $\sigma = 5.3(9.7)$ Å, a reference mass $m = 42.3(364)$ g/mol, and a reference energy $\varepsilon/k_B = 443(490)$ K. From m, σ, ε one immediately obtains reference values for any other quantity such as viscosity, time, stress etc. by dimension analysis: $\sqrt{m\varepsilon}/\sigma^2 = 0.07(0.07)$ mPas, $\sigma\sqrt{m/\varepsilon} = 1.8(9) \times 10^{-12}$ s, 40 (7.5) MPa, 0.46 (0.67) g/cm^3, 553 (109) GHz. Corresponding reference values for other polymers are obtained along this procedure. Care has to be taken when predicting quantities which are sensitive to the ratio between the systems longest and shortest relaxation time (τ_{N_c}/τ_1) such as the shear viscosity (proportional) and the shear rate at the onset of shear thinning (inversely proportional). To illustrate this, for polystyrene the simulation predicts the correct zero shear viscosity $\eta_0 = \sqrt{m\varepsilon}/\sigma^2 \Gamma \eta_0^* = 68$ Pas (at $N = N_c$) for a factor $\Gamma = 10^4$ which happens to be equal to the ratio of relaxation times $\tau_{N_c}/\tau_1 = 10^4$. Accordingly, from the onset of shear thinning at shear rate $\dot{\gamma} = 10^{-4}$ obtained for the FENE chain melt at $N = N_c$ (see Fig. 4.1) we predict for the real shear rate (for polystyrene) $\dot{\gamma}_c = \dot{\gamma}^*(\sigma/\Gamma)^{-1}\sqrt{\varepsilon/m} = 1100$ s^{-1} which is again in agreement with experimental findings [164]. As a result, the shear stress at onset of shear thinning is correctly reproduced without adjustment by Γ, i.e., $(\eta_c\dot{\gamma}_c)/(\eta_c^*\dot{\gamma}_c^*) = 7.5$ MPa for polystyrene.

5

Chain Models for Transient and Semiflexible Structures

In order to be prepared for the analysis of the flexible FENE-C (FENE model which allows for scission/recombination), FENE-B (which allows for bending stiffness) and FENE-CB fluids (both bending stiffness and scission) to be discussed below, we summarize results for the configurational statistics of wormlike chains (WLC) in external fields by using the method of functional integrals (FI) in quasimomentum space. From the correlation functions, statistical properties of WLCs, such as gyration radius and scattering functions can be obtained. By varying the bending rigidity the WLC exhibits a crossover from an ideal Gaussian chain to a rodlike chain. Simulations on the WLC model are widely available, see e.g. [185–188].

In 1960 Edwards [189] proposed a continuum model for polymer chains. For the ideal Gaussian chain, the FI can be solved exactly, and after taking excluded volume into account, a perturbation expansion as well as the renormalization group method are used to study the configurational statistics of polymer solutions [190–194].

5.1 Conformational Statistics of Wormlike Chains (WLC)

The wormlike chain (WLC) model was first proposed by Kratky and Porod [195] and extended to the continuum level in [190, 191]. It is described by a statistical weighting factor p for a polymer contour path $\boldsymbol{r}(s)$ with contour position s (imaged as time) $0 \leq s \leq L$:

$$p^{\mathrm{WLC}}(\boldsymbol{r}(s)) \propto \exp\left(-\frac{3}{2l}\int_0^L \boldsymbol{u}^2(s)ds - \frac{\kappa}{2}\int_0^L \dot{\boldsymbol{u}}^2(s)ds\right), \tag{5.1}$$

where L is the contour length of the chain, κ the bending elastic coefficient, $\boldsymbol{u}(s) \equiv \partial\boldsymbol{r}(s)/\partial s$ the differential (tangent) of the curve, and $\dot{\boldsymbol{u}} \equiv \partial\boldsymbol{u}/\partial s$. Using the constraint

$|\boldsymbol{u}(s)| = 1$ a series solution for the tangent distribution (Green's function) $G(\boldsymbol{u},\boldsymbol{u}';L,0)$ has been derived in [190]. Releasing the constraint and considering stretchable chains, end-to-end distances and the tangent distribution have been derived by using the method of Feynman [191, 196]. Later it turned out that functionals in momentum space often used in field theories are a convenient method of studying properties of WLCs [31]. For a uniform system, the configurational statistics of WLCs can be accessed by considering the correlation function

$$\begin{aligned} C(\boldsymbol{R}_1,\boldsymbol{r}_2;s_1,s_2) &\propto \langle \delta(\boldsymbol{r}(s_1)-\boldsymbol{R}_1)\delta(\boldsymbol{r}(s_2)-\boldsymbol{R}_2)\rangle \\ &\propto \langle \delta(\boldsymbol{r}(s_1)-\boldsymbol{r}(s_2)-\boldsymbol{R})\rangle \propto C(\boldsymbol{R};s) \end{aligned} \tag{5.2}$$

where $\boldsymbol{R} = \boldsymbol{R}_1 - \boldsymbol{R}_2$, $s = s_1 - s_2$, $0 \le s_1, s_2 \le L$ and $\langle..\rangle$ denotes a statistical average over various configurations of the chain by FI. The correlation function (5.2) is actually more fundamental than the end-to-end functions for WLCs [197], caused by chain end effects, except in the limit of Gaussian chains ($\kappa \to 0$).

5.1.1 Functional Integrals for WLCs

We consider a polymer chain which is described by a three-dimensional curve $\boldsymbol{r}(s)$ with $0 \le s \le L$. For convenience, the infinite long chain limit is taken then the normal mode coordinate, i.e., the Fourier transformation of $\boldsymbol{r}(s)$ is obtained as [198] $\boldsymbol{r}(s) = 1/\sqrt{2\pi}\,\hat{\boldsymbol{r}}(k)e^{iks}\,dk$, satisfying $\hat{\boldsymbol{r}}(k) = \hat{\boldsymbol{r}}^*(-k)$ because $\boldsymbol{r}(s)$ is real. The statistical weighting factor $p^{\mathrm{WLC}}[\hat{\boldsymbol{r}}(k)]$ for the WLC is, according to (5.1),

$$p^{\mathrm{WLC}}[\hat{\boldsymbol{r}}(k)] \propto \exp\left(-\frac{3}{2l}\int k^2\hat{\boldsymbol{r}}^2(k)dk - \frac{\kappa}{2}\int k^4\hat{\boldsymbol{r}}^2(k)dk\right) . \tag{5.3}$$

Physical properties X are obtained by FI in the quasimomentum space:

$$X = \int \mathcal{D}[\hat{\boldsymbol{r}}(k)]X[\hat{\boldsymbol{r}}(k)]\,p[\hat{\boldsymbol{r}}(k)] , \tag{5.4}$$

where $\int \mathcal{D}[\hat{\boldsymbol{r}}(k)]$ denotes the FI [199]. With regard to the correlation function (5.2) one has

$$\begin{aligned} &\delta((\boldsymbol{r}(s)-\boldsymbol{r}(0))-\boldsymbol{R}) = \\ &\left(\frac{1}{2\pi}\right)^{\frac{3}{2}}\int_{-\infty}^{\infty}\exp\left(i\boldsymbol{w}\cdot\left[\frac{1}{\sqrt{2\pi}}\int_{-\infty}^{\infty}\hat{\boldsymbol{r}}(k)(e^{iks}-1)dk-\boldsymbol{R}\right]\right)\mathrm{d}^3w \end{aligned} \tag{5.5}$$

and the tangent of the curve at contour position s reads

$$\boldsymbol{u}(s) = (\sqrt{2\pi})^{-1}\int_{-\infty}^{\infty} ik\hat{\boldsymbol{r}}(k)\exp^{iks}dk . \tag{5.6}$$

Using standard methods [198, 200], one obtains for the correlation function (5.2) for WLC from (5.3)

$$C(\boldsymbol{R},\boldsymbol{0};s,0) = \int \mathcal{D}[\hat{\boldsymbol{r}}(k)]\delta((\boldsymbol{r}(s)-\boldsymbol{r}(0))-\boldsymbol{R})\, p^{\text{WLC}} \propto \exp(-R^2/4\Gamma_1)\,, \tag{5.7}$$

where $\Gamma_1 = l\{s-\alpha^{-1}(1-e^{-s\alpha})\}/6$, $\alpha^2 \equiv 3/(\kappa l)$, and therefore (5.7) simplifies to the expression $\exp\{-3R^2/(2ls)\}$ for ideal Gaussian chains. There is a variety of related correlation functions which have been discussed [31]. For example, one may consider the adsorption on a surface where the polymer has a fixed orientation $\boldsymbol{U}_0$ at $\boldsymbol{r}(0)$. The orientation distribution function of the tangent $\boldsymbol{U}$ at position position s becomes

$$C(\boldsymbol{U};s) = \int \mathcal{D}[\hat{\boldsymbol{r}}(k)]\delta(\boldsymbol{U}(s)-\boldsymbol{U})\, p^{\text{WLC}}[\hat{\boldsymbol{r}}(k)] \propto \exp[-U^2/4\Gamma 2]\,, \tag{5.8}$$

independent of s due to translational invariance.

5.1.2 Properties of WLCs

From (5.7) we calculate the average monomer-monomer distance (MMD) separated by the contour distance s for WLC

$$\langle R^2\rangle(s) = l\left(s-\frac{1}{\alpha}[1-e^{-s\alpha}]\right), \quad \langle R^4\rangle(s) = \frac{5}{3}l^2\left(s-\frac{1}{\alpha}[1-e^{-s\alpha}]\right)^2. \tag{5.9}$$

Equation (5.9) are also obtained in [197, 198] and differentiate from the average end-to-end distance obtained in [191]: $\langle R^2\rangle(L) = l\{L-(2\alpha)^{-1}\tanh(L\alpha)\}$, which demonstrates the difference between basic end-to-end and correlation functions through an end-effect. In order to patch up the difference, an additional term describing the end effect has been added to the Hamiltonian in [197]. For Gaussian chains, i.e., $\alpha \to \infty$, one recovers from (5.9): $\langle R^2\rangle = lL$, and for the opposite limit of rodlike chains, i.e., $\kappa \to \infty, \alpha -> \infty$ the WLC at first glance give incorrect results and in order to make the model valid, an additional condition of the average length of the chain being L should be used, i.e., as discussed in detail by Freed [191], let $\int_0^L d\tilde{s} = \int_0^L \langle(\boldsymbol{u}(s)\cdot\boldsymbol{u}(s))^{1/2}\rangle\, ds = L$, where $d\tilde{s}$ is differential arc length. Then we will obtain constraint on the parameters, l and κ, by (5.8)

$$\langle u^2\rangle = \frac{\int u^2 G(\boldsymbol{u};L)d\boldsymbol{u}}{\int G(\boldsymbol{u};L)d\boldsymbol{u}} = 6\Gamma_2 = \frac{3l}{4\kappa} = 1\,, \tag{5.10}$$

being equivalent to $l = 4\kappa/3$. For example, if κ is selected as the independent parameter l will depend on κ and will have a meaning of an effective monomer length Kuhn length!. Another reasonable constraint can be obtained from $\langle|\boldsymbol{u}|\rangle = 1$ which leads to $l = 3\pi^2\kappa/16$. A different is derived by Freed [191] ($l = \kappa/3$ obtained from the end to end tangent distribution function, and in [197], $l = (4/3)\kappa$ is derived by taking a limit on (5.7). Substituting $l = 4\kappa/3$ into (5.9) we have

$$\langle R^2\rangle = l\{L-l(1-e^{-2L/l})/2\}\,, \tag{5.11}$$

and l is proportional to persistence length (see below). For $\kappa \to \infty$ we now properly obtain the result for a rodlike polymer $\langle R^2\rangle = L^2$.

Persistence Length

The persistence length l_p for finite contour length is obtained along the same line using the definition:

$$l_p \equiv \frac{\int R \cos\vartheta\, C(\boldsymbol{R},\boldsymbol{0},\boldsymbol{U}_0;s,0)\mathrm{d}^3R\,\mathrm{d}^3U_0}{(\int C(\boldsymbol{R},\boldsymbol{0},\boldsymbol{U}_0;s,0)\mathrm{d}^3R\,\mathrm{d}^3U_0)}\,, \tag{5.12}$$

i.e., $l_p = \phi_1 \langle|\boldsymbol{U}_0|\rangle \Gamma_2^{-1}$ and therefore $l_p = [1-\exp(-s\alpha)]/\alpha$, where α is given after (5.7), which is similar to the result of Porod-Kratky [190]. For $s \to \infty$ one has $l_p = \alpha^{-1} = (2/3)\kappa = l/2$.

Radius of Gyration

For the radius of gyration, defined as $R_G^2 = \frac{1}{2L^2}\int_0^L ds \int_0^L ds' \langle (R(s)-R(s'))^2\rangle$, we obtain, by making use of (5.9)

$$R_G^2 = \frac{lL}{6} - \frac{l^2}{4} + \frac{l^3}{4L^2}\left[L - \frac{l}{2}(1-e^{-2L/l})\right]. \tag{5.13}$$

For $\alpha \to \infty$ (5.13) becomes $R_G^2 = lL/6$, which is just the ideal Gaussian chain radius of gyration. When $\alpha \to 0$, using $l = 4\kappa/3$ we have $R_G^2 = \alpha l L^2/24 = L^2/12$ which is just the expected result for a rodlike polymer. But there is notable peculiarity in the statistics when approaching the rodlike limit, as will be seen from the scattering function.

Scattering Function

In order to compare the result for the WLC with the ones for ideal Gaussian chains and rodlike chain, let us write down the corresponding isotropic scattering functions, for the Gaussian chain

$$I(x) = N\frac{2}{x^2}(x-1+e^{-x})\,, \tag{5.14}$$

where $x \equiv k^2 R_G^2$, and for the rodlike polymer,

$$I(x) = L^2\frac{1}{6x}\{2\sqrt{3x}\,\mathrm{Si}(2\sqrt{3x}) + \cos(2\sqrt{3x}) - 1\}\,, \tag{5.15}$$

where $\mathrm{Si}(x) = \int_0^x (\sin(t)/t)dt$.

The scattering function for WLC is obtained from the Fourier transform of the correlation function $C(\boldsymbol{k},\boldsymbol{U},\boldsymbol{U}_0;L)$ and gives

$$I(k) = 2N/(L^2)\int_0^L (L-s)\exp\{-k^2 l[s-(1-\exp(-s\alpha)/\alpha]/6\}ds\,. \tag{5.16}$$

If we let $\kappa \to 0$, we see that the Gaussian limit is reobtained. But if we let $\kappa \to \infty$, this doesn't lead to the above $I(x)$ for rodlike chains. For that reason, the demonstrated approach leads to a so called Gaussian rodlike polymer' for $\kappa \to \infty$. Properties of the presented model have been also worked out for the case of WLC in external fields [31]. Finally, we mention a difference between the approaches discussed here and the one by Saito et al. [190]. We obtain

$$\langle \boldsymbol{u}(s) \cdot \boldsymbol{u}(s') \rangle \approx \begin{cases} 1 - a^{-1} \approx \mathrm{e}^{-\frac{2}{3}\frac{|s-s'|}{l_p}}, |s-s'| \ll l_p \\ a^{-1}/3 \approx \mathrm{e}^{-\frac{|s-s'|}{l_p}}, \quad |s-s'| \gg l_p, \end{cases} \tag{5.17}$$

which means, that for two segments far from each other these two models are consistent.

For molecules whose intrinsic rigidity against twist is important to interpret results the statistics to be presented for WLC had been extended to chiral ribbons [201].

5.2 FENE-C Wormlike Micelles

Aqueous surfactant solutions are known to form wormlike micelles under certain thermodynamic conditions, characterized by surfactant concentration, salinity or temperature In the semi-dilute solution regime these linear and flexible particles, with persistence lengths varying from 15 to 150 nm form an entangled viscoelastic network. In equilibrium their behavior is analogous to that of polymer solutions and their properties obey the scaling laws predicted for the semi-dilute range [202]. See [203] for the prediction of more general surfactant microstructures (such as bilayers), their shapes, and shape fluctuations. In contrast to ordinary polymers, wormlike micelles can break and recombine within a characteristic time (breaking time) and their length distribution is strongly affected by flow. Phenomena such as shear banding structures, the variety of phase transitions and thixotropy are not completely understood [179]. This section contributes to this debate with a mesoscopic concept. There is huge number of both macroscopic and microscopic models available which deals with the prediction of the wormlike micellar phase, or a full phase diagram, changes in topology, etc. To summarize these works is certainly outside the scope of this monograph (see, e.g. the book by Gelbart, Ben-Shaul and Roux [27]). For a review on simulations of self-assembly see [28].

Wormlike micelles, with certain similarities to equilibrium polymers [204] can be modeled on a mesoscopic scale which disregards amphiphilic molecules and their chemistry by a modified version of the FENE potential which allows for scissions and recombinations of worms, the so called 'FENE-C'(ut) for which the ıPotential!FENE-C connector force between adjacent beads is parameterized by Q_C:

$$\boldsymbol{F}^{\text{FENE-C}}(r) = \boldsymbol{F}^{\text{FENE}}(r), \quad r \le Q_C\,, \tag{5.18}$$

and $\boldsymbol{F}^{\text{FENE-C}} = 0$ for $r \ge Q_C$ with a rather irrelevant smooth interpolation at Q_C [205–207]. FENE-C reduces to FENE for $Q_C = Q_0$ and Q_C is trivially related to the scission energy (energy barrier for scission). In this section we will analyze this model both numerically (via NEMD) and analytically. The analytic model is based on an expression for the free energy of Gaussian chains, modified by a term which takes into account a finite scission energy in order to describe micelles, and extended to flow situations. In equilibrium, the length distribution then depends on two parameters, namely the micellar concentration and the scission energy. The shape of this distribution has a significant influence on flow alignment and the rheological behavior of linear micelles. The analytic approach to be discussed first exhibits similarities to the calculation of products in polymerization kinetics and to association theory [4, 208, 209]. Results will be compared with the exact numerical solution in Sect. 5.2.3. The example in the next section has been chosen for illustrative purpose. Shear thickening rather than thinning occurs for a wide range of micellar systems, cf. [210, 211] which is also obtained via a modified FENE-C which includes bending stiffness (FENE-CB models) and allows for the formation of networks. Recently, shear-thickening has been observed for the FENE-Cx model (implemented as transient polymer network, ie. 'FENE-C∞'), where instead of bending additional FENE-C cross-linkers were used [212].

5.2.1 Flow-Induced Orientation and Degradation

Consider an ideal solution of linear chains (micelles) being modeled as bead-spring chains. We assume that each bead can have two bonds and we exclude ring formation. We consider a total number of N_b beads at (micellar) concentration c, where a bead represents a number of chemical units as already discussed in this monograph. Let $N_M \equiv cN_b$ denote the number of beads able to form linear chains ('M-beads') and which can associate and dissociate, and $N_S \equiv (1-c)N_b$ the number of solvent particles ('S-beads'). The system is then characterized by the number n_i of micellar chains made of i beads and c. At equilibrium the distribution of chains results from the grand canonical partition function

$$\Xi = \sum_{n_1=0}^{\infty} \cdots \sum_{n_N=0}^{\infty} (q_1\lambda_1)^{n_1}(n_1!)^{-1} \ldots (q_N\lambda_N)^{n_N}(n_N!)^{-1} = \prod_{i=1}^{N} e^{q_i\lambda_i}, \qquad (5.19)$$

where q_i and μ_i are the partition function and activity, respectively, of an i-chain ('subsystem' i), $\lambda_i = \exp(\beta\mu_i)$, and $\beta = 1/(k_BT)$. For the average number of i-chains one has $\langle n_i\rangle = \lambda_i \partial(\ln\Xi)/\partial\lambda_i = \lambda_i q_i$. Let us require that the various subsystems are in a chemical equilibrium with each other, i.e., $\mu_i = i\mu_1$. Thus, with $\lambda \equiv \lambda_1$, we have $\langle n_i\rangle = \lambda^i q_i$. For an i-chain the Hamiltonian $\mathcal{H}$ is formulated in terms of momentum and space coordinate of the center of mass, $\boldsymbol{p}_c$ and $\boldsymbol{r}_c$, respectively, and $i-1$ internal momenta and coordinates $\boldsymbol{P}_k, \boldsymbol{Q}_k$ with $(k = 1,\ldots,i-1)$. We choose the internal coordinates such that $\boldsymbol{Q}_k$ denotes the kth bond vector between beads k and $k+1$. Carrying out the integration over momenta (Maxwell distribution) and coordinates yields $\int \exp(-\beta\mathcal{H})\, d\boldsymbol{p}_c d\boldsymbol{P}^{i-1} d\boldsymbol{r}_c d\boldsymbol{Q}^{i-1} = (2\pi m k_B T)^{3i/2} V q_i^{\rm int}$, where m is the mass of a single bead and V is the total volume of the solution, $q_i^{\rm int}$ denotes the internal configurational integral, and we can write $q_i = V q_i^{\rm int} \Lambda^{-3i}$, with the thermal de Broglie wavelength of a bead Λ. In order to simplify the structure of the following equations we equal the masses of M- and S-beads. For the calculation of the configurational integral we introduce a configurational distribution function f. The configurational integral is related to the free energy via $q_i^{\rm int} = \exp(-\beta A_i^{\rm int})$, with $A_i^{\rm int} = \int d\boldsymbol{Q}^{i-1} f_i (k_B T \ln f_i + U_i)$, where U_i denotes the internal energy of an i-chain.

In order to keep this example simple, we assume Gaussian distributions, i.e.

$$f_i(\boldsymbol{Q}^{[i-1]}) = \frac{1}{(2\pi)^{3(i-1)/2}} \frac{1}{|\boldsymbol{C}_{i-1}^{-1}|^{1/2}} \times \exp\left(-\frac{1}{2} \boldsymbol{Q}^{[i-1]} \cdot \boldsymbol{C}_{i-1}^T \cdot \boldsymbol{Q}^{[i-1]} \right), \qquad (5.20)$$

with $\boldsymbol{Q}^{[i-1]} \equiv (\boldsymbol{Q}_1, \boldsymbol{Q}_2, \ldots, \boldsymbol{Q}_{i-1})$. The $3(i-1) \times 3(i-1)$ matrix of covariances is given by $\boldsymbol{C}_{i-1}^{-1} = \langle \boldsymbol{B}_i \rangle$ with $(\boldsymbol{B}_i)_{\mu\nu} \equiv \boldsymbol{Q}_\mu \boldsymbol{Q}_\nu$ $(\mu, \nu = 1, .., i-1)$ and $|\cdots|$ denoting a determinant. The tensor $\boldsymbol{B}_i$ becomes anisotropic under flow conditions. In the 'slow reaction limit' in which changes in micellar size occur on a time scale long compared to orientational diffusion of the segments in presence of flow, one can assert that the deformation energy can be added to the micellar free energy [213]. The internal energy of i-chains is then given by

$$U = -(i-1)E_{\rm sc} + \frac{1}{2} \sum_{j=1}^{i-1} H \langle Q_j^2 \rangle , \qquad (5.21)$$

where $E_{\rm sc}$ is the scission energy, i.e. $E_{\rm sc}$ is the energy required to break a chain (independent of its length, for a more general case see [214]). For the moment we consider in (5.21) the FENE-regime where bond stretching is not relevant which is especially reasonable for FENE-C chains for which $Q_{\rm C}$ is considerably smaller than Q_0. Inserting (5.20, 5.21) into the above integral expression for the free energy and performing the integration yields

$$A_i^{\rm int} = -\frac{3}{2}(i-1)k_B T(1+\ln(2\pi)) - \frac{1}{2} k_B T \ln|\langle \boldsymbol{B}_i \rangle| - (i-1)E_{\rm sc} + \frac{H}{2} \sum_{j=1}^{i-1} \langle Q_j^2 \rangle \, . \qquad (5.22)$$

and, as such, is similar to an expression given by Booij [215]. Note that the last term on the rhs is proportional to the trace of the pressure tensor for an i-chain within the Rouse model, $H \sum_{j=1}^{i-1} \langle Q_j^2 \rangle = V \mathrm{Tr}(\boldsymbol{P}_i)$. Strict usage of the above relationships leads to

$$\langle n_i \rangle = V \left(\frac{\lambda}{\Lambda^3} \right)^i (2\pi)^{3(i-1)/2} |\langle \boldsymbol{B}_i \rangle|^{1/2} \, e^{\left((i-1)(\beta E_{\rm sc} + \frac{3}{2}) - \beta \frac{V}{2} \mathrm{Tr}(\boldsymbol{P}_i)\right)} \, . \qquad (5.23)$$

This expression provides a basis to analyze the length distribution for both equilibrium and nonequilibrium states. One can evaluate (5.23) in equilibrium by making use of expressions resulting from the Rouse model [4, 216–218]. The number of i-chains is then given by $\langle n_i \rangle_0 = V(\lambda/\Lambda^3)^i z^{i-1}$, where

$$z \equiv (2\pi)^{3/2} \, | \langle \boldsymbol{Q}\boldsymbol{Q} \rangle_0 \, |^{1/2} \, e^{\beta E_{\rm sc}} = \left(\frac{2\pi a^2}{3} \right)^{3/2} e^{\beta E_{\rm sc}}, \qquad (5.24)$$

inherits the scission energy and represents an apparent volume of a bead. For the number density of micellar i-chains $\rho_i \equiv \langle n_i \rangle_0 / V$ we arrive at $\rho_i = \rho_1^i z^{i-1}$. Through the constraint of conserved total density of beads $\rho = N_b/V$ the density ρ_1 of 1-chains can be expressed in terms of the concentration c and z in (5.24) by using rules for geometric series as

$$\rho c \equiv \sum_{i=1}^{N} i\rho_i = \frac{\rho_1}{(1-\rho_1 z)^2} \,. \tag{5.25}$$

5.2.2 Length Distribution

The length distribution in equilibrium is thus determined by the scission energy and concentration. and may also be rewritten in exponential form, $\langle n_i\rangle_0 / \langle n_{i-1}\rangle_0 = \rho_i/\rho_{i-1} = \rho_1 z$. The normalized equilibrium distribution function $C_0(L)$ of L-chains is then equivalent to the expression derived by Cates [219] and reads

$$C_0(L) = \frac{1}{\langle L\rangle_0} \exp\left(-\frac{L}{\langle L\rangle_0}\right) . \tag{5.26}$$

From $\rho_i = \rho_1^i z^{i-1}$ we obtain the average equilibrium length (number of beads) of the micelles $\langle L\rangle_0 \equiv \sum_{i=1}^N i\rho_i/(\sum_{i=1}^N \rho_i) = (1-\rho_1 z)^{-1}$. Solving for ρ_1 leads to the relation $\langle L\rangle_0^2 - \langle L\rangle_0 = z\rho c$, which – itself – is solved (for positive lengths L) by

$$\langle L\rangle_0 = \frac{1}{2} + \left(\frac{1}{4} + z\rho c\right)^{1/2} . \tag{5.27}$$

For a simple fluid which is, within this framework, modeled by an infinitely large negative scission energy (FENE limit) we obtain the correct result $\langle L\rangle_0 = 1$ which we call a generalization of the square root dependence obtained earlier. The generalization is important in reproducing the results from the microscopic model as well as to describe experimental results, for which at low concentrations the dependence of the micellar length on concentration seems to be quite weak.

For the case of FENE-C chains with Q_c close to Q_0 expressions become slightly more complicated, cf. [30]. More precisely, the ratio ρ_i/ρ_{i-1} increases weakly with i and therefore the length distribution $C_0(L)$ decreases weaker than exponentially with L. The concentration dependence of the average micellar length $\langle L\rangle_0$ is more pronounced than the square root behavior given in (5.27). The formalism presented also allows, for FENE-C chains, to calculate the variation of the length distribution with the flow rate, but the treatment becomes considerably more lengthy due to correlations between the bond vectors and the dependence of the pressure tensor on flow parameters.

Results presented in the figures have been obtained numerically using the above 'algorithm' (an extended version can be found in [30]). The second moment $\langle \boldsymbol{QQ}\rangle$ becomes anisotropic, the covariance matrix $|\langle \boldsymbol{B}_i\rangle|$ represents the shear induced orientation of segments. The concentration c is obtained numerically by the summation in (5.25). Varying the shear rate a maximum in the distribution of micellar lengths $C(L)$ occurs, which shifts to shorter chain length with increasing shear rate. Additionally the distribution becomes less broad with increasing rate. The flow alignment angle χ is expressed through the viscosities (assuming validity of the SOR) by

$\chi = \pi/4 + \tan^{-1}[\dot{\gamma}\Psi_1/(2\eta)]$. We evaluate the material quantities such as the shear viscosity η from expressions involving $|\langle \boldsymbol{B}_i \rangle|/|\langle \boldsymbol{B}_i \rangle_0|$ and $\rho_1(\dot{\gamma})$ [30]. It turns out that even for high scission energies the alignment angle does not decrease with increasing shear rate towards zero, because, in opposite to 'classical' polymers, here the average length of chains decreases implying a flow alignment angle which is just moderatly decreasing.

A simplified approach to the analytic treatment of the FENE-C model subjected to flow may neglect the variation of the determinant of the covariance matrix with the shear rate, as done in [4, 215] for (classical=non-breakable) polymers, The approximation is justified by the fact that the determinant is of the order of $\dot{\gamma}^{1/2}$ which is small compared with the exponential of the trace of the pressure tensor. From the approximation follows an increase of the scalar pressure $p = \mathrm{Tr}(\boldsymbol{P}_i)/3$ with shear rate $\dot{\gamma}$, i.e. $\partial p/\partial \dot{\gamma} > 0$ which influences the given result (5.23) as if one would decrease the scission energy (see 5.22, 5.23). A decrease of that energy is connected with a decrease of the average length according to (5.24, 5.27) and hence with a decrease of the viscosity [4].

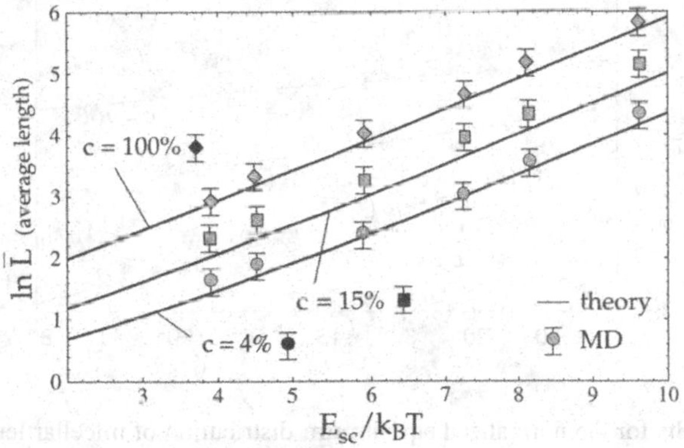

Fig. 5.1. MD results for average micellar length $\langle L \rangle_0$ vs. the scission energy βE_{sc} for FENE-C micellar solutions (from 4% to 100%) in equilibrium. *Lines*: the mesoscopic result (5.27). The fit is parameter-free

5.2.3 FENE-C Theory vs Simulation, Rheology, Flow Alignment

Let us now compare the predictions of the non-simplified analytic model described in the foregoing sections with NEMD simulation results for the full FENE-C model (temperature is kept constant at $k_BT/\varepsilon = 1$, cutoff radius of the FENE-C potential chosen as $R_0 = 1.13\sigma$ implying $E_{sc} = 8.09$, bead number density $n = 0.84$). Results can be compared without any remaining adjustable parameter, see Figs. 5.1, 5.2, and 5.3 As can be seen clearly from Fig. 5.2 only the dependence of average length $\langle L \rangle$ divided by E_{sc} (representation motivated by (5.27)) on concentration is not in

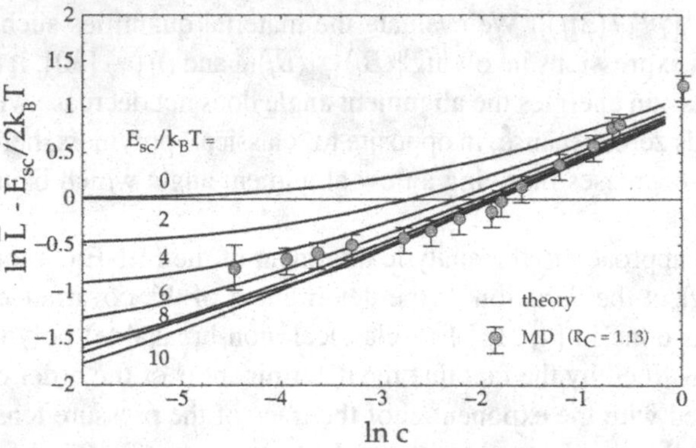

Fig. 5.2. MD results for the average micellar length $\langle L \rangle_0$ (reduced form) vs. concentration c as compared with the mesoscopic result (5.27). The expression of Cates [219] predicts a constant slope in this representation

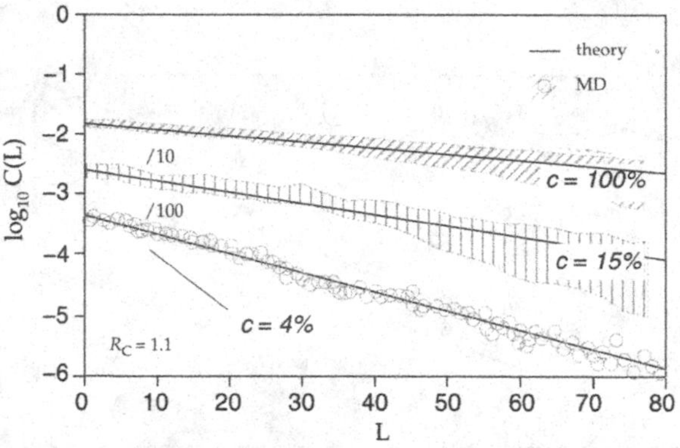

Fig. 5.3. MD results for the normalized equilibrium distribution of micellar length $C_0(L)$ for three samples at different concentrations c. *Lines*: the mesoscopic result (5.26) with same parameters as for the microscopic model

ideal agreement, but a tendency to a small slope at low concentrations is obvious. The slope at high concentrations is around 0.8 for the systems studied here. All other – nonequilibrium – quantities shown in Figs. 5.4–5.7 are described well.

Through (5.23) a phase separation between the short chain and long chain systems can be expected if the sign of $\partial p/\partial\dot{\gamma}$ depends on the length of chains as it has been detected for the microscopic FENE chain melt in [17]. Various hints for such a phase separation exist, e.g., under shear, a shear banding structure has been observed by one of us [220]. Theoretical studies on the latter phenomenon have been already performed [221–223].

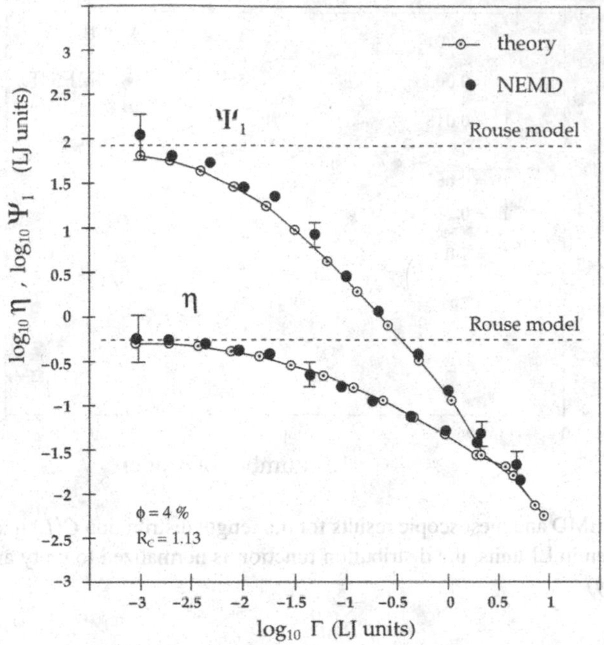

Fig. 5.4. Both NEMD and mesoscopic results for the non-Newtonian shear viscosity η, the viscometric function Ψ_1 vs the dimensionless shear rate Γ. All quantities are given in Lennard–Jones (LJ) units

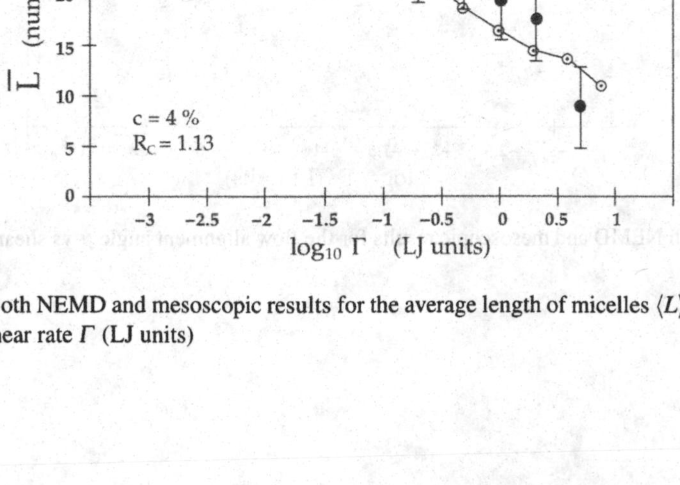

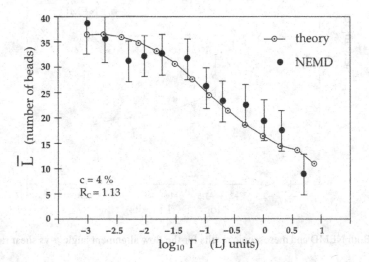

Fig. 5.5. Both NEMD and mesoscopic results for the average length of micelles $\langle L \rangle$ vs dimensionless shear rate Γ (LJ units)

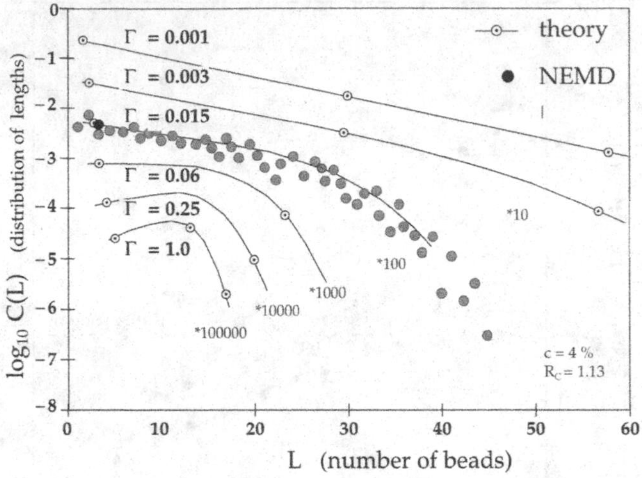

Fig. 5.6. Both NEMD and mesoscopic results for the length distribution $C(L)$ under shear (the shear rate is given in LJ units, the distribution function is normalized to unity and shifted for reasons of clarity)

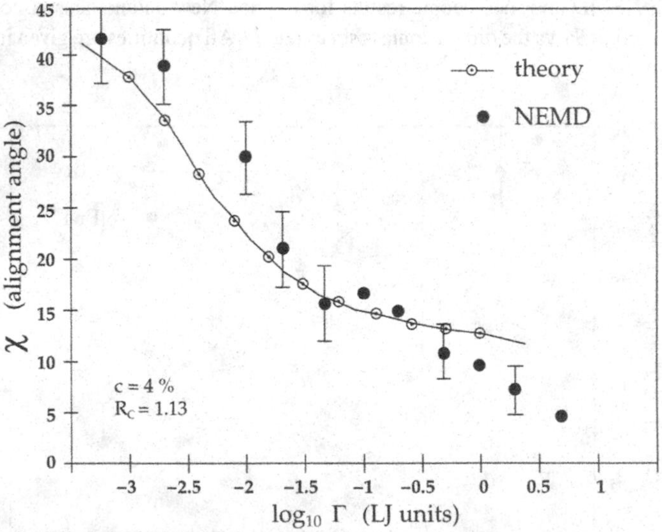

Fig. 5.7. Both NEMD and mesoscopic results for the flow alignment angle χ vs shear rate (LJ units)

5.3 FENE-B Semiflexible Chains

Polymerized actin (F-actin) plays an essential role in cell mechanics and cell mobility, and is an attractive model for studying the fundamental physical properties of semiflexible polymers. Monomeric actin (G-actin, relative molecular mass $M_r = 42,000$) polymerizes in physiological salt solutions (pH 7.5, 2 mM $MgCL_2$, 150 mM KCl) to double-stranded filaments (F-actin). The F-actin solutions usually exhibit a polydisperse length distribution of 4–70 μm with a mean length of 22 μm. F-actin filaments have been extensively studied by Sackmann et al. Details about its physics and biological function can be obtained from [224, 225], its role as model polymer for semiflexible chains in dilute, semidilute, liquid crystalline solutions [226] and also gels [227] has been recently discussed. Bio-molecular dynamics simulations have been also reviewed by Berendsen [228].

Our goal is to demonstrate, that the simple FENE-B model defined through its intramolecular bending (5.32) and FENE (2.1) potential (with $R_C = R_0$ in order to prevent chain breaking) plus the WCA potential for interactions between all beads allows for a rather efficient study of semiflexible model actin filaments at arbitrary concentrations and subjected to external fields on a coarse-grained level, i.e. in particular simple compared with dynamic rigid-rod models and atomistic MD. This is so since it is impossible to keep constraints exactly within a numerical approach, and approximative methods are 'expensive'. Moreover, even actin filaments are stretchable, and conformations of FENE chains share a fractal dimension $d_f = 1$ with non-stretchable (line) models. Gaussian chains and random walk conformation, in the opposite, are inappropriate models for actin since they belong to a class of fractal dimension $d_f = 2$.

If the model is restricted to the formation of linear molecules, the model serves to study linear actin filaments, if this restriction is released, we are going to model semiflexible networks. Notice also similarities with the case of flexible (linear and branched) micelles, for which FENE-C and FENE-CB models are used in the study of linear and branched micelles, respectively. For reviews discussing the relevant aspects in the formation of flexible and stiff networks and their mechanical properties we refer to [226, 229–231]. Semiflexible block copolymers have been studied for a FENE model in [232].

5.3.1 Actin Filaments

Actin filaments can be regarded as classical wormlike chains which are shorter or comparable in length with their persistence length. Further to Sect. 5.1 we mention the result for the radial distribution function $C(\boldsymbol{R};L)$ of the end-to-end vector [31, 185, 233] in the extreme limit of relatively stiff filaments:

$$C(\boldsymbol{R};L) \approx \ell_p (AL^2)^{-1} f(\ell_p \{1 - \|\boldsymbol{R}\|/L\}/L) , \tag{5.28}$$

with

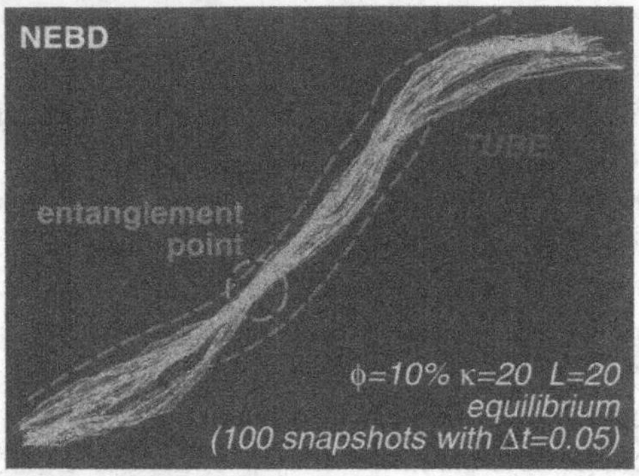

Fig. 5.8. Transient contours of a single FENE-B actin filament with 100 beads embedded in a semidilute solution

$$f(x) = \begin{cases} \frac{\pi}{2}\,\mathrm{e}^{-\pi^2 x}, & x > 0.2 \\ \frac{x^{-1}-2}{8(\pi x)^{\frac{3}{2}}}\,\mathrm{e}^{-1/4x}], & x \leq 0.2 \end{cases} \tag{5.29}$$

and a normalization factor A close to 1 according to [185]. The result is valid for $L \leq 2\ell_p, x \leq 0.5$ and (space dimension) $d = 3$.

For actin filaments, concentration c is usually given in units of mass per volume, whereas theoretical and simulation studies prefer to deal with concentrations $\tilde{c}$ in units of length per volume. The relevant regime is $c \approx 1\,\mathrm{mg/ml}$. Since for the weight of actin one has $370 \times 43\,\mathrm{kD/\mu m} = 2.64 \times 10^{-11}\,\mathrm{\mu g/\mu m}$, a solution with the desired concentration c contains $3.8 \times 10^{10}\,\mathrm{\mu m/ml} = 38\,\mathrm{\mu m/\mu m^3}$, i.e., we are interested in systems with $\tilde{c} \approx 10 - 100\,\mathrm{\mu m/\mu m^3}$. For simplicity, considering a cubic (equidistant) lattice with lattice spacing ξ_l we have: $\xi_l = \sqrt{3/\tilde{c}} \approx \sqrt{0.1} \approx 0.3\,\mathrm{\mu m}$. A minimum estimate for the length of a segment of the multibead FENE-B chain a should be $\xi_l \approx 5a$, and the segment (or bead number) concentration n to be used in the simulation of FENE-B filaments is $n = \tilde{c}/a \approx 5\tilde{c}/\xi_l = 5\tilde{c}^{3/2}/\sqrt{3}$. Concerning the system size, if we need to study a realistic regime, where the length L of filaments is $L \approx 5\,\mathrm{\mu m}$, and the box size is twice the contour length, the total number of beads is $40L^3\tilde{c}^{3/2}/\sqrt{3}$. For the desired concentration of about $1\,\mathrm{mg/ml}$, we arrive at a large number. The system should contain $8 \times 125 \times 5 \times 135 \approx 7 \times 10^5$ beads. The situation is better – from the viewpoint of number of particles – for a minimum (still revelevant) concentration of $0.1\,\mathrm{mg/ml}$, for which 20000 beads are sufficient.

Restrictions for the chain dynamics within an entangled polymer solution can be demonstrated by comparing the transient contours of a free actin filament with the ones of an actin filament embedded in semidilute solution. A decrease of the amplitudes for the thermally excited undulations is measured for the embedded filament, see Fig. 5.8 for an animation of our NEMD computer simulation result. The restricted chain motion can be understood in terms of of the undulations of a filament in a tube formed by the surrounding entangled filaments, and allows to determine its local diameter by measuring the maximum flicker amplitudes: Let y_i denote the local axes of the tube at the two ends ($i = 1,2$). The reptation diffusion coefficient along the tube, $D_{\|}$, according to [225], can be determined by evaluating the random fingering motion of the chain ends. If the chain end positions (x_i, y_i), with respect to a local coordinate system with y-axes parallel to the tube axes at the ends are recorded at fixed time intervals Δt, $D_{\|}$ is determined as the arithmetic mean of the diffusion coefficients parallel to the tube at both ends according to $D_{\|} = (A\Delta t)^{-1} \sum_{i=2}^{N_{\rm steps}} (y_1^i - y_1^{i-1})^2 + (y_2^i - y_2^{i-1})^2$, where $A \equiv 4(N_{\rm steps} - 1)$, and $N_{\rm steps}$ is the number of steps. In [225], projections of the filament contour to a plane $(x - y)$ were analyzed from experiment.

In order to extract the corresponding reptation diffusion coefficient from the bead trajectories of the FENE-B model, embedded in 3D space, one has to precise the above definition, i.e., we hereby define the orientation of a tube on the basis of the temporary end-to-end vector of the semiflexible chain: $\boldsymbol{R}(T) \equiv T^{-1} \int_0^T [\boldsymbol{r}_N(t) - \boldsymbol{r}_1(t)]\,\mathrm{d}t$, which depends on the chosen time interval T. Let $\boldsymbol{n}_T$ denote the normalized quantity $\boldsymbol{n}_T \equiv \boldsymbol{R}(T)/\|\boldsymbol{R}(T)\|$, then the diffusion coefficient of a single bead parallel to 'its' tube is $D_T^{\rm k} \equiv (2T)^{-1} \left\langle (\boldsymbol{n}_T \cdot [\boldsymbol{r}_k(T) - \boldsymbol{r}_k(0)])^2 \right\rangle$, where $\langle \ldots \rangle$ represents a time average. The reptation diffusion coefficient along the tube of the polymer with N beads is then expressed as $D_{\|} \equiv (D_T^1 + D_T^{\rm N})/2$. For rods the expected result is $D_{\|} = k_B T \ln(L/b)/(2\pi\eta_s L)$, where L is the contour length, b the diameter of the filament, k_B is Boltzmann constant, T is the temperature and η_s is the viscosity of the solvent. In addition, we need to have a formula to extract the orientation diffusion coefficient $D_{\rm or}$. and a tube width a, based on the time evolution of the end bead coordinates of the semiflexible chain. The concept has physical meaning for semiflexible or stiff chains, but is obviously meaningless for ideal chains. Now, let $\boldsymbol{r}_1(t)$ and $\boldsymbol{r}_N(t)$ denote the coordinates of the end beads of a representative chain, separated by $\boldsymbol{R} \equiv \boldsymbol{r}_N(t) - \boldsymbol{r}_1(t)$. The natural choice for a definition of the orientational diffusion coefficient is $D_{\rm or}(T) \equiv (4T)^{-1}(\boldsymbol{n}_T - \boldsymbol{n}_0)^2$, to be extracted in a range where $D_{\rm or} \ll 1/T$. In this range, $D_{\rm or}(T)$ should be independent of T. For rods the theoretical result is $D_{\rm or} = 3k_B T\,(\ln(L/b) - \gamma)/(\pi\eta_s L^3)$, where $\gamma \approx 0.8$, but slightly dependent on L/b [216]. Finally, based on the trajectories of all the three beads we estimate a perpendicular diffusion coefficient as follows

$$D_{\perp}(T) \equiv \frac{1}{2T} \int_0^T \left(\frac{\boldsymbol{R}(T)}{\|\boldsymbol{R}(T)\|} \times \frac{d\boldsymbol{r}_C(t)}{dt} \right)^2 \mathrm{d}t\,. \tag{5.30}$$

For rods, the theoretical result is $D_{\perp} = D_{\|}/2$, and the so called 'disentanglement time' can be related to $D_{\|}$ through $\tau_d = L^2/D_{\|}$, a 'tube radius' a can be defined by $a^2 \equiv L^2 D_{\rm or} \tau_d = L^4 D_{\rm or}/D_{\|}$, and the center of mass diffusion is obtained via $D_{\rm cm} = (D_{\|} + 2D_{\perp})/3$. Experimentally, thermal undulations of the filament (visible by microscopy) have been used to define the tube diameter; it is estimated as the maximum deflection along the contour, at sufficiently large concentrations, within a limited time interval.

Figures 5.9, 5.10 provide snapshots of FENE-B model actin filaments in equilibrium as well as in a nonequilibrium situation. Our preliminary results (which should

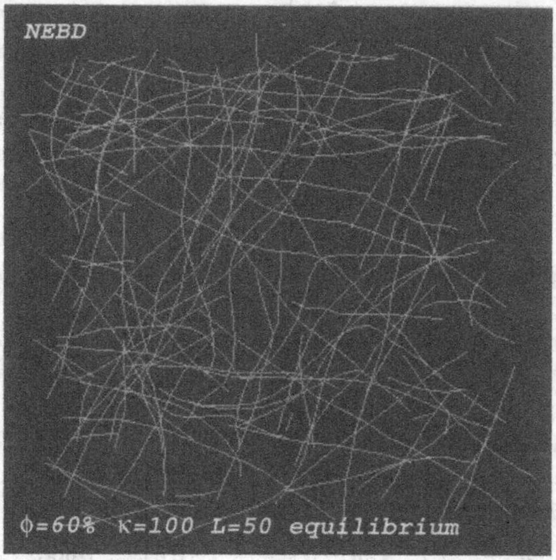

Fig. 5.9. Equilibrium high density semiflexible FENE-B chains (5.33) for system parameters given in the figure

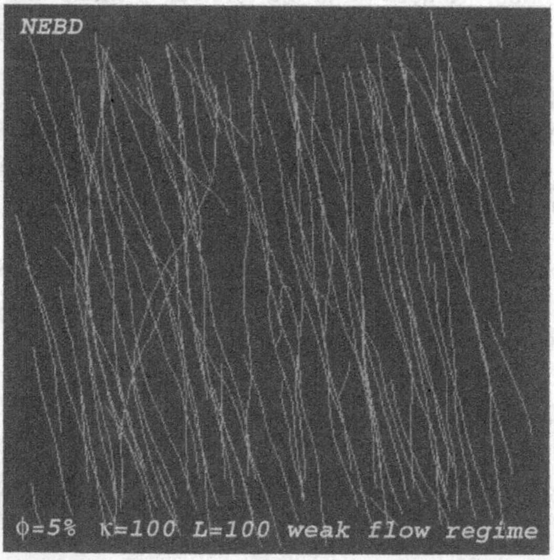

Fig. 5.10. Flow-aligned FENE-B chains for system parameters given in the figure

Table 5.1. Preliminary simulation result for the scaling behavior of various diffusion coefficients (see text part) for semidilute solutions of the FENE-B model actin filaments. The scaling exponents have been estimated in the concentration regime (5-60%), relative bending rigidities $\kappa/L = 0.5 - 2$

		D_{mon}	D_{or}	$D_{\parallel}$	$D_{\perp}$
$\propto c^{-\alpha}$	with $\alpha =$	0.6(1)	0.7(2)	0.5(1)	0.6(1)
$\propto \kappa^{-\beta}$	with $\beta =$	0.3(1)	0.3(1)	0.3(1)	0.3(1)
$\propto L^{-\gamma}$	with $\gamma =$	0.5(4)	2.1(2)	0.5(5)	0.3(2)

be improved in the near future) for the reptation and orientational diffusion coefficients defined in the previous section are summarized in Table 5.1. The effect of concentration on the end-to-end distribution function of FENE-B actin filaments is demonstrated by Fig. 5.11, for the diffusion coefficient $D_{\parallel}$ vs chain length see Fig. 5.12. A solutions of actin filaments exhibits pronounced shear thinning, non-newtonian rheological behavior of the FENE-B model is reported in Fig. 5.13. The simulation of dilute and semidilute solutions of actin filaments remains a challenge for computer simulation due to the stiffness of filaments which requires large samples in order to prevent finite size effects.

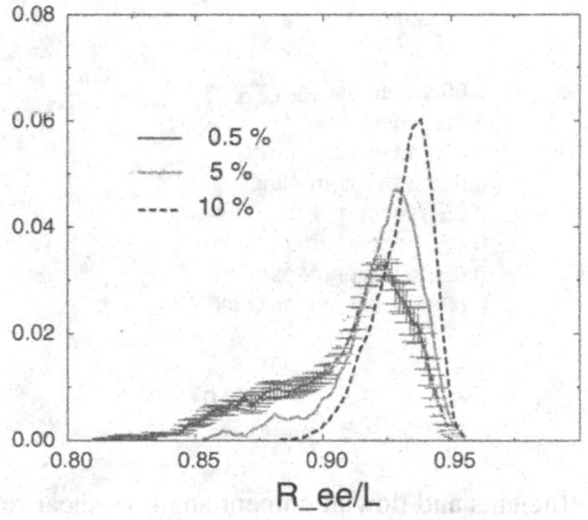

Fig. 5.11. Effect of concentration c on the end-to-end distribution function $f(R_{\text{ee}})$ vs R_{ee}/L of FENE-B actin filaments ($\kappa = 200$, $L = 100$). For the curve with $c = 0.5\%$, error bars are shown

To give an impression for possible further applications of the presented FENE-C and FENE-CB models, we end up this section with few snapshots. Figures 5.14, 5.15 show FENE-CB3 networks with different rigidities, whereas Fig. 5.16 has been obtained for an extended version of the FENE-CB∞ model, for which the bending potential (5.32) has been modified such that in-plance scissions between more than three beads (at branching points) are prefered (see Table 14.1 conc. nomenclature).

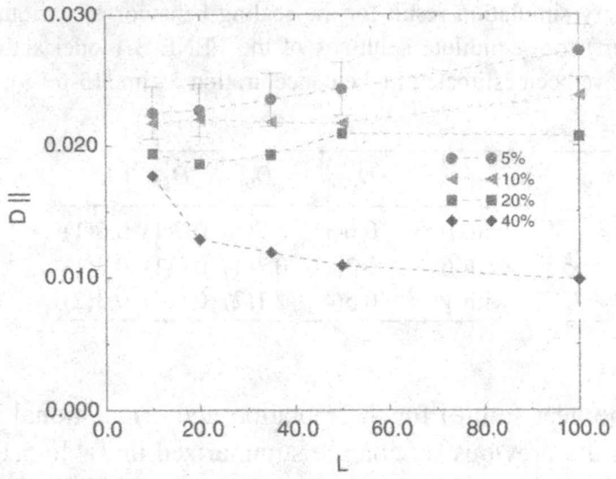

Fig. 5.12. Diffusion coefficient parallel to the tube vs chain length L for the FENE-B model actin filaments at various concentrations

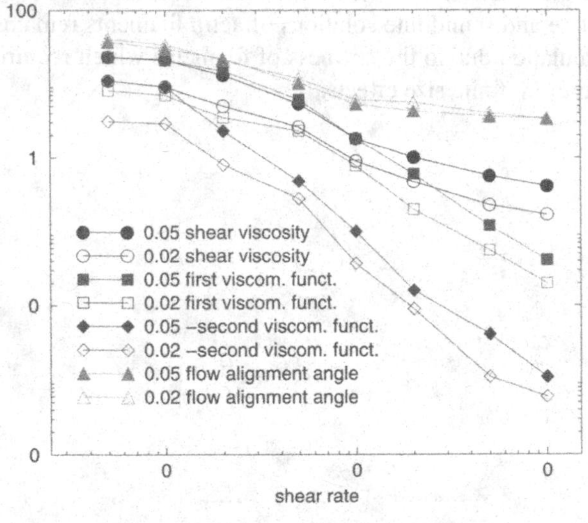

Fig. 5.13. Viscosity coefficients and flow alignment angle vs shear rate for both, 2% and 5% solutions of FENE-B actin filaments ($\kappa = 100$, $L = 100$)

The incorporation of f-branching into the FENE-C model, which carries a single scission energy E_{sc} (since $f = 1$ in its simplest form) generally introduces f independent paramaters characterizing scissions and recombinations.

5.4 FENE-B Liquid Crystalline Polymers

Thermotropic liquid crystals form mesophases intermediate between a solid phase at low temperatures and an isotropic liquid phase at high temperatures [234–236].

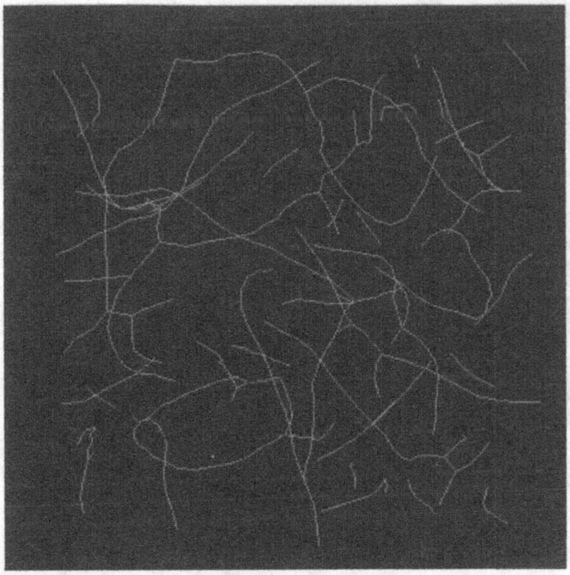

Fig. 5.14. Sample snapshot of a realization of a system made of FENE-CB6 chains (5.32). Beside scissions/recombinations of chains (parameterized through a scission energy E_{sc}) the model allows for the formation of branchings and carries a parameter for the (in plane) stiffness of chains. The concentration is $c = 5\%$. Results obtained via BD

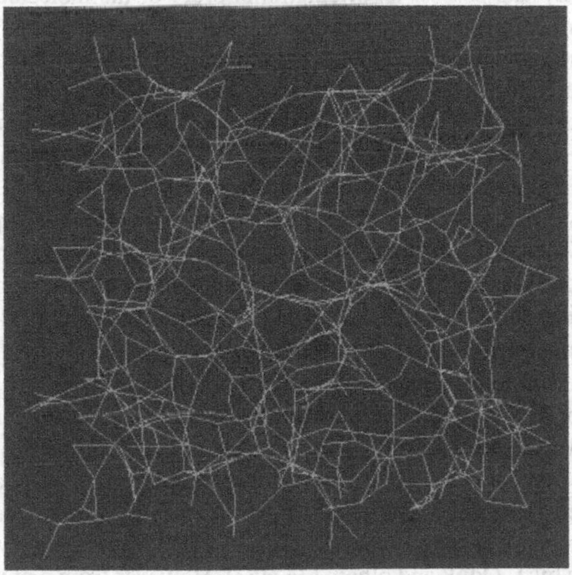

Fig. 5.15. Same system as in Fig. 5.14 at concentration $c = 20\%$. Results obtained via BD

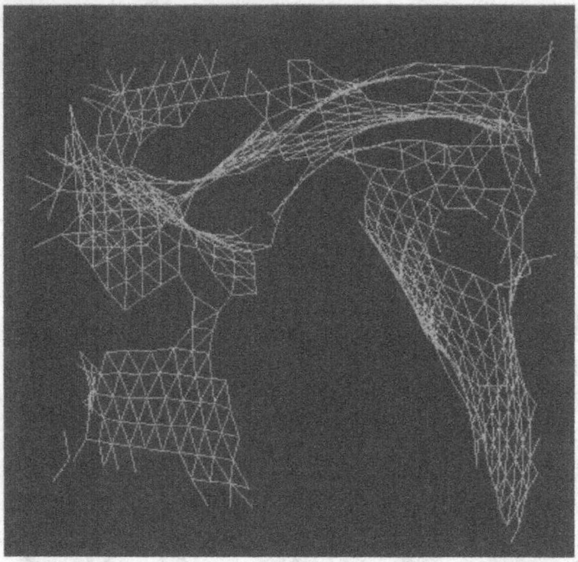

Fig. 5.16. Sample snapshot of a realization of a system made of semiflexible FENE-CB chains. Beside scissions/recombinations of chains (parameterized through a scission energy E_{sc}) the model potential naturally allows for the formation of branchings and carries a parameter for the stiffness of chains

Nematic liquid crystals possess an orientational order of the molecular axes but no long range positional order. Smectic liquid crystals, in particular those referred to as SmA and SmC have a nematic like orientational order and in addition their centers of mass are confined to layers. Previous computational studies on the phase behavior of model liquid crystals by MD and Monte Carlo (MC) simulations have been performed on various levels of simplification of the molecular interactions [207, 237, 238]. Simulations of the Lebwohl-Lasher lattice model [239, 240] gave hints on the basic features of the phase transitions. The simplest approach where the dynamics of the centers of mass of the particles are properly taken into account is to treat molecules as stiff non-spherical particles like ellipsoids or spherocylinders, or to consider particles interacting by a Gay-Berne potential [241–243]. Going further the internal configuration has been taken into account by treating the molecules as being composed of interaction sites (monomers) connected by formulating constraints or binding forces. Both Monte Carlo [207, 244–247] and MD methods [248–250] were applied to study the static and dynamic properties, respectively. Extremely huge compounds such as lipids in the liquid crystalline phase have been simulated as well [251, 252]. The effect of semiflexibility and stiffness of macromolecules on the phase behavior of liquid crystals has been extensively discussed on analytic grounds by T. Odijk and others [34, 253]. However from a physical point of view the construction of model interactions remains in question [254, 255], and from the technical point of view, the development of efficient implementations [256–259]

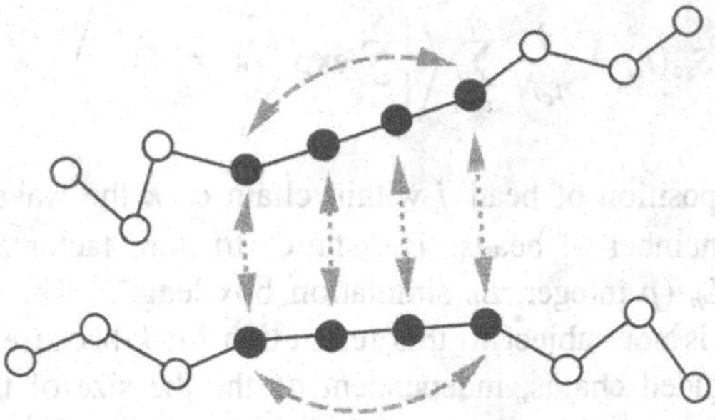

Fig. 5.17. The bead-bead interactions. In addition to the interactions indicated in this figure, there are also a FENE interaction between all connected beads in chains and a repulsive Lennard–Jones between all beads of the system [33]

is challenging due the complexity of detailed models which involve long range electrostatic forces or many body potentials.

This section reviews a simple microscopic model for a 'representative' thermotropic liquid crystals composed of partially stiff, partially flexible molecules. Our system is composed of intramolecularly inhomogeneous FENE-B chains, interacting via a Lennard–Jones potential, and the attractive part of the Lennard–Jones potential is taken into account only between their stiff parts. This model has been introduced in [33]. The model system is composed of n_c multibead chains with N beads per chain. Each chain, as shown in Fig. 5.17 is made of two identical terminal flexible parts (N_{flex} beads) and a central stiff part (N_{stiff} beads) where $N_{\text{stiff}} + 2 \times N_{\text{flex}} = N$. The notation ($N_{\text{flex}} - N_{\text{stiff}} - N_{\text{flex}}$) had been used to characterize the different systems. For example, (*3*–4–*3*) means that the chains in this system are composed of a central stiff part of 4 beads and two terminal flexible parts of 3 beads. Simulations are performed in the NVT ensemble. Results to be reported below were obtained for a system of $n_c = 288$ chains of length $N = 10$ at bead number density $n = 0.8$. All beads are interacting with a WCA potential. Adjacent (connected) beads within chains interact via a FENE force. The central part of each chain is kept stiff with a strong (large κ) FENE-B interaction. Additionally, corresponding beads within the stiff parts of different chains interact via the attractive part of the Lennard–Jones potential ('smectic' biased) producing an effectively anisotropic interaction between stiff parts. The strength of the attractive interaction is adjustable by a depth parameter ε_{att}.

5.4.1 Static Structure Factor

The static structure factor of the multibead fluid where each bead is assumed to act as a 'scatterer' can be written as a product between inter- and intramolecular structure factors $S(\boldsymbol{k}) = S_{\text{sc}}(\boldsymbol{k})\, S_{\text{inter}}(\boldsymbol{k})$. The single chain static structure factor representing the intramolecular correlations is defined as:

$$S_{\rm sc}(\boldsymbol{k}) = \frac{1}{n_c N} \sum_{\alpha=1}^{n_c} \left\langle \left| \sum_{j=1}^{N} \exp\left(i\boldsymbol{k} \cdot \boldsymbol{x}_j^{(\alpha)} \right) \right|^2 \right\rangle . \tag{5.31}$$

Here $\boldsymbol{x}_j^{(\alpha)}$ denotes position of bead j within chain α, $\boldsymbol{k}$ the wave vector transfer, and n_cN the total number of beads. The static structure factor $S(\boldsymbol{k})$ is restricted to $k = |\boldsymbol{k}| = 2\pi p/L_b$ (p integer, L_b simulation box length). The single chain static structure $S_{\rm sc}(\boldsymbol{k})$ is not subject to this restriction for k because it can be calculated from the unfolded chains, independent of the the size of the basic simulation box. A long range positional is revealed by Bragg like peaks in another static structure factor $S_{\rm cm}(\boldsymbol{k})$ where the centers of mass of the molecules are taken as scatterer. For ideal crystals the height of the Bragg peaks approaches n_c, the number of molecules in the scattering volume. For a layered (smectic) structure with a separaion distance d between layers a peak occurs at $k = 2\pi/d$. Its height divided by n_c provides a convenient measure for the degree of positional order σ, i.e., we have $\sigma \equiv \left|\left\langle n_c^{-1} \sum_{\alpha=1}^{n_c} \exp(2i\pi z^{(\alpha)}/d) \right\rangle\right|$, where $z^{(\alpha)}$ is a center of mass coordinate of chain α with respect to a symmetry-adapted coordinate system, and $\langle .. \rangle$ denotes a time average.

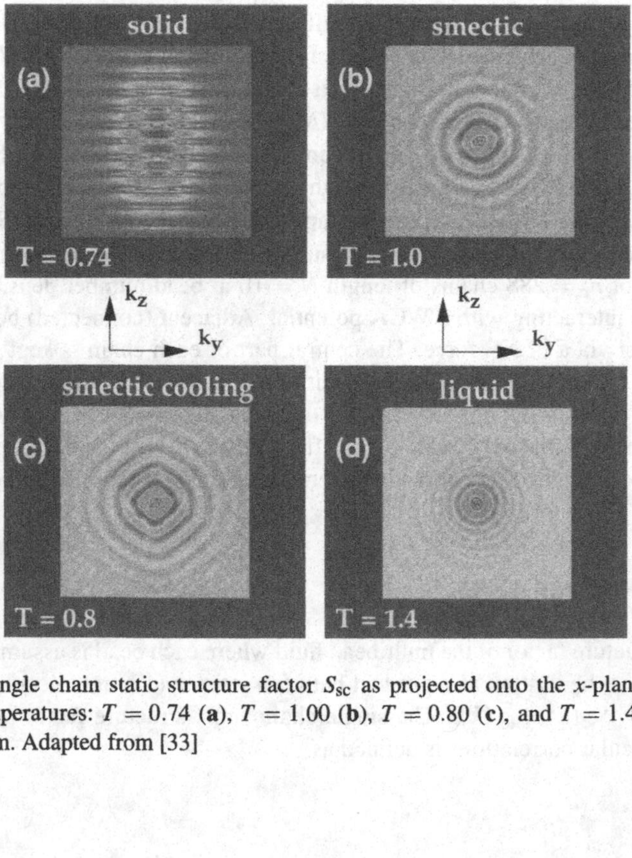

Fig. 5.18. Single chain static structure factor $S_{\rm sc}$ as projected onto the x-plane ($k_x = 0$) at different temperatures: $T = 0.74$ **(a)**, $T = 1.00$ **(b)**, $T = 0.80$ **(c)**, and $T = 1.40$ **(d)** for the *3–4–3* system. Adapted from [33]

For a number of these semiflexible systems it had been observed that a smectic phase is well defined over a wide range of temperatures whereas the nematic phase is too narrow in temperature to be seen clearly. The smectic phase becomes increasingly disordered upon decreasing the strength of attraction (parameter ε_{att}). The effect of architecture (amount stiff/flexible) has been studied to a certain extend in [33]. According to Table 5.2 clearing temperatures as well as melting temperatures increase for this model upon increasing the length of the stiff part. This in qualitative agreement with experiments. Some snapshots and results for order parameters are given in Figs. 5.19–5.20.

Table 5.2. Influence of the ratio between stiff and overall length of the special FENE-B molecules on their melting and clearing temperatures [33]

$n_{flex}-n_{stiff}-n_{flex}$	*3−4−3*	*3−5−3*	*0−10−0*
$(n_{stiff}-1)/n_b$	0.34	0.40	1.00
melting Temperature	0.75	0.90	3.0
clearing Temperature	1.2	2.0	>5.0

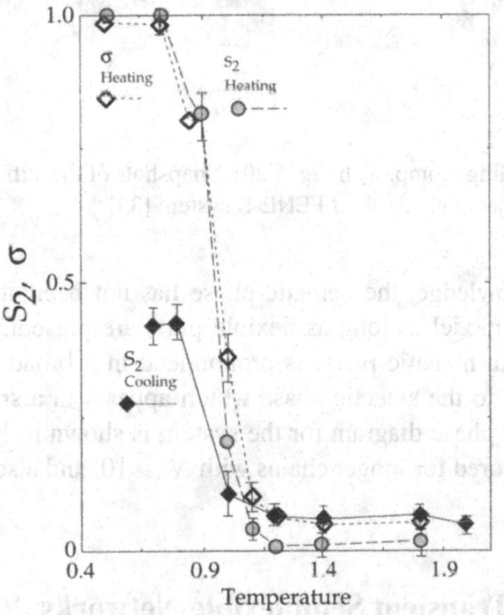

Fig. 5.19. Orientational order parameter S_2 and positional order parameter σ at different temperatures, for the *3−4−3* FENE-B system, observed in heating (from an ideal fcc structure) and (subsequent) cooling [33]

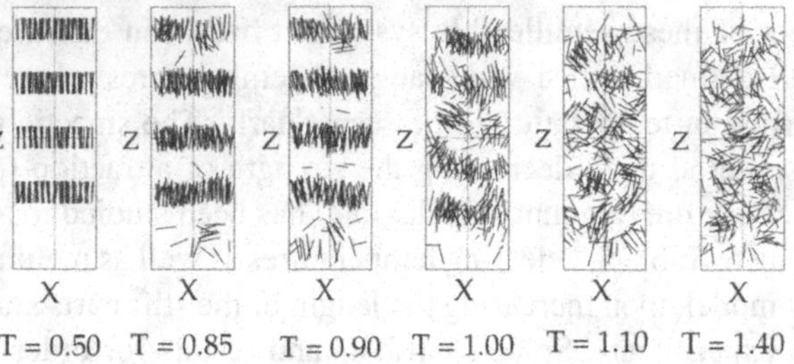

Fig. 5.20. During heating: Snapshots of the stiff central parts of molecules at different temperatures T (increasing from *left* to *right*). for the *3–4–3* FENE-B system [33]

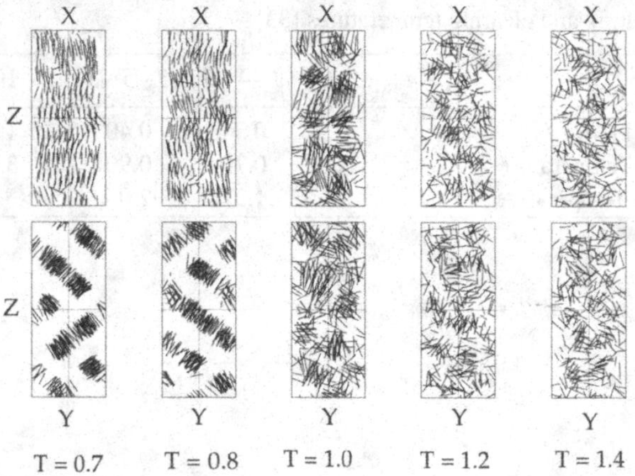

Fig. 5.21. During cooling (comp. with Fig. 5.20): Snapshots of the stiff parts of the molecules in two orthogonal projections) *3–4–3* FENE-B system [33]

To our best knowledge, the nematic phase has not been studied via computer simulation for this model as long as flexible parts are present. Of course, for stiff molecules [260], the nematic phase is pronounced in a broad temperature regime in contradistinction to the smectic phase which appears in a small temperature interval. An expected phase diagram for the system is shown in Fig. 5.23. A nematic phase should be favored for longer chains with $N \gg 10$, and also for non-symmetric molecules.

5.5 FENE-CB Transient Semiflexible Networks, Ring Formation

Both the anxalytic and numerical tools for linear wormlike micelles reviewed in the foregoing sections can be used to predict the extent of loop formation as function of the micellar concentration, the end-cap energy and the flexibility of linear micelles. As a matter of fact, even if loop formation is unfavorable under many conditions, e.g., for stiff micelles and low end-cap energies, they have to be treated correctly

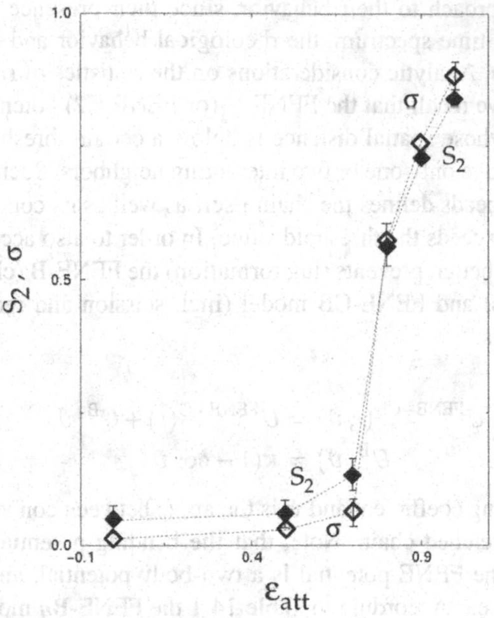

Fig. 5.22. The order parameters S_2 (nematic) and σ (smectic) as function of ε_{att} at the temperature $T = 0.8$ for the *3–4–3* FENE-B system [33]

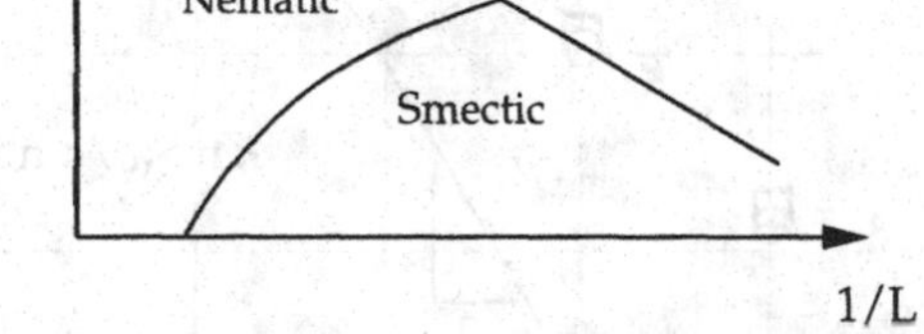

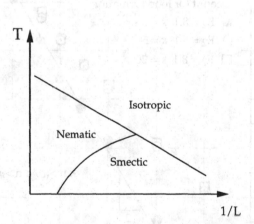

Fig. 5.23. Typical experimental phase diagram where L is length of the chains and T denotes temperature [261]

in any statistical approach to their behavior, since their presence can significantly affect the relaxation time spectrum, the rheological behavior and correlation function of various types. Analytic considerations on the statistics of ring formation are available in [262]. We recall that the FENE-C (or FENE-C2) potential acts between all pairs of beads (whose spatial distance is below a certain threshold value Q_C) as long as both beads have only one or two interacting neighbors. Such a transient bond between connected beads defines the chain itself as well as its contour and it breaks if any bond length exceeds the threshold value. In order to also account for stiffness (which disfavors, or better, prevents ring formation) the FENE-B (classical semiflexible linear polymers) and FENE-CB model (incl. scission and recombination) are introduced as follows:

$$U^{\text{FENE-CB}}(r,\vartheta) = U^{\text{FENE-C}}(r) + U^{\text{B}}(\vartheta) \tag{5.32}$$

$$U^{\text{B}}(\vartheta) = \kappa(1-\cos\vartheta)\,, \tag{5.33}$$

where κ is the bending coefficient and ϑ is the angle between connected bonds, such that $\vartheta = 0$ for a stretched chain. Note, that the bending potential is a three-body potential, whereas the FENE potential is a two-body potential, and the notation in (5.32) is a formal one. According to Table 14.1 the FENE-Bn model is the natural

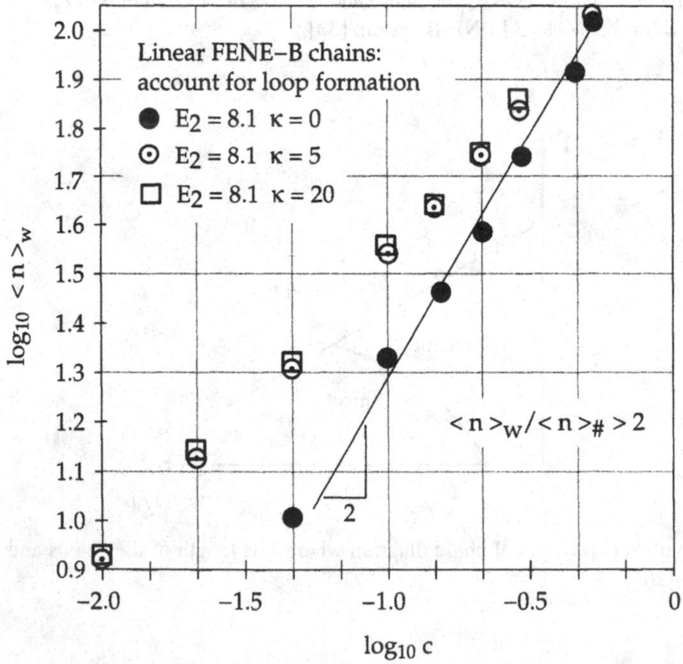

Fig. 5.24. The average weight size for linear FENE-CB chains vs concentration for different bending coefficients κ. Results obtained via BD

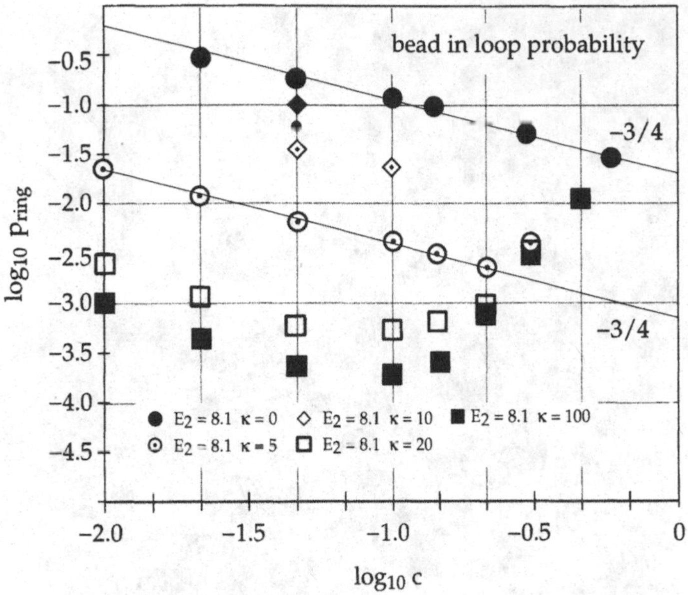

Fig. 5.25. Probability to find a bead inside a loop for different bending coefficients κ and concentrations c. Model system as for Fig. 5.24

extension of the FENE-B allowing for maximum functionality n (classical saturated and unsaturated networks for small and large bending stiffness, respectively).

The FENE-CB and FENE-B models have not yet been characterized in an exhaustive fashion. Flexible FENE-n networks also known as 'finitely extensible network strand (FENS)' [263] models have been used to investigate strain hardening behavior for associating polymeric systems in [264], overshoot in the shear stress growth function and strand extensibility in [265]. Remarkable progress has been made in the understanding of polymer gels [266] where 'equilibrium' properties of a FENE-C type network model were studied in detail via MC. The authors artificially prohibit asssociation of direct neighbors but it seems that agreement between experiments and FENE model predictions can be further improved by taking bending stiffness into account (FENE-CB). At the same this article provides an excellent review on continuum and molecular theories of stress-strain relations for networks (incl. classicial network theory, nonaffine deformation theory, scaling model, rod and coil model). To get a feeling on the power of FENE-CB network models and their range of application we present a tiny result obtained in a preliminary study. The model exhibits characteristic behaviors as those shown in Figs. 5.24, 5.25 when solving the FENE-CB model via BD. With increasing concentration the probability of loop formation decreases resulting from the increase of average length of micelles. With increasing scission energy loop formation becomes favorable, but increasing

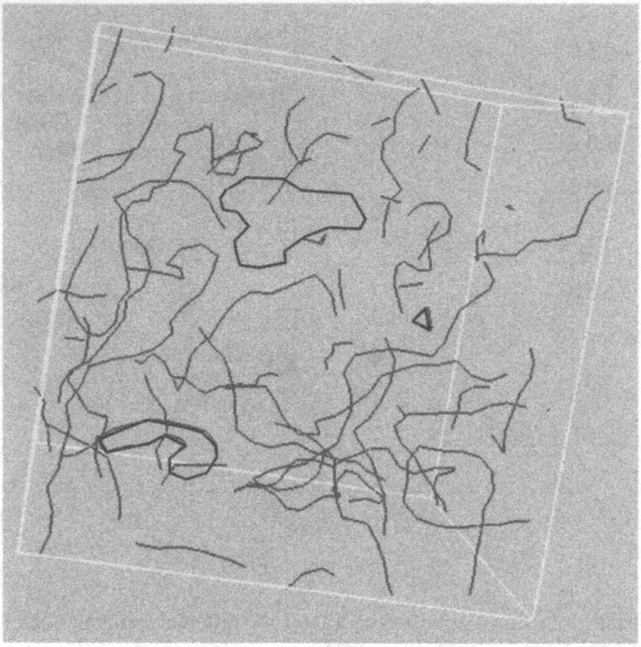

Fig. 5.26. Snapshot of a BD computer simulation configuration of FENE-C wormlike micelles with parameters $c = 0.02$, $\kappa = 5$ and $E_2 = 4$. Here, a small system size, containing 1000 beads, was chosen for reasons of clarity

stiffness decreases the tendency of ring formation. At large concentrations and large values for the bending stiffness parameter κ there are deviations from the square root behavior $\langle n \rangle_{\#} \propto \sqrt{c}$ which are expected when a mean-field approach is used to describe the effect of concentration. A snapshot is given in Fig. 5.26.

6

Primitive Path Models

Having discussed the range of applicability for various FENE chain models mostly listed in the upper part of Table 14.1 one may expect that we also review the FENE models in the lower part of this table. Forttunately, several reviews exist summarizing the constitutive equations following from the approximations involved in the FENE-P.. and FENE-L. models, cf. [4, 35, 66] such there is no need to summarize them – and there usefulness in micro-macro applications – here. Rather, we turn to simple low dimensional models depicted in the upper part of Fig. 1.2, i.e. tube models and elongated particle models for the dexcription of complex fluids. One may ask how these levels of descriptions are related. This will be discussed in Sect. 8.10.1.

6.1 Doi-Edwards Tube Model and Improvements

A molecular model for polymer melts was elaborated by Doi and Edwards (DE) [216] who extended the reptation idea introduced by de Gennes [153] to a tube idea in order to describe the viscoelastic behavior of entangled polymers in the presence of 'obstacles'. Within the tube and reptation pictures, the complex entanglement interaction between polymer chains has been treated in a rather direct approach, i.e. each chain in the polymer system is equivalent to a chain restricted to one dimensional motion (so called 'reptation') in a confining tube, except for its two ends which can move in any possible direction. In addition to the reptation mechanism, DE originally assumed instantaneous and complete chain retraction, affine tube deformation by the flow, and independent alignment of tube segments. By doing so, they obtained a closed-form constitutive equation which only involves the second moment of the orientation vector for a tube segment. For highly entangled, linear polymers, the original DE model has been extended to incorporate chain contour length fluctuations [267, 268] and constraint release due to the motion of the surrounding chains (so called 'double reptation') [269,270]. The combination of these two effects lead to a refined description of the linear viscoelastic properties [271], however, the model is much less successful for the nonlinear properties. The major experimental observations that the original DE theory fails to describe in the nonlinear regime are the following [175, 272]: A) There exist irreversible effects in double-step strain experiments with flow reversal, B) Over a wide range of shear rates $\dot{\gamma}$ above the inverse disentanglement time $1/\tau_d$ the steady shear stress is nearly constant for very highly entangled melts or solutions or increases slowly with shear rate for less highly entangled ones. The first normal stress difference N_1 increases more rapidly with shear rate than does the shear stress over the same range of shear rates. The slope of N_1 versus $\dot{\gamma}$ increases as the molecular weight decreases, C) The steady-state shear viscosity of different molecular weights merge into a single curve in the high shear rate, power-law regime, D) The shear stress shows transient overshoots in the start-up of steady shear flow at low shear rates. The strain at which the maximum in the overshoot occurs increases with shear rate at high rates, E) The first normal stress difference exhibits transient overshoots in the start-up of steady shear flow at moderate shear rates, F)The rate of stress relaxation following cessation of steady shear flow is shear rate dependent, G) The steady-state extinction angle decreases more gradually with shear rate than predicted by the DE model, H) The transient extinction angle shows an undershoot at the start-up of steady shear at high shear rates; it also shows an immediate undershoot when the shear rate is suddenly decreased after a steady state has been reached, finally it reaches a higher steady-state value [273], I) Steady-state values of the dimensionless uniaxial extensional viscosity are non-monotonic functions of extension rate.

In order to improve the situation, many attempts of modifying the original DE model have been made during the last years and been reviewed in [5]. Several physical effects have been found to be important for more realistic modeling of nonlinear properties of entangled polymers. Upon these the most important are avoiding the independent alignment (IA) approximation, double reptation, chain stretching, convective constraint release (CCR), and anisotropic tube cross sections. For a review on these effects, their influence on the quality of predictions for rheological quantities a good reference might be [175]. Recently, reptation models incorporating all the well established phenomena (except for anisotropic tube cross sections) have been formulated based on a full-chain stochastic approach suitable for computer simulations [148, 274–276]; on a full-chain, temporary network model with sliplinks, chain-length fluctuations, chain connectiviy and chain stretching [277]; on coupled integral-differential equations [278]; and a reptation model including anisotropic tube cross sections, chain stretching, double reptation, and CCR, while avoiding the IA approximation [175, 180]. The predicitive power of the Jacobi idenity has been demonstrated for the latter model which is thermodynamiccaly admissible, i.e., compatible with the GENERIC framework (Sect. 8.3). It is encouraging that these reptation models can quite successfully reproduce the experimentally observed rheological behavior in a large number of flow situations. Very recently, Doi merged together the network model of Green & Tobolsky, and the tube model of Edwards and de Gennes. The resulting model, called the dual slip-link model, can be handled by computer simulation, and it can predict the linear and nonlinear rheological behaviours of linear and star polymers with arbitrary molecular weight distribution [279]. Unified stress-strain models for polymers, including polymer networks have been presented by Wagner [280, 281]. Rather than going into further detail with these models for polymer melts, and in order to go into detail with any of the established models, we take an illustrative example from our own research, where the original tube model is subject to a very minor modification. This will allow us to discuss an analytic expression for the dynamic viscosities, a decoupling approximation used to evaluate nonlinear elastic behaviors, and Galerkin's method to solve the underlying Fokker–Planck equation efficiently.

6.2 Refined Tube Model with Anisotropic Flow-Induced Tube Renewal

Point of departure are classical kinetic equations for the orientational distribution function of polymer segments in melts. In the DE tube model the macromolecules of a polymeric liquid are idealized as freely jointed primitive paths characterized by the orientation of a segment $\boldsymbol{u}$ at contour label s (we use $0 \leq s \leq 1$). The orientation of the segment at the 'position' s is determined by the orientational distribution function $f = f(t,s,\boldsymbol{u})$ which, in general, also depends on the time. The kinetic equation for $f = f(t,s,\boldsymbol{u})$ is written as

$$\frac{\partial f}{\partial t} = -\boldsymbol{\omega}\cdot \mathcal{L}f - \mathcal{L}\cdot(\boldsymbol{T}_{\text{flow}} f) + \mathcal{D}_{\text{rep}}(f) + \mathcal{D}_{\text{or}}(f)\,,$$
$$\boldsymbol{T}_{\text{flow}} \equiv \frac{1}{2} B\mathcal{L}(\boldsymbol{u}_{[2]} : \boldsymbol{\gamma})\,, \qquad (6.1)$$

with angular operator $\mathcal{L} \equiv \boldsymbol{u} \times \partial/\partial\boldsymbol{u}$, cf. Chap. 10, vorticity $\boldsymbol{\omega} \equiv (\nabla \times \boldsymbol{v})/2$ associated with the (macroscopic) flow field $\boldsymbol{v}$, $\boldsymbol{\gamma} \equiv (\boldsymbol{\kappa} + \boldsymbol{\kappa}^\dagger)/2$ with $\boldsymbol{\kappa} \equiv (\nabla\boldsymbol{v})^\dagger$, and $\boldsymbol{T}_{\text{flow}}$ is the orienting torque exerted by the flow. The kinetic equation of Peterlin and Stuart [282] for solutions of rod-like particles (where the variable s is not needed) is of the form (6.1) with $\mathcal{D}_{\text{or}}(f) \equiv \mathrm{w}\mathcal{L}^2 f$, where w stands for the orientational diffusion coefficient. Often the corresponding relaxation time $\tau \equiv (6\mathrm{w})^{-1}$ is used to discuss results. The (reptation) diffusion term of DE can be written as $\mathcal{D}_{\text{rep}} \equiv \lambda^{-1}\partial^2/\partial s^2 f$, with a relaxation time $\lambda = L^2/D$, which is connected with a disentanglement time via $\tau_d = \lambda\pi^{-2}$. The $\mathcal{D}$-terms describe the 'damping', which guarantees that f approaches the isotropic distribution $f_0 = (4\pi)^{-1}$ in the absence of orienting torques. With an additional torque caused by a mean field taken into account in (6.1), such a kinetic equation will be applied below to the flow alignment of liquid crystals [74]. Here we consider both diffusion mechanisms. For the case of rodlike segments ($B = 1$) the Fokker–Planck equation (6.1) is equivalent with the diffusion equation in [4, 283].

With the normalization $\int f \mathrm{d}^2 \boldsymbol{u} = 1$ for the orientational distribution function $f = f(t,s,\boldsymbol{u})$ (time t) the average $\langle \psi \rangle$ of a function $\Psi = \Psi(\boldsymbol{u})$ is given $\langle \Psi \rangle = \int \Psi f \mathrm{d}^2 \boldsymbol{u}$ and depends on t and s. Here, the (2nd rank) alignment tensor (4.8) $\boldsymbol{a}_{[2]} = \boldsymbol{a}(t,s) = \langle \overline{\boldsymbol{uu}} \rangle = \langle \boldsymbol{u}_{[2]} \rangle$ is once more of particular importance. The symbol $\overline{\cdots}$ denotes the symmetric traceless part of a tensor, and $\mathbf{1}$ is the unit tensor. Considering a planar Couette flow in x–direction, gradient in y-direction, the shear rate $\dot{\gamma}$ for the macroscopic velocity profile $\boldsymbol{v}$ is $\dot{\gamma} \equiv \partial v_x / \partial y$. For this geometry, only 3 of the 5 independent components of the alignment tensor do not vanish. In the spirit of Sect. 2.1 we abbreviate – for the present purpose – as follows:

$$
\begin{aligned}
a_+ &\equiv \langle u_x u_y \rangle \,, \\
a_- &\equiv \frac{1}{2} \langle u_x^2 - u_y^2 \rangle \,, \\
a_0 &\equiv \frac{3}{4} \left\langle u_z^2 - \frac{1}{3} \right\rangle \,, \\
\tilde{\boldsymbol{a}} &\equiv (a_+, a_-, a_0)^T \,.
\end{aligned} \tag{6.2}
$$

A viscous flow gives rise to a flow alignment [282, 284] which can be detected optically via its ensuing birefringence. The alignment, in turn, affects the viscous flow [284, 285] and consequently the stress tensor $\boldsymbol{\sigma}$ contains a contribution associated with the alignment, more specifically, $\boldsymbol{\sigma} = 2\eta_{iso} \boldsymbol{\gamma} + \boldsymbol{\sigma}_a$, and $\boldsymbol{\sigma}_a = 3 n_p k_B T R \int_0^1 \boldsymbol{a}(t,s) \mathrm{d}s$, where η_{iso} is the 'isotropic' viscosity for $\boldsymbol{a} = \tilde{\boldsymbol{a}} = 0$. n_p and T are the molecule number density and the temperature of the liquid. The relation between $\boldsymbol{\sigma}_a$ and $\boldsymbol{a}_{[2]}$ (SOR, discussed in Sect. 4.6) which has been derived by Giesekus [285] and used by DE is a limiting expression for long and thin segments corresponding to $B = 1$. In general the factor B is the ratio of two transport coefficients [74, 284]. Curtiss and Bird [4] replaced $3B$ by 1 and presented additional viscous contributions associated with the 'link tension'. These terms are disregarded here.

6.2.1 Linear Viscoelasticity of Melts and Concentrated Solutions

Multiplication of (6.1) with $\overline{\boldsymbol{uu}}$ and integration over the unit sphere yields

$$
\left(\frac{\partial}{\partial t} + \tau^{-1} - \lambda^{-1} \frac{\partial^2}{\partial s^2} \right) \boldsymbol{a}_{[2]} = \frac{2}{5} B \boldsymbol{\gamma} + \cdots \,, \tag{6.3}
$$

with $\tau = (6\mathrm{w})^{-1}$. The dots stand for terms involving products of $\boldsymbol{a}_{[2]}$ with the vorticity $\boldsymbol{\omega}$ and $\boldsymbol{\gamma}$, as well a term which couples $\boldsymbol{a}_{[2]}$ with an alignment tensors of rank 4. These terms can be inferred from [74], they are of importance for the non-newtonian viscosity and the normal pressure differences (see next section). For an analysis of the frequency dependence of the viscosity in the Newtonian regime, these terms can be disregarded, i.e. we consider the only nonvanishing component a_+ of $\boldsymbol{a}_{[2]}$.

The complex viscosity $\eta^* = \eta' - i\eta''$ of a viscoelastic medium can be determined by measurements under oscillatory shear flow (or deformation) $\gamma \sim e^{-i\omega t}$. The relaxation of the material causes a phase-shift δ between (complex) stress and deformation which is related to the complex viscosity ($\tan\delta = \eta'/\eta''$), or alternatively, to the storage G' and loss modulus G'' via $G^* = G' + iG'' \equiv i\omega\eta^*$. With the ansatz $a_+ = 2B\gamma C/5$ the scalar function $C(\omega,s)$ with dimension of time obeys $(\tau^{-1} - i\omega)C - \lambda^{-1}(\partial^2/\partial s^2)C = 1$. The boundary condition proposed by DE are random orientations for the chain ends, $\forall_t f(s=0,\boldsymbol{u}) =$ const. This implies $\forall_\omega C(\omega, s=0) = 0$. We wish to take into account the property of chain ends to participate in the flow alignment of the complete chain, or equivalently, anisotropic (flow-induced) tube-renewal. Working out this modification, we set $C(\omega, s=0) = \tau_{\text{end}}$, in order to introduce an additional relaxation time τ_{end} for this process. The solution reads

$$C(\omega,s) = \lambda\left[\frac{1}{z^2} + \left(\frac{1}{z^2} - g\right)\left(\frac{\tanh(z/2)}{\sinh^{-1}(sz)} - \cosh(sz)\right)\right], \quad z \equiv \sqrt{\tau^{-1}\lambda - i\omega\lambda}\,, \tag{6.4}$$

with $g \equiv \tau_{\text{end}}\lambda^{-1}$ being a dimensionless 'order' parameter for the chain ends. From the above relations alone we immediately obtain an analytic expression for the complex viscosity:

$$\eta^*(\omega) = G_a\lambda\left[\frac{1}{z^2} + \left(g - \frac{1}{z^2}\right)\left(\frac{2\tanh(\frac{z}{2})}{z}\right)\right], \tag{6.5}$$

with a shear modulus $G_a = 3B^2 n_p k_B T/5$. A Maxwell model type expression is obtained if $\tau \ll \lambda$. For polymer melts and highly concentrated solutions where the reorientational motion is strongly hindered, one expects the opposite situation, viz. $\tau \gg \lambda$. The pure reptation model considered by DE corresponds to $\tau^{-1}\lambda \to 0$ and consequently $z \to y$ with $y \equiv (-i\omega\lambda)^{1/2} = (1-i)\Omega^{1/2}$, and $\Omega \equiv \omega\lambda/2$. In this case (6.5) reduces to $\eta^*(\omega) = \eta_{\text{DE}}[H^*_{\text{DE}}(\omega) + H^*_{\text{end}}(\omega)]$ with the DE viscosity $\eta_{\text{DE}} = G_a\lambda/12 = n_p k_B T\lambda/20$, and dimensionless (complex) damping functions $H^*_{\text{DE}} = 12y^{-2}\{1 - 2y^{-1}\tanh(y/2)\}$, and $H^*_{\text{end}} = g\,24y^{-1}\tanh(y/2)$. The index 'end' labels a term, which vanishes for $g = 0$ and represents the influence of anisotropic tube renewal on the frequency behavior of the viscosity.

A more convenient expression for the dynamic viscosities G' and G'' in the DE limit $\lambda \ll \tau$ is immediately obtained from (6.5) and reads:

$$\frac{G'}{G_a} = \frac{1}{C}\{C - \sin(A) - B\sin(A) + (B-1)\sinh(A)\}\,, \tag{6.6}$$

$$\frac{G''}{G_a} = \frac{1}{C}\{(B-1)\sin(A) + (B+1)\sinh(A)\}\,, \tag{6.7}$$

with

$$\begin{aligned} A &\equiv (\lambda\omega)^{1/2} \\ B &\equiv 2gA^2\,, \\ C &\equiv A(\cos(A) + \cosh(A))\,. \end{aligned} \tag{6.8}$$

Some of deficiencies of the DE model (now recovered from (6.8)) with $B = 0$) have been overcome by inclusion of anisotropic chain ends. By NEMD simulation of a FENE melt in [283] we found strong support for implementing this modification. Moreover, the expected scalings $\tau_{\mathrm{end}} \propto \eta_{\mathrm{Rouse}} \propto L$ and $\lambda \propto \eta_{\mathrm{DE}} \propto L^{3.4}$ and therefore $g \sim L^{-2.4}$ (L is proportional to the molecular weight) allow to predict – in good agreement with experiments – the effect of chain length on the dynamics viscosities, and in particular on the width of the plateau regime.

In distinction to the DE theory ($g = 0$), for high frequencies the presented modification predicts one region, where both moduli display the same characteristics, independent of g, and another (plateau) region, where the storage modulus is nearly constant within a g-dependent frequency range. For a plot of the dynamic viscosities see Fig. 6.1. Notice that the moduli tend to overlap with increasing values for the shear frequency. The positive slope of G' and G'' at high frequencies ω follows here without the recourse to 'glassy relaxation modes', as suggested by Ferry [164]. To complete the discussion we mention the explicit result for the shear relaxation modulus $G(t) \equiv \int_0^\infty \eta'(\omega)\cos(\omega t)\,d\omega$. We obtain

$$G(t) = 8G_a \sum_{\alpha,\mathrm{odd}} ((\pi\alpha)^{-2} + g)\mathrm{e}^{-t/\lambda_\alpha}\,, \tag{6.9}$$

with $\lambda_\alpha = \lambda/(\pi\alpha)^2 = \tau_d\alpha^{-2}$, thus reducing to the DE result for vanishing g. For short chains, i.e., large g one obtains an expression $G_R(t)$ – by the way quite similar to the one of the Rouse model – which satisfies $G_R(t) = -g\lambda\tau^{-1}dG/dt$.

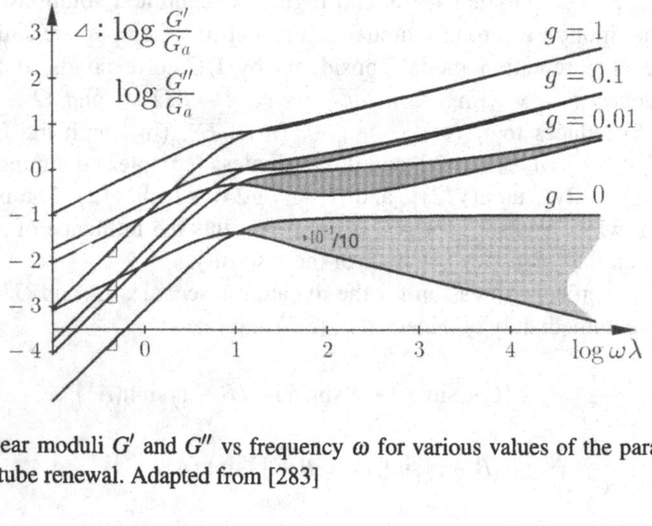

Fig. 6.1. Shear moduli G' and G'' vs frequency ω for various values of the parameter g for anisotropic tube renewal. Adapted from [283]

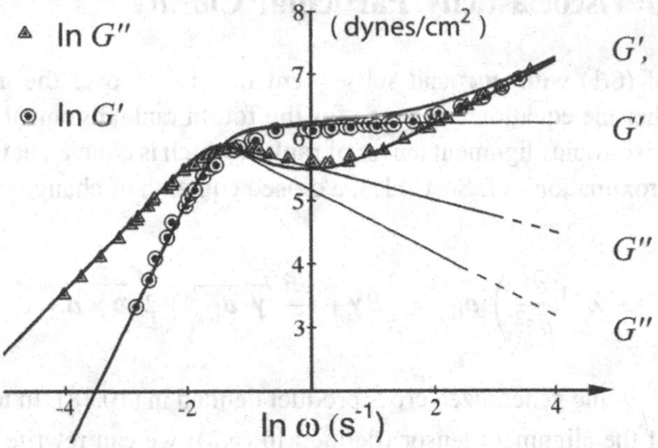

Fig. 6.2. Comparison between theory and experiment for the loss and storage moduli, (6.5). Experiments (symbols) are for on a monodisperse polysterene melt ($M_w = 215000$) [286]. The moduli are functions of shear rate reduced to a reference temperature of $T^{red.} = 160°C$ by a factor a_T. The two upper *solid lines* (for G' and G'') pertain to the theoretical parameters $G_a = 1.7 * 10^6$ dynes cm^{-2}, $\lambda = 260$s and $\tau_{\text{end}} = g\lambda = 1$s. The theoretical curves for $g = 0$ corresponding to the results of Doi and Edwards [216], Curtiss and Bird [4], de Gennes [202] are also shown. The calculation of Doi [267] takes into account fluctuations in the length of the 'primitive chain'. Adapted from [283]

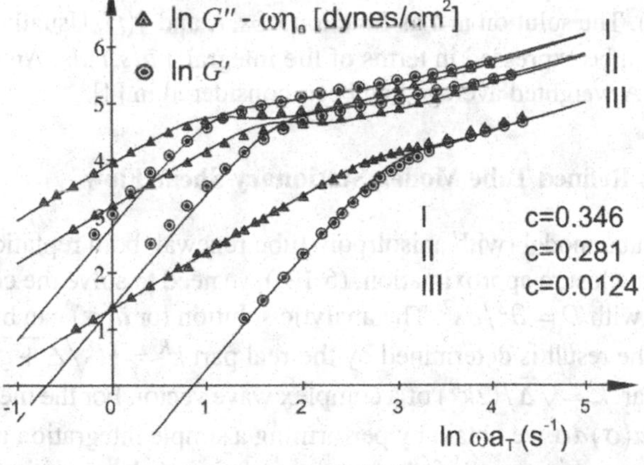

Fig. 6.3. Comparison between theory (6.5) and experiment (symbols) for the loss and storage moduli vs frequency for polysterene of molecular weight 267,000 dissolved in chlorinated diphenyl at the concentrations c shown (in g/cm^3) [17,287]

6.3 Nonlinear Viscoelasticity, Particular Closure

Multiplication of (6.1) with $\overline{\boldsymbol{uu}}$ and subsequent integration over the unit sphere, considering further the equation of change for the fourth rank alignment tensor and neglecting the anisotropic alignment tensor of rank 6, which is equivalent to a specific 'decoupling approximation', cf. Sect. 11.3, a closed equation of change is obtained,

$$\left(\frac{\partial}{\partial t}+\tau^{-1}-\lambda^{-1}\frac{\partial^2}{\partial s^2}\right)\boldsymbol{a}_{[2]}=\frac{2}{5}B\boldsymbol{\gamma}+\frac{6B}{7}\overline{\boldsymbol{\gamma}\cdot\boldsymbol{a}_{[2]}}+2\,\overline{\boldsymbol{\omega}\times\boldsymbol{a}_{[2]}}\,, \tag{6.10}$$

where $\overline{\boldsymbol{\omega}\times\boldsymbol{a}_{[2]}}_{ij}$ is the generalized cross product defined in (10.18). In terms of the components $\tilde{\boldsymbol{a}}$ of the alignment tensor (defined in (6.2)) we can rewrite (6.10) for stationary of time-dependent simple shear flow as

$$\mathcal{D}\underbrace{\begin{pmatrix}a_+\\a_-\\a_0\end{pmatrix}}_{\equiv\tilde{\boldsymbol{a}}}=\underbrace{\begin{pmatrix}\varsigma&\Gamma&\Upsilon\\-\Gamma&\varsigma&0\\\Xi&0&\varsigma\end{pmatrix}}_{\equiv\boldsymbol{M}}\cdot\tilde{\boldsymbol{a}}+\underbrace{\begin{pmatrix}-\Theta\\0\\0\end{pmatrix}}_{\equiv\boldsymbol{\Theta}},\quad\text{with}\quad\begin{pmatrix}\Upsilon\\\Xi\\\Theta\end{pmatrix}\equiv B\Gamma\begin{pmatrix}2/7\\3/14\\1/5\end{pmatrix}, \tag{6.11}$$

i.e., $\mathcal{D}\tilde{\boldsymbol{a}}=\boldsymbol{M}\cdot\tilde{\boldsymbol{a}}+\boldsymbol{\Theta}$ with the differential operator $\mathcal{D}$, a matrix $\boldsymbol{M}$ and inhomogeneity (vector) $\boldsymbol{\Theta}$, dimensionless shear rate $\Gamma=\dot{\gamma}\lambda$, ratio between reptation and orientational relaxation times $\varsigma=\lambda/\tau=\pi^2\tau_d/\tau$ and dimensionless coefficients Υ,Ξ,Θ defined in (6.11). The solution is $\tilde{\boldsymbol{a}}$ as function of s, t, and $\dot{\gamma}(t)$. Usually the rheological quantities can be expressed in terms of the integral $\int\tilde{\boldsymbol{a}}(s,t)ds$. An example will be given below. A weighted average had been considered in [4].

6.3.1 Example: Refined Tube Model, Stationary Shear Flow

For the refined tube model (with anisotrpoiv tube renewal, both reptation and orientational damping, closure approximation, (6.10)) we need to solve the corresponding matrix equation with $\mathcal{D}=\partial^2/\partial s^2$. The analytic solution for $\tilde{\boldsymbol{a}}(\sigma)$ can be immediatly written down. The result is determined by the real part $k^R=\{(\sqrt{\Delta+\varsigma^2}-\varsigma)/2\}^{1/2}$ and imaginary part $k^I=\sqrt{\Delta}/(2k^R)$ of a complex wave vector. For the mean alignment (vector) $\tilde{\boldsymbol{a}}\equiv\int_0^1\tilde{\boldsymbol{a}}(\sigma)\,d\sigma$ we obtain by performing a simple integration an explicit result for the alignment in terms of shear rate, reptation and orientational relaxation times, shape factor B, and parameterized tube renewal:

$$\begin{pmatrix}a_+\\a_-\\a_0\end{pmatrix}=\frac{1}{\sqrt{\Delta(\Delta+\varsigma^2)}}\begin{pmatrix}\sqrt{\Delta}\cap&\sqrt{\Delta}\cup\\\Gamma\cup&-\Gamma\cap\\-\Xi\cup&\Xi\cap\end{pmatrix}\cdot\begin{pmatrix}k^R&k^I\\k^I&-k^R\end{pmatrix}\cdot\begin{pmatrix}\sin\frac{k^R}{2}\cosh\frac{k^I}{2}\\\cos\frac{k^R}{2}\sinh\frac{k^I}{2}\end{pmatrix}$$
$$+\frac{\Theta}{(\Delta+\varsigma^2)}\begin{pmatrix}\varsigma\\\Gamma\\-\Xi\end{pmatrix}. \tag{6.12}$$

where all symbols except $\cap, \cup, V, \Lambda$ being related to the parametric (tube renewal) boundary conditions $a_{\pm}^{\text{end}} \equiv a_{\pm}(s=0)$ were introduced in terms of dimensionless shear rate, shape factor B, and ratio ς just above. We have $(\cap,\cup) \equiv (\cos k^R + \cosh k^I)^{-1}\,((\Lambda,-V),(V,\Lambda))\;(\cos\frac{k^R}{2}\cosh\frac{k^I}{2}, \sin\frac{k^R}{2}\sinh\frac{k^I}{2})$. with $\Lambda \equiv a_{+}^{\text{end}} - \varsigma\Theta/(\Delta+\varsigma^2)$ and $V \equiv \Gamma^{-1}\sqrt{\Delta}(a_{-}^{\text{end}} - \Gamma\Theta/(\Delta+\varsigma^2))$. Assuming the SOR, the non-newtonian shear viscosity η is obtained from $\tilde{\boldsymbol{a}}$ through $\eta = 2C^{-1}\dot{\gamma}^{-1}a_{+}$ with a stress-optic coefficient C discussed earlier. The same applies to the normal stress differences (captured by a_{-}, a_0).

6.3.2 Example: Transient Viscosities for Rigid Polymers

For this example we evaluate (6.10) without reptation ($\lambda^{-1}=0$) and the differential operator is identified to be $\mathcal{D} = -\lambda\partial/\partial t$ (just formally, λ drops out in the result). The analytic solution for the time-dependent alignment vector reads $\tilde{\boldsymbol{a}}(t) = \boldsymbol{B}\cdot[\tilde{\boldsymbol{a}}(t_0)+\boldsymbol{c}] - \boldsymbol{c}$ with $\boldsymbol{B} = \exp\{-\boldsymbol{M}(t-t_0)/\lambda\}$ and $\boldsymbol{c} = \boldsymbol{M}^{-1}\cdot\boldsymbol{\Theta}$. The solution can be rewritten in terms of the eigensystem of $\boldsymbol{M}$. For a case of isotropic rods, $B=1$ at time $t_0=0$, the time evolution of $\tilde{\boldsymbol{a}}(t)$ is plotted in Fig. 6.4.

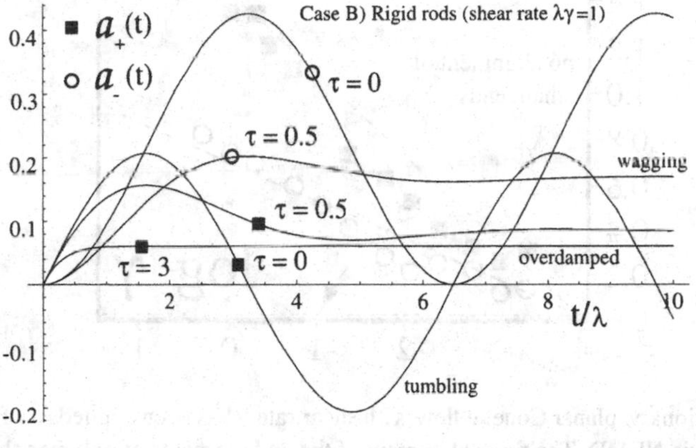

Fig. 6.4. A particular case of the presented analytical solution of (6.10) for the alignment tensor components $\mathbf{a}_{\pm}(t)$ of initial isotropically distributed rigid rods subjected to shear

6.3.3 Example: Doi-Edwards Model as a Special Case

We should notice, that the analytic solution (6.12) for isotropic chain ends ($a_{\pm}^{\text{end}}=0$) and without orientational damping ($\tau^{-1}=0$) provides an analytical approximation for the numerical result of the DE model [58, 216]. Using (6.12) we arrive – for steady shear – at

$$a_{+} = \frac{1}{10}\dot{\gamma}\lambda B\left(\frac{\sinh x - \sin x}{\cosh x + \cos x}\right)x^{-3}, \quad x \equiv \frac{1}{\sqrt{2}}(\dot{\gamma}\lambda)^{\frac{1}{2}}\left(1 - \frac{3}{49}B^2\right)^{\frac{1}{4}}, \tag{6.13}$$

As can be seen from this expression, for low shear rates the shear alignment a_+ increases linearly with shear rate $\dot{\gamma}$, for high rates $a_+ \sim \dot{\gamma}^{-1/2}$ in agreement with [216]. Using the SOR, $\eta \sim a_+\dot{\gamma}^{-1}$ is the shear viscosity, and $\Psi_1 \sim -2a_-\dot{\gamma}^{-2}$ and $\Psi_2 \sim (2a_0 + a_-)\dot{\gamma}^{-2}$ are the viscometric functions [4]. In the DE limit our approximate model yields $\Psi_2 = \Psi_1 \lim_{\dot{\gamma}\to 0} \Psi_2/\Psi_1$ and $\lim_{\dot{\gamma}\to 0} \Psi_2/\Psi_1 = 3B/14 - 1/2$, showing that Ψ_1 and Ψ_2 possess the same characteristic dependence on shear rate. The original DE model considers rod-like segments, i.e. $B = 1$, for which recover the expected and famous result $\Psi_2/\Psi_1 = -2/7$. If both the orientational diffusion constant and anisotropic tube renewal are taken into account, different power laws appear which can be used to classify the systems rheological behavior [4, 216, 283]. A consistent procedure is still missing to calculate the tube renewal parameter $a_\pm^{\text{end}}$. Figure 6.5 suggests $a_\pm^{\text{end}}/a_\pm^{center} \propto \dot{\gamma}$.

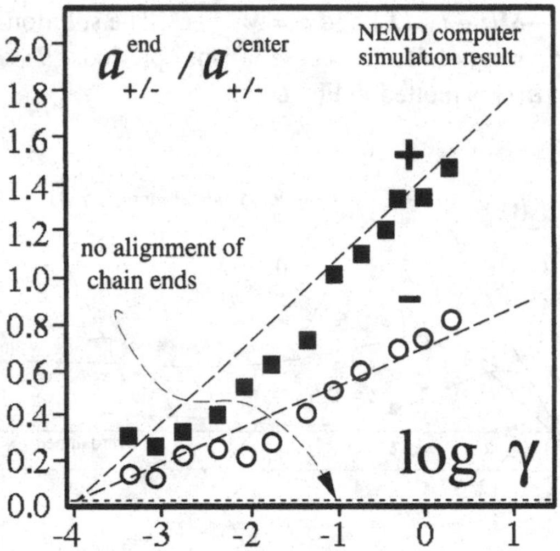

Fig. 6.5. A stationary, planar Couette flow with shear rate $\dot{\gamma}$ has been applied to a FENE model polymer melt via NEMD. The finite alignment of the end segments of polymer chains relative to the alignment of the centers of chains is shown for two components of the alignment tensor. Note that for a_+ (being closely related to the shear viscosity) the alignment of the chain ends is more pronounced than the alignment of the centers of chains at sufficiently high shear rates. The effect on the rheological quantities is important, and quantified in this paragraph

6.4 Nonlinear Viscoelasticity without Closure

For the three examples just discussed we started from a closed, approximate equation of change for the second rank alignment tensor, (6.10). We want to shortly summarize on how the underlying Fokker–Planck equation (6.1) incl. the effect of anisotropic tube renewal had been solved to within given precision in [288] using Galerkin's principle [4]. The same methodology had been recently used in [289] to improve on an efficient realization of the micro-macro CONNFFESSIT [290] approach for the case where a low-dimensional Fokker–Planck equation carrying the recommended ingredients (double reptation, convective constraint release etc., cf. Sect. 6.1) is available. There are several alternative strategies. One of them is BD which we already used in the first chapters (see also Sect. 8.5), and which should be the prefered method for solving non-trivial high dimensional Fokker–Planck equations [58].

6.4.1 Galerkin's Principle

Galerkin's principle The idea is to solve the Fokker–Planck equation (6.1) by expanding $f(\boldsymbol{u},\sigma)$ in spherical harmonics ψ and even Euler polynomials E [291]

$$f^{(M,I)}(\boldsymbol{u},\sigma) = \sum_{k=0}^{1}\sum_{n=0}^{M}\sum_{m=0}^{n}\sum_{i=0}^{I} A^i_{knm}\,\psi^m_{kn}(\boldsymbol{u})E_{2i}(\sigma)\,, \tag{6.14}$$

with $\psi^m_{0n} = P^m_n(\cos\theta)\cos\phi$, $\psi^m_{1n} = P^m_n(\sin\theta)\sin\phi$. Inserting the series $f^{(M,I)}$ into (6.1) and applying Galerkin's principle $\int d\phi \int d\theta \int d\sigma\, \hat{D}[f^{(M,I)}]\,\psi^p_{lq}E_j \sin\theta = 0$, for $l=0..1$, $q=0..M$, $p=0..q$, $j=0..I$ leads to coupled linear equations for the coefficients A^i_{knm} as function of the dimensionless ratio $\varsigma = \lambda/(6\tau)$ and the dimensionless shear rate $\Gamma = \dot{\gamma}\tau$. These equations were derived in [288]. A finite bending of f at the chain ends (anisotropic tube renewal) is captured through a coefficient

$$x \equiv \partial^2/\partial\sigma^2 \int f(\boldsymbol{u},\sigma)\,d^2\boldsymbol{u}\,|_{\sigma=0,\sigma=1} = A^2_{000}\,, \tag{6.15}$$

while we allow the integral $\int f(\boldsymbol{u},\sigma)\,d^2\boldsymbol{u}$ to depend on σ. The normalization for f reads $\sum_{i=0}^{I} A^i_{000}\,N_E(i,0) = 1$, with $N_E(i,f) \equiv \int_0^1 d\sigma E_i E_f = \alpha_{if}((i+f+2)!)^{-1}B_{i+f+2}$, involving the Bernoulli numbers B [291] and $\alpha_{if} \equiv 4(-1)^i(2^{i+f+2}-1)i!f!$. The coefficients $\forall_{n,i}A^i_{10n}$ are left undetermined in the ansatz (6.14). Finally there is an equal number of $(M/2+1)^2(I/2+1)$ nontrivial equations and unknowns to solve for given parameters Γ, ς and x.

The rheological behavior is infered from the moments (or weighted moments, cf. the parameter ε used by Bird et al. [4] for additional 'viscous' contributions) of f, and had been also discussed in [288]. The effect of ratio of relaxation times ς on the alignment tensor components $a_\pm$ (for a fixed value for x), together with the corrresponding components of the viscous contribution proposed by Bird et al. [4] and denoted as $k_\pm$ are shown in Fig. 6.6. A plateau (undershoot) in a_+ appears with decreasing ς, and $k_\pm$ dominates at very high rates. The latter term can bbe actually used to predict a wide range of power law behaviors for the shear viscosity vs rate by varying ς. The influence of the finite bending of f at the chain ends, i.e. $x \neq 0$, on the alignment of segments is shown in Fig. 6.7. Perhaps surprising is the result for the dependence of a_+ on the contour position. At vanishing shear rates the components

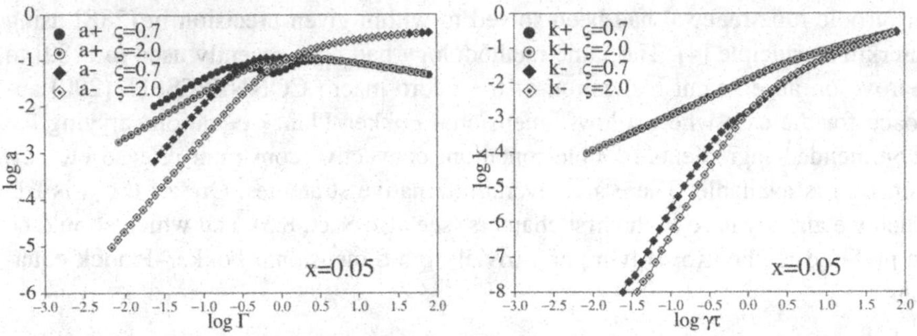

Fig. 6.6. Components of the dimensionless alignment quantitites (*left*) $a_\pm = \langle \boldsymbol{uu} \rangle_\pm$ (relevant for rheological properties if the SOR is valid) and (*right*) $k_\pm$ denoting a corresponding contributions to the viscous part [4] of the stress tensor $\boldsymbol{\kappa} : \langle \boldsymbol{uuuu} \rangle_\pm$. for selected ratios ς between orientational and reptatkon diffusion coefficients, and with boundary condition for the chain end $x = 0.05$ vs. dimensionless shear rate $\Gamma = \dot{\gamma}\tau$

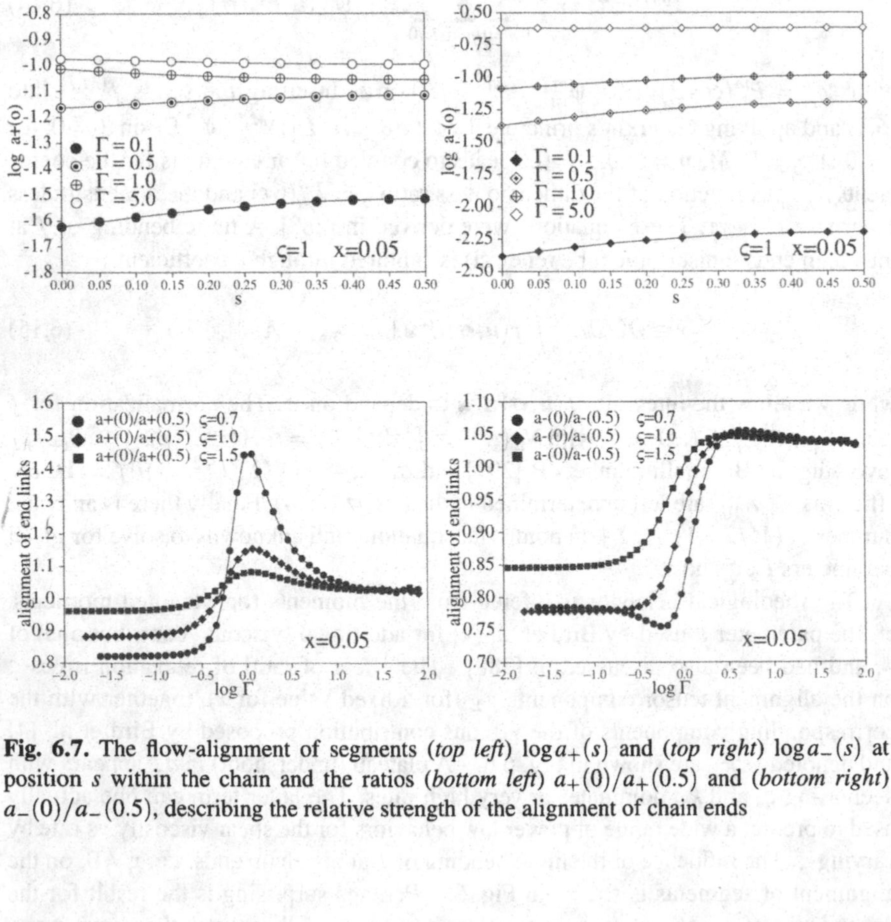

Fig. 6.7. The flow-alignment of segments (*top left*) $\log a_+(s)$ and (*top right*) $\log a_-(s)$ at position s within the chain and the ratios (*bottom left*) $a_+(0)/a_+(0.5)$ and (*bottom right*) $a_-(0)/a_-(0.5)$, describing the relative strength of the alignment of chain ends

$a_{\pm}$ of the symmetric traceless 2nd rank alignment tensor vanish. At high rates the component a_+ at the ends is larger than a_+ at the chain's center, while the component a_- monotoneously increases with rate – for all contour positions. The centers of the chains are more aligned in direction of flow (characterized by a_-) than the outer parts. Since the a_+-component must rise and fall with shear rate and has a maximum at a certain characteristic shear rate, the chain end will follow this behavior - just shifted to larger rates. These predictions are in very good agreement with results from NEMD of polymer melts described in Chaps. 4, 6 and also illustrate why the effect of anisotropic tube renewal has an important effect on the shear viscosity (which is connected with a_+ but not with a_-).

7

Elongated Particle Models

In [292] we provided a statistical interpretation of the director theory of Ericksen and Leslie (EL) [293–295] for nematic liquid crystals. Starting from a Fokker–Planck equation of the type (6.1) supplemented by a mean-field plus external potential, and using an expression for the stress tensor derived for structural theories of suspensions, we interpreted the EL viscosity coefficients and molecular fields in terms of the parameters characterizing a suspension, i.e., particle geometry, particle concentration, degree of alignment, solvent viscosity, and the potential. It turned out that the theory of Kuzuu and Doi [73] for concentrated suspensions of rod-like polymers, the affine transformation model by Hess and Baalss [296], the results by Hand [297] and Sin-Doo Lee [298], were contained as special cases. In distinction to Kuzuu and Doi in [292] we also obtained an expression for the tumbling parameter in terms of order parameters and particle shape, which had been confirmed independently by Archer and Larson [299]. Here, in order to review the highly coarse-grained models depicted at the top of Fig. 1.2 we summarize the macroscopic framework developed by EL. We give an example on how the microscopic quantities such as an anistropic gyration tensor for polymeric chains, or the shape of suspended ellipsoidal (colloidal) particles enter the anisotropic viscosities.

There are various approaches in the literature to modeling fluids with microstructure. For example, equations for suspensions of rigid particles have been calculated by averaging the detailed motion of the individual particles in a Newtonian fluid. In particular, the solution for the motion of a single ellipsoid of revolution in a steady shear [49] can be used to determine the governing equations for the slow flow of a dilute suspension of non-interacting particles. For more concentrated systems, various approximations to the particle motions have been used. This approach, based upon a detailed analysis of the microstructure, has been called 'structural' by Hinch and Leal (HL) [50]. Alternatively, 'phenomenological' continuum theories for anisotropic fluids have been postulated. They tend to be quite general, being based upon a small number of assumptions about invariance, perhaps the most successful and well known example being the EL director theory for uniaxial nematic liquid crystals [293, 294], Additionally, numerous models have been developed and discussed in terms of symmetric second- and higher-order tensorial measures of the alignment [216, 234, 235, 300–305].

Given these diverse methods of derivation and apparently diverse domains of application, one may ask, however, if and how such diverse approaches may be interrelated. Several comparisons have already been made. In particular, Hand [297] obtained the governing equations for dilute suspensions of ellipsoids of revolution without rotary diffusion and subject to no potential (thus perfectly aligned), showed that they could be modeled also by the simpler EL director theory for transversely isotropic fluids [306], and calculated the corresponding viscosities in terms of the suspension parameters. Furthermore Marrucci [307], Semonov [308], and Kuzuu and Doi [73] related the EL theory to a dynamical mean-field theory for concentrated suspensions of rigid rods and thereby calculated the Leslie and Miesowicz viscosity coefficients in terms of the suspension parameters.

7.1 Director Theory

The traditional EL theory of anisotropic fluids [293, 309] assumes that there is a unit vector field $\boldsymbol{n}(\boldsymbol{x},t)$ (called the director) representing the average alignment at each point of the fluid. The extension [295] also introduces a variable degree of alignment represented by the scalar field $S(\mathbf{x},t)$, where $-1/2 \leq S \leq 1$. The extended EL (also denoted by EL in the following) constitutive relation for the hydrodynamic stress tensor $\boldsymbol{\sigma}$ of an incompressible anisotropic fluid with velocity $\boldsymbol{v}$ is given by the following expression linear in the nonequilibrium variables $\dot{S}$, $\boldsymbol{\gamma}$, and $\boldsymbol{N}$:

$$\boldsymbol{\sigma} = (\alpha_1 \boldsymbol{n}_{(2)} : \boldsymbol{\gamma} + \beta_1 \dot{S})\boldsymbol{n}_{(2)} + \alpha_2 \boldsymbol{n}\boldsymbol{N} + \alpha_3 \boldsymbol{N}\boldsymbol{n} + \alpha_4 \boldsymbol{\gamma} + \alpha_5 \boldsymbol{n}_{(2)} \cdot \boldsymbol{\gamma} + \alpha_6 \boldsymbol{\gamma} \cdot \boldsymbol{n}_{(2)} , \quad (7.1)$$

where $\boldsymbol{n}_{(2)} \equiv \boldsymbol{n}\boldsymbol{n}$, and

$$\boldsymbol{N} \equiv \dot{\boldsymbol{n}} - \boldsymbol{\Omega} \cdot \boldsymbol{n} , \quad (7.2)$$

$$\boldsymbol{\gamma} \equiv (\boldsymbol{\kappa} + \boldsymbol{\kappa}^\dagger)/2 = \boldsymbol{\gamma}^{\mathrm{T}} , \quad (7.3)$$

and

$$\boldsymbol{\Omega} \equiv (\boldsymbol{\kappa} - \boldsymbol{\kappa}^\dagger)/2 = -\boldsymbol{\Omega}^{\mathrm{T}} , \quad (7.4)$$

with $\boldsymbol{\kappa} = (\nabla \boldsymbol{v})^\dagger$. In addition to the usual balance of momentum, $\rho \dot{\boldsymbol{v}} = -\nabla p + \nabla \cdot \boldsymbol{\sigma}^{\mathrm{T}}$, there are two additional equations governing the microstructure: *i)* a vector equation for the director $\boldsymbol{n}$ (here we neglect director inertia)

$$\boldsymbol{0} = \boldsymbol{n} \times (\boldsymbol{h}_n - \gamma_1 \boldsymbol{N} - \gamma_2 \boldsymbol{\gamma} \cdot \boldsymbol{n}) , \quad (7.5)$$

or equivalently, $\boldsymbol{0} = (\boldsymbol{1} - \boldsymbol{n}_{(2)}) \cdot (\boldsymbol{h}_n - \gamma_1 \boldsymbol{N} - \gamma_2 \boldsymbol{\gamma} \cdot \boldsymbol{n})$, where $\boldsymbol{h}_n$ is the vector molecular field (which is indeterminate to a scalar multiple of $\boldsymbol{n}$); *ii)* a scalar equation for the degree of alignment S (again neglecting inertia)

$$0 = h_S - \beta_2 \dot{S} - \beta_3 \boldsymbol{n}_{(2)} : \boldsymbol{\gamma} \,, \tag{7.6}$$

where h_S is the scalar molecular field.[1] The α_i are commonly called Leslie viscosity coefficients. The β_i were recently introduced in by Ericksen [295] for the case of variable degree of alignment. Furthermore the coefficients γ_i are related to the α_i by $\gamma_1 = \alpha_3 - \alpha_2$, $\gamma_2 = \alpha_6 - \alpha_5$. There are also two restrictions (Onsager relations) that follow from the existence of a dissipation potential: $\alpha_2 + \alpha_3 = \alpha_6 - \alpha_5$ (Parodi's relation [234]), and $\beta_1 = \beta_3$ (proposed by Ericksen). Dissipation arguments lead to the following restrictions on the coefficients [295]: $\alpha_4 \geq 0$, $\gamma_1 \geq 0$, $\beta_2 \geq 0$, $\alpha_1 + 3\alpha_4/2 + \alpha_5 + \alpha_6 - \beta_1^2/\beta_2 \geq 0$, $2\alpha_4 + \alpha_5 + \alpha_6 - \gamma_2^2/\gamma_1 \geq 0$. Particular micro-based realizations of the 'macroscopic' equations will be presented next.

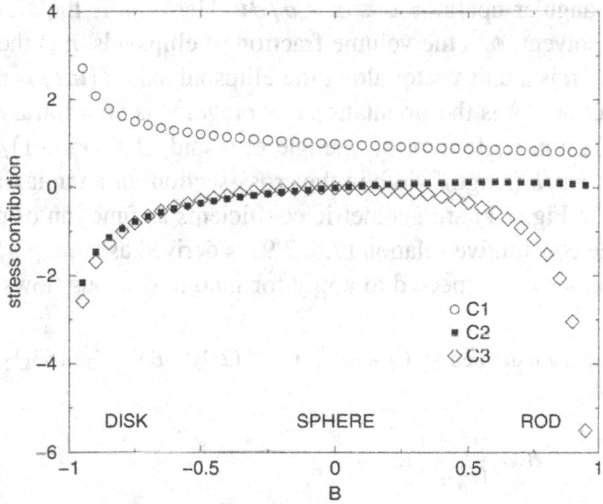

Fig. 7.1. Effect of particle shape on the relevance of the stress contributions for suspensions of ellipsoids of revolution, (7.8)

7.2 Structural Theories of Suspensions

Consider a dilute suspension of neutrally buoyant, rigid ellipsoids of revolution dispersed in an incompressible Newtonian fluid at thermal equilibrium. The governing equations can be determined from Jeffery's [49] solution for the motion of a single ellipsoid in a homogeneous shear flow. In terms of the notation of Brenner and Condiff [310], we have for the dynamic stress tensor

[1] In terms of a free energy $F(S, \nabla S, \boldsymbol{n}, \nabla \boldsymbol{n})$, the molecular fields are given by

$$\boldsymbol{h}_n = \nabla \cdot \frac{\partial F}{\partial \nabla \boldsymbol{n}} - \frac{\partial F}{\partial \boldsymbol{n}}, \quad h_S = \nabla \cdot \frac{\partial F}{\partial \nabla S} - \frac{\partial F}{\partial S} \,. \tag{7.7}$$

$$\boldsymbol{\sigma} = 2\mu_0\boldsymbol{\gamma} + 5\mu_0\phi\,\langle\boldsymbol{A}\rangle - \frac{n}{2}\boldsymbol{\varepsilon}\cdot\langle\boldsymbol{L}\rangle + O(\phi^2)\,, \tag{7.8}$$

where $\boldsymbol{A}$ is the stresslet and $\boldsymbol{L}$ is the applied couple on each particle. They are given by

$$\begin{aligned}\langle\boldsymbol{A}\rangle &= C_1\boldsymbol{\gamma} + C_2(\boldsymbol{\gamma}\cdot\boldsymbol{a}_{(2)} + \boldsymbol{a}_{(2)}\cdot\boldsymbol{\gamma}) - C_3\boldsymbol{\gamma}:\boldsymbol{a}_{(4)}\\ &\quad + ND_r\left[\left\langle \boldsymbol{u}\frac{\partial}{\partial\boldsymbol{u}}U\right\rangle + \left\langle\left(\frac{\partial}{\partial\boldsymbol{u}}U\right)\boldsymbol{u}\right\rangle\right],\\ \langle\boldsymbol{L}\rangle &= -\left\langle \boldsymbol{u}\times\frac{\partial}{\partial\boldsymbol{u}}V\right\rangle = -\langle\mathcal{L}V\rangle\,,\end{aligned} \tag{7.9}$$

with the 'brownian potential' $U \equiv \ln f + V/k_BT$, cf. Chap. 11 and footnote on Page 175, and angular operator $\mathcal{L} \equiv \boldsymbol{u}\times\partial/\partial\boldsymbol{u}$. Here, μ_0 is the Newtonian shear viscosity of the solvent, ϕ is the volume fraction of ellipsoids, n is the number density of ellipsoids, $\boldsymbol{u}$ is a unit vector along the ellipsoid axis, $f(\boldsymbol{u},t)$ is the orientation distribution function, $\langle\cdot\rangle$ is the orientational average, V is an arbitrary potential, D_r is the rotary diffusion coefficient of a single ellipsoid, $B \equiv (r_p^2-1)/(r_p^2+1)$ with the axis ratio $r_p = a/b$ (length/width in the cross-section) of an uniaxial ellipsoid, N and $C_{..}$ (plotted in Fig. 7.1) are geometric coefficients as function of particle shape, cf. footnote[2] The constitutive relation (7.8, 7.9) is derived assuming a homogeneous shear flow. It can also be expected to apply for inhomogeneous flows [311]. There

[2] Shape coefficients are defined by $C_1 \equiv 2Q_1, C_2 \equiv (2Q_3 - BN), C_3 \equiv (3Q_2 + 4Q_3 - 2BN)$. with

$$\begin{aligned} B &= \frac{(r_p^2-1)}{(r_p^2+1)},\\ N &= \frac{2(r_p^2-1)^2}{5r_p^2[2r_p^2\beta - \beta - 1]},\\ Q_1 &= \frac{4(r_p^2-1)^2}{5r_p^2(3\beta + 2r_p^2 - 5)},\\ Q_2 &= \frac{2Q_1}{3}\left[1 - \frac{2r_p^2 + 1 - (4r_p^2-1)\beta}{4(2r_p^2+1)\beta - 12}\right],\\ Q_3 &= Q_1\left[\frac{[r_p^2(\beta+1)-2](3\beta+2r_p^2-5)}{4[\beta(2r_p^2-1)-1](r_p^2+2-3r_p^2\beta)} - 1\right],\end{aligned} \tag{7.10}$$

where $r_p \equiv a/b$ is the axis ratio of ellipsoid and

$$\beta = \frac{\cosh^{-1} r_p}{r_p(r_p^2-1)^{\frac{1}{2}}} \quad \text{(for } r_p > 1\text{, i.e., prolate spheroids)}$$

$$\beta = \frac{\cos^{-1} r_p}{r_p(1-r_p^2)^{\frac{1}{2}}} \quad \text{(for } r_p < 1\text{, i.e., oblate spheroids)}$$

is also a convection-diffusion equation (of the Fokker–Planck type) for the orientation distribution function f, which allows for the calculation of the evolution of the moments of the alignment, cf. Chap. 11, i.e., (6.1)) with an orienting torque due to external fields (flow plus potential)

$$\boldsymbol{T} = \boldsymbol{T}_{\text{flow}} - \frac{D_r}{k_B T}\mathcal{L}V\,. \tag{7.11}$$

In equilibrium the orientation distribution function becomes the canonical distribution, cf. Chap. 11. We will make use only of the equation for the rate of change of the second-moment of the alignment $\boldsymbol{a}_{(2)} \equiv \langle \mathbf{uu} \rangle$. It follows directly from the Fokker–Planck equation:

$$\begin{aligned} \frac{\partial}{\partial t}\boldsymbol{a}_{(2)} &= 2B\boldsymbol{\gamma} : \boldsymbol{a}_{(4)} + \boldsymbol{\Omega}\cdot\boldsymbol{a}_{(2)} - \boldsymbol{a}_{(2)}\cdot\boldsymbol{\Omega} + B(\boldsymbol{\gamma}\cdot\boldsymbol{a}_{(2)} + \boldsymbol{a}_{(2)}\cdot\boldsymbol{\gamma}) \\ &\quad - D_r\left[\left\langle \boldsymbol{u}\frac{\partial}{\partial\boldsymbol{u}}U + (\frac{\partial}{\partial\boldsymbol{u}}U)\boldsymbol{u}\right\rangle\right]. \end{aligned} \tag{7.12}$$

Furthermore, we have the following relations between the coefficients [310]:

$$Bck_BT = 10\mu_0\phi N D_r, \quad \phi = n v_p\,, \tag{7.13}$$

where $v_p = 4\pi ab^2/3$ is the volume of an ellipsoid. The correspondence between micro- and macroscopic equations will be presented for a special case in Sect. 7.2.2. A more general case had been discussed in [292].

7.2.1 Semi-Dilute Suspensions of Elongated Particles

Batchelor [312] has calculated the effect of hydrodynamic interaction of parallel elongated particles (without brownian motion) in a pure steady straining motion ($\boldsymbol{\Omega} = \mathbf{0}$) on the bulk stress tensor. For elongated particles of length a on which no external force or couple acts and taking up the same preferred orientation, Batchelor gave the approximate relation for the stress tensor which can be compared immediately to those of the EL theory with $\alpha_1 = 4\pi/(3V)\,\sum(a/2)^3/(\ln h/R_0)$, $\alpha_4 = 2\mu_0$, $\alpha_{2,3,5,6} = 0$, $S = 1$. where $\boldsymbol{n}$ is the direction of the particle axes, the sum is over the particles in the volume V, R_0 is the effective radius of the cross-section of the particle, and $h = (na)^{-1/2}$.

7.2.2 Concentrated Suspensions of Rod-Like Polymers

Doi [313] has presented a dynamical mean field theory for concentrated solutions of rod-like polymers. We follow here the version by Kuzuu and Doi [73]. Viscous contributions to the stress tensor are generally assumed negligible, but we include the viscosity μ_0 of the solvent. The stress tensor of this model formally equals expression (7.8) with $C_1 = C_2 = C_3 = 0$ in (7.9). The potential is composed of two contributions

$$V = V_m + V_e, \quad V_e = -\frac{1}{2}\chi_a \boldsymbol{HH} : \boldsymbol{u}_{[2]}, \quad V_m = -\frac{3}{2} U_m k_B T \boldsymbol{a}_{[2]} : \boldsymbol{u}_{[2]}, \tag{7.14}$$

V_e denotes the contribution due to an induced dipole by an external field $\boldsymbol{H}$, χ_a being the anisotropic susceptibility of a rod, and V_m denotes the mean-field contribution, U_m being a constant reflecting the energy intensity of the mean field. A similar equation was also presented by Hess [314].

7.3 Uniaxial Fluids, Micro-Macro Correspondence

It is common to classify the types of alignment according to the eigenvalues of the second moment of the alignment:

$$\boldsymbol{a}_{(2)} \equiv \langle \boldsymbol{uu} \rangle = A_1 \boldsymbol{ll} + A_2 \boldsymbol{mm} + (1 - A_1 - A_2) \boldsymbol{n}_{(2)}, \tag{7.15}$$

where $\boldsymbol{l}$, $\boldsymbol{m}$, and $\boldsymbol{n}$ form a triad of orthogonal unit vectors. In the special case in which the distribution of particles of the suspension in a given flow is *uniaxial*, e.g., $f_{uni} = f(|\boldsymbol{u} \cdot \boldsymbol{n}|)$, $\boldsymbol{n}(\boldsymbol{x},t)$ denoting the axis of symmetry, one obtains that $A_1 = A_2$. Traditionally, the parameter $S_2 \equiv 1 - 3A_1$ is used. In this case, we have the following explicit relations for the second and fourth moments of the alignment [73, 82]:

$$\boldsymbol{a}_{[2]_{\mathrm{uni}}} = S_2 \boldsymbol{n}_{[2]} \Leftrightarrow \boldsymbol{a}_{(2)_{\mathrm{uni}}} = S_2 \boldsymbol{n}_{(2)} + \frac{1}{3}(1 - S_2)\boldsymbol{1}, \tag{7.16}$$

and (in cartesian coordinates) $\langle u_i u_j u_k u_l \rangle_{\mathrm{uni}} = S_4 n_i n_j n_k n_l + (S_2 - S_4)(\delta_{ij} n_k n_l + \delta_{ik} n_j n_l + \delta_{kj} n_i n_l + \delta_{il} n_j n_k + \delta_{jl} n_i n_k + \delta_{kl} n_i n_j)/7 + (7 - 10S_2 + 3S_4)(\delta_{ij}\delta_{kl} + \delta_{ik}\delta_{jl} + \delta_{il}\delta_{jk})/105$, where S_2 and S_4 are scalar measures of the degree of orientation related to Legendre polynomials: $S_2 = \langle P_2(\boldsymbol{u} \cdot \boldsymbol{n}) \rangle$, $S_4 = \langle P_4(\boldsymbol{u} \cdot \boldsymbol{n}) \rangle$, cf. Chap. 10. They must satisfy $-\frac{1}{2} \leq S_2, S_4 \leq 1$. In the case of perfect alignment $S_2 = S_4 = 1$, and in the case of random alignment $S_2 = S_4 = 0$. Note that the odd moments vanish identically due to symmetry of the distribution function f. Similar relations hold for the higher moments, but we refrain from writing them.

The uniaxial assumption is not valid for all flows of the suspension. More generally, the alignment will be biaxial, i.e., $A_1 \neq A_2$. The biaxial case requires the use of multiple directors plus additional biaxial scalar measures (see [315] and refs. cited herein). For this monograph we are however concerned only with those flows for which this assumption holds since we want to make a comparison to the EL theory, which assumes uniaxial symmetry. In this case we need only a single unit vector plus the set $\{S_{2i}\}$ of scalars to completely describe the alignment. Furthermore, note that each even-order moment of the alignment introduces a new scalar measure of the alignment S_{2i}. The EL theory assumes that there is a closure relation so that all higher-order parameters can be expressed as a function of S_2. Such an assumption is consistent, for example, with a Gaussian distribution about the symmetry axis $\boldsymbol{n}$. However, it will not be necessary to specify any particular closure relation.

7.3.1 Concentrated Suspensions of Disks, Spheres, Rods

Comparing micro- (7.8) with macroscopic (7.1) stress tensors and also comparing the equation of change for the alignment tensor (7.12) with (7.5) one obtains for the particular case of concentrated suspensions of rod-like polymers, cf. Sect. 7.2.2, upon extending from rods ($B=1$) to uniaxial ellipsoids also including disks ($B=-1$) and spheres ($B=0$) the following microscopic interpretation of the EL parameters [292], with $\chi \equiv nk_BT/(2D_r)$

$$\begin{aligned}
\alpha_1 &= -2\chi B^2 S_4\,,\\
\alpha_2 &= -\chi B\left(1+\lambda^{-1}\right) S_2\,,\\
\alpha_3 &= -\chi B\left(1-\lambda^{-1}\right) S_2\,,\\
\alpha_4 &= 2\mu_0 + 2B^2\left(\frac{1}{5}+\frac{1}{7}S_2\right) - \eta\frac{4}{35}B^2 S_4\,,\\
\alpha_5 &= \frac{3}{7}\chi B^2\left(S_2+\frac{4}{3}S_4\right) + \chi B S_2\,,\\
\alpha_6 &= \frac{3}{7}\chi B^2\left(S_2+\frac{4}{3}S_4\right) - \eta B S_2\,,\\
\beta_1 &= -\chi B\,,\\
\beta_2 &= 35\chi(21+15S_2-36S_4)^{-1}\,,\\
\beta_3 &= \beta_1\\
\gamma_1 &= \alpha_3-\alpha_2 = 2\chi B\lambda^{-1} S_2\,,\\
\gamma_2 &= \alpha_3+\alpha_2 = -2\chi B S_2\,,\\
\lambda &\equiv -\frac{\gamma_2}{\gamma_1} = \frac{\alpha_3-\alpha_2}{\alpha_3+\alpha_2} = \frac{(14+5S_2+16S_4)B}{35 S_2}\,,
\end{aligned} \tag{7.17}$$

where λ is the 'tumbling parameter'. Vector and scalar molecular fields are given by

$$\begin{aligned}
\boldsymbol{n}\times\boldsymbol{h}_n &= -n\left\langle\left(\boldsymbol{u}\times\frac{\partial}{\partial\boldsymbol{u}}\right)V\right\rangle_{\text{uni}},\\
h_S &= 35\,nk_BT\left\langle \boldsymbol{u}\frac{\partial}{\partial\boldsymbol{u}}U\right\rangle_{\text{uni}} (24S_4-10S_2-14)^{-1}\,.
\end{aligned} \tag{7.18}$$

One easily confirms that Parodi's relation and all other relationships known from the director theory (summarized in Sect. 7.1) are in full agreement with our micro-based expressions (7.17). Carlsson's conjecture [316,317] on the signs of α_2 and α_3 provided that S_2 is positive is also confirmed by (7.17).

7.3.2 Example: Tumbling

One way to characterize materials is according to the behavior of the director in a steady shear flow. As discussed by Chandrasekhar [236] and de Gennes [234],

$|\lambda| < 1$ implies that the director always tumbles in steady shear flow, whereas $|\lambda| \geq 1$ implies that the director has a steady solution. The above expression for the tumbling parameter λ (not provided by Kuzuu and Doi [73]) has been confirmed by Archer and Larson [299] who also took into account numerically the flow-induced biaxiality showing that there can be a modest but significant effect on the coefficient λ. Predictions (7.17) have been already compared with experiments [318–320], and extended to biaxial fluids [315]. A very similar expression for λ (using $S_4 \propto S_2^2$) had been derived early by Hess [74] for the case of uniaxial symmetry, cf. Chap. 10, based on a truncation approximation to the Fokker–Planck equation, obtaining $\gamma_1 \propto S_2^2(1-c_1 S_2^2)$, and $\gamma_2 \propto -BS_2(1+c_2S_2-c_3S_2^2)$, where $c_{1,2,3}$ are temperature dependent constants. A typical relaxation time [321] for reorientations of the director is given by $\tau = 1/(\dot{\gamma}\sqrt{\lambda^2-1})$, where $\dot{\gamma}$ is the shear rate. Thus τ is seen to be a function of the order parameters and the axis ratio.

Also the coefficients α_2 and α_3 determine the type of flow via λ. For a negative product $\alpha_2\alpha_3$ (i.e., $|\lambda| < 1$) there is no steady state solution in simple shearing, for positive $\alpha_2\alpha_3$ the molecules will be aligned under shear flow, with a flow angle χ given by $\cos 2\chi = \lambda^{-1}$. In Fig. 7.2, we can see how the sign of $\alpha_2\alpha_3$ varies with order parameter S_2 and geometry B (using the closure relation [322] $S_4 = S_2 - S_2(1-S_2)^{\nu}$ where $\nu = 3/5$, again there is no qualitative difference in the choice of the exponent ν). According to (7.17), $\lambda \longrightarrow B$ when both $S_2, S_4 \longrightarrow 1$. Also $\lambda \longrightarrow \infty$ when both $S_2, S_4 \longrightarrow 0$. Thus we will always have tumbling in the case of suspensions of almost perfectly aligned (i.e., $S_2, S_4 \approx 1$) rigid ellipsoids of revolution but steady solutions for suspensions with small degree of alignment ($S_2, S_4 \approx 0$). The transition between the two regimes is given by $|\lambda| = 1$. Note that in the case of perfect alignment (i.e., $S_2 = S_4 = 1$), (7.17) reduces to $\lambda = B$ and for ellipsoids of revolution we always have $|B| < 1$, which is the classical result that a single ellipsoid of revolution tumbles in steady shear flow [49, 323]. Figure 7.4 indicates the dependence of the tumbling of the director on the degrees of alignment S_2 and S_4. These results are independent of the particular potential, thus apply also to mean-field theory.

The calculated viscosity coefficients in (7.17) are subject to the restrictions given in Sect. 7.1. From (7.17) it follows that $\beta_2 \geq 0$ if and only if $S_4 \leq (5S_2+7)/12$. which excludes arbitrary pairs of values for S_2 and S_4. The excluded region is shown in Fig. 7.4. The remaining inequalities are automatically satisfied when $\beta_2 \geq 0$.

7.3.3 Example: Miesowicz Viscosities

It is common to measure the three Miesowicz viscosities η_i, $i = 1,2,3$ defined as the ratio of the yx-component of the stress tensor and the shear rate $\dot{\gamma}$. The label $i = 1,2,3$ refers to the cases where the director $\boldsymbol{n}$ is parallel to the x-,y-,z-axis, respectively. An orienting (magnetic) field has to be strong enough to overcome the flow induced orientation. A fourth coefficient η_4 with $\boldsymbol{n}$ parallel to the bisector between the x- and y-axes is needed to characterize the shear viscosity completely. Instead of η_4, the Helfrich viscosity coefficient $\eta_{12} = 4\eta_4 - 2(\eta_1+\eta_2)$ is used in addition to the Miesowicz coefficients. The 'rotational' viscosity γ_1 can be measured via the torque exerted on a nematic liquid crystal in the presence of a rotating magnetic field

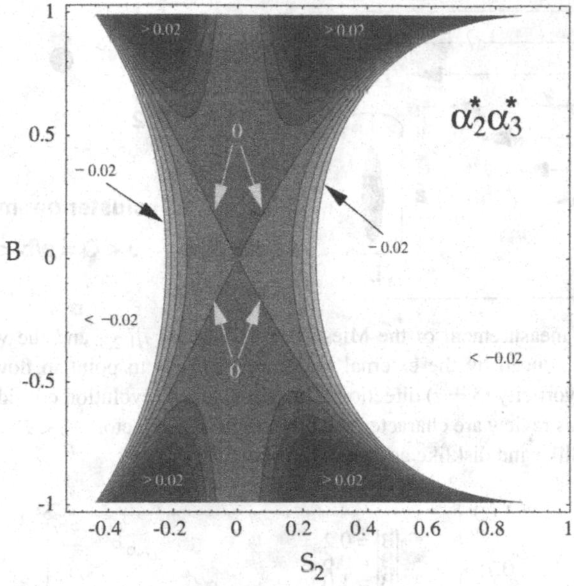

Fig. 7.2. Contour plot of $\alpha_2\alpha_3$ as a function of geometry B and order parameter S_2. Positive region corresponds to tumbling regimes, negative region to steady flow alignment. The dimensionless viscosities are defined by $\alpha_i^* := \alpha_i D_r/(ck_BT)$

(Tsvetkov effect). The four effective viscosities measurable in a flow experiment, cf. Fig. 7.3, are related to the EL viscosity coefficients by

$$\begin{aligned}\eta_1 &= (\alpha_4+\alpha_6+\alpha_3)/2\,,\\ \eta_2 &= (\alpha_4+\alpha_5-\alpha_2)/2\,,\\ \eta_3 &= \alpha_4/2\,,\\ \eta_{12} &= \alpha_1\,.\end{aligned} \tag{7.19}$$

Explicit expression for these quantities are obtained by inserting the viscosity coefficients from (7.17).

7.4 Uniaxial Fluids: Decoupling Approximations

In this section we briefly comment on the validity of closure schemes often used in the literature, in particular the so called Hinch and Leal (HL) closures. They have been used to close the infinite number of coupled equations of motion for alignment tensors, derived from the Fokker–Planck equation such as (6.1). Here we wish to point out that for the case of uniaxial symmetry there is a single possible closure which requires the knowledge of a scalar function $S_4(S_2)$ rather than a full tensorial relationship, and we will show, that this closure is different from the HL closures.

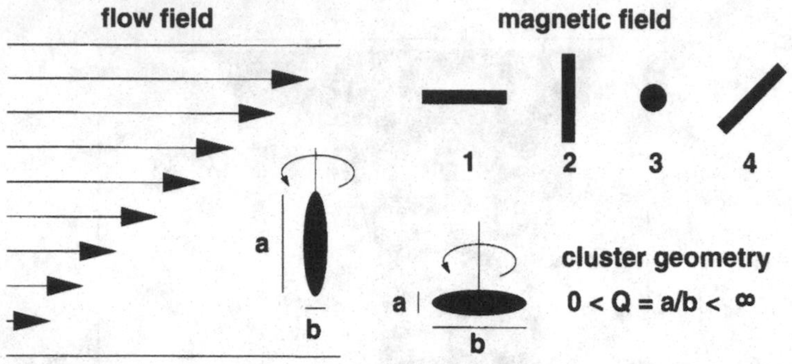

Fig. 7.3. For the measurement of the Miesowicz viscosities $\eta_{1,2,3}$ and the viscosity η_4 the magnetization – induced by the external magnetic field has to point in flow ($1 = x$), flow gradient ($2 = y$), vorticity ($3 = z$) direction. The ellipsoids of revolution considered within the FP approach in this review are characterized by a single shape factor $-1 < B < 1$ where $B > 0$ and $B < 0$ for rodlike and disklike aggregates, respectively

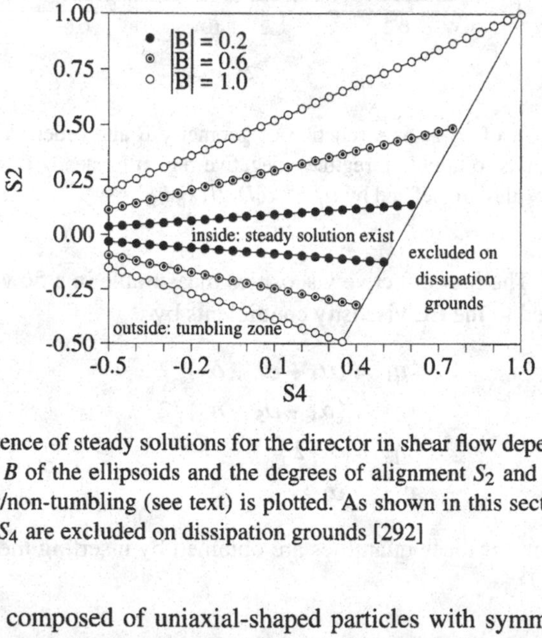

Fig. 7.4. The existence of steady solutions for the director in shear flow depends upon the geometric coefficient B of the ellipsoids and the degrees of alignment S_2 and S_4. The boundary between tumbling/non-tumbling (see text) is plotted. As shown in this section, some combinations of S_2 and S_4 are excluded on dissipation grounds [292]

For systems composed of uniaxial-shaped particles with symmetry axis $\boldsymbol{u}$, the tensorial second- and fourth-order moments of the (non-anisotropic) alignment are denoted by $\boldsymbol{a}_{(2)} = \langle \boldsymbol{uu} \rangle$, $\boldsymbol{a}_{(4)} = \langle \boldsymbol{uuuu} \rangle$, where $\langle \cdot \rangle$ is an orientational average. As shown before in this monograph it is often convenient to use alternative but equivalent tensorial measures that are symmetric in all indices and traceless when contracted over any pair of indices. We denoted such alignment tensors with the '$\overline{\ \ }$' symbol. For the second and fourth order anisotropic alignment tensor $\boldsymbol{a}_{[2]}$ and $\boldsymbol{a}_{[4]}$ one explicitly has $\boldsymbol{a}_{[2]} = \boldsymbol{a}_{(2)} - \mathbf{1}/3$, and

$$\boldsymbol{a}_{[4]} = \boldsymbol{a}_{(4)} - \frac{6}{7}\{\boldsymbol{a}_{(2)}\mathbf{1}\}^{\mathrm{sym}} + \frac{3}{35}\{\mathbf{1}\,\mathbf{1}\}^{\mathrm{sym}}, \tag{7.20}$$

respectively, where $\{\cdot\}^{\rm sym}$ denotes a symmetrized expression defined by

$$\{X_{\mu\nu}Y_{\kappa\lambda}\}^{\rm sym} \equiv \frac{1}{6}(X_{\mu\nu}Y_{\kappa\lambda}+X_{\mu\kappa}Y_{\nu\lambda}+X_{\mu\lambda}Y_{\kappa\lambda}+X_{\nu\kappa}Y_{\mu\lambda}+X_{\nu\lambda}Y_{\mu\kappa}+X_{\kappa\lambda}Y_{\mu\nu}) \tag{7.21}$$

for the dyadic product of symmetric tensors $\boldsymbol{X}$ and $\boldsymbol{Y}$. For a more detailed treatment see Chap. 11, and in particular Sect. 10.2. There are orthogonal unit vectors $\boldsymbol{n}, \boldsymbol{m}, \boldsymbol{l}$ such that

$$\boldsymbol{a}_{(2)} = \lambda_1 \boldsymbol{n}_{(2)} + \lambda_2 \boldsymbol{mm} + \lambda_3 \boldsymbol{ll}\,, \tag{7.22}$$

or equivalently[3]

$$\boldsymbol{a}_{[2]} = \left(\lambda_1 - \frac{1}{3}\right)\boldsymbol{n}_{(2)} + \left(\lambda_2 - \frac{1}{3}\right)\boldsymbol{mm} + \left(\lambda_3 - \frac{1}{3}\right)\boldsymbol{ll}\,. \tag{7.23}$$

The λ_i are the principal values of $\boldsymbol{a}_{(2)}$, and the unit vectors $\boldsymbol{n}$, $\boldsymbol{m}$ and $\boldsymbol{l}$ are the principal directions. The λ_i are subject to the constraint ${\rm Tr}\,\boldsymbol{a}_{[2]} = 0$, i.e. $\sum_i \lambda_i = 1$. Similar relations hold for alignment tensors of arbitrary orders. The symmetry of orientational distribution f which defines the moments (alignment tensors) is directly reflected by the number of distinct principal values. For example, for the second-order moment $\boldsymbol{a}_{[2]}$, we have 1, 2 and 3 distinct principal value(s) for isotropic, uniaxial, and biaxial symmetry, respectively. Let us summarize some trivial implications.

Isotropic Symmetry

All eigenvalues are identical, $\lambda_{1,2,3} = 1/3$, implying $\boldsymbol{a}_{(2)} = \boldsymbol{1}/3$, $\boldsymbol{a}_{[2]} = \boldsymbol{0}$, $\boldsymbol{a}_{(4)} = \{\boldsymbol{1}\boldsymbol{1}\}^{\rm sym}/5$, $\boldsymbol{a}_{[4]} = \boldsymbol{0}$. Any closure relation for $\boldsymbol{a}_{[4]}$ in terms of $\boldsymbol{a}_{[2]}$ which should at least be non-violated close to equilibrium (if isotropic) must therefore fulfill the relationship $\boldsymbol{a}_{(4)} = (9/5)\{\boldsymbol{a}_{(2)}\boldsymbol{a}_{(2)}\}^{\rm sym}$.

Uniaxial Symmetry

Two of the principal values of the second-order alignment tensor are equal (say $\lambda_2 = \lambda_3$). In this case we can write $\boldsymbol{a}_{(2)} = S_2\boldsymbol{n}_{(2)} + (1-S_2)\boldsymbol{1}/3$, $\boldsymbol{a}_{[2]} = S_2\boldsymbol{n}_{[2]}$, with an order parameter $S_2 \equiv (3\lambda_1 - 1)/2$. The fourth-order moments are given by $\boldsymbol{a}_{[4]} = S_4\boldsymbol{n}_{[4]}$, and

$$\boldsymbol{a}_{(4)} = S_4\,\boldsymbol{n}_{(4)} + \frac{6}{7}(S_2 - S_4)\{\boldsymbol{1}\,\boldsymbol{n}_{[2]}\}^{\rm sym} + \frac{1}{35}(7 - 10S_2 + 3S_4)\{\boldsymbol{1}\,\boldsymbol{1}\}^{\rm sym}\,. \tag{7.24}$$

Here S_2 and S_4 are the uniaxial scalar order parameters. They are related to the particle orientations by averages of Legendre polynomials: $S_2 = \langle P_2(\boldsymbol{u}\cdot\boldsymbol{n})\rangle$, $S_4 = \langle P_4(\boldsymbol{u}\cdot\boldsymbol{n})\rangle$, and range in value by $-1/2 \le S_2 \le 1$, $-3/8 \le S_4 \le 1$. The principal direction $\mathbf{n}$ is called the uniaxial director. As for the isotropic case, both anisotropic moments $\boldsymbol{a}_{[2]}$ and $\boldsymbol{a}_{[4]}$ are simpler to handle than $\boldsymbol{a}_{(2)}$ and $\boldsymbol{a}_{(4)}$.

[3] Using $\boldsymbol{1} = \boldsymbol{nn} + \boldsymbol{mm} + \boldsymbol{ll}$, one can also eliminate one of the eigenvectors from these equations. From a representation of the type (7.22) we can most easily read of the inverse of a matrix, here $\boldsymbol{a}_{(2)}^{-1} = \lambda_1^{-1}\boldsymbol{n}_{(2)} + \lambda_2^{-1}\boldsymbol{mm} + \lambda_3^{-1}\boldsymbol{ll}$ such that $\boldsymbol{a}_{(2)}\cdot\boldsymbol{a}_{(2)}^{-1} = \boldsymbol{1}$ due to the orthonormality relations between $\boldsymbol{n}$, $\boldsymbol{m}$, and $\boldsymbol{l}$.

Biaxial Symmetry

For the case of biaxial symmetry see Sect. 10.6.2.

7.4.1 Decoupling with Correct Tensorial Symmetry

Substitution of $\boldsymbol{n}_{(2)}$ in terms of $\boldsymbol{a}_{[2]}$ and S_2 into (7.24) yields

$$S_2^2 \boldsymbol{a}_{[4]} = S_4 \, \overline{\boldsymbol{a}_{[2]} \boldsymbol{a}_{[2]}} \, . \tag{7.25}$$

No assumption has been made other than uniaxial symmetry, so that this tensorial closure relationship is *exact* for uniaxial and isotropic symmetry, but carries still unspecified scalar order parameters S_2, S_4. The generalization of (7.25) is

$$\forall_{n,l} S_n S_l \boldsymbol{a}_{[n+l]} = S_{n+l} \, \overline{\boldsymbol{a}_{[n]} \boldsymbol{a}_{[l]}} \, . \tag{7.26}$$

Obviously, there is not such a simple analog for the biaxial case.

Based on the above representations of the second- and fourth-order alignment tensors, we now consider possible closure schemes for $\boldsymbol{B} : \boldsymbol{a}_{(4)}$ with $\boldsymbol{B}$ an arbitrary symmetric and traceless tensor. such a closure is needed, e.g. in (7.12) to derive a closed form nonlinear equation for the second moment.

Two more commonly cited closures, motivated by HL [324], are the HL1 and HL2 closures (beside closures which are either invalid in the isotropic or aligned state, cf. [325], for example).

HL1 Closure

$$\boldsymbol{B} : \boldsymbol{a}_{(4)} = \frac{1}{5} \left(6 \boldsymbol{a}_{(2)} \cdot \boldsymbol{B} \cdot \boldsymbol{a}_{(2)} - \boldsymbol{B} : \boldsymbol{a}_{(2)} \boldsymbol{a}_{(2)} + 21 \left(\boldsymbol{a}_{(2)} - \boldsymbol{a}_{(2)} \cdot \boldsymbol{a}_{(2)} \right) : \boldsymbol{B} \right) , \tag{7.27}$$

HL2 Closure

$$\begin{aligned} \boldsymbol{B} : \boldsymbol{a}_{(4)} &= \boldsymbol{B} : \boldsymbol{a}_{(2)} \boldsymbol{a}_{(2)} + 2 \boldsymbol{a}_{(2)} \cdot \boldsymbol{B} \cdot \boldsymbol{a}_{(2)} - \frac{2 \boldsymbol{a}_{(2)} \cdot \boldsymbol{a}_{(2)} : \boldsymbol{B}}{\boldsymbol{a}_{(2)} : \boldsymbol{a}_{(2)}} \boldsymbol{a}_{(2)} \cdot \boldsymbol{a}_{(2)} \qquad (7.28) \\ &\quad + \mathrm{e}^{\frac{(2-6 \boldsymbol{a}_{(2)} : \boldsymbol{a}_{(2)})}{(1-\boldsymbol{a}_{(2)} : \boldsymbol{a}_{(2)})}} \left[\frac{52 \boldsymbol{B}}{315} - \frac{8}{21} \left(\boldsymbol{B} \cdot \boldsymbol{a}_{(2)} + \boldsymbol{a}_{(2)} \cdot \boldsymbol{B} - \frac{2}{3} \boldsymbol{B} : \boldsymbol{a}_{(2)} \right) \mathbf{1} \right] . \end{aligned}$$

These are based on interpolation between weak and strong flow limits in a brownian suspension of rods.

For the closure (7.25), which is exact for the case of uniaxial symmetry, and relies only on an approximation between scalar quantities S_4 and S_2, we obtain by straightforward calculation, for the special case $S_4 = S_2^2$, which fulfills $S_4 = 0 \leftrightarrow S_2 = 0$ and $S_4 = 1 \leftrightarrow S_2 = 1$ and is the only consistent one which is parameter-free:

KS1 Closure

$$
\begin{aligned}
\boldsymbol{B}:\boldsymbol{a}_{(4)} = \frac{1}{105}\{ & 2\boldsymbol{B} - 10(\boldsymbol{B}\cdot\boldsymbol{a}_{(2)} + \boldsymbol{a}_{(2)}\cdot\boldsymbol{B}) + 35\boldsymbol{B}:\boldsymbol{a}_{(2)}\boldsymbol{a}_{(2)} \\
& -20(\boldsymbol{B}\cdot\boldsymbol{a}_{(2)}\cdot\boldsymbol{a}_{(2)} + \boldsymbol{a}_{(2)}\cdot\boldsymbol{a}_{(2)}\cdot\boldsymbol{B}) \\
& +70\boldsymbol{a}_{(2)}\cdot\boldsymbol{B}\cdot\boldsymbol{a}_{(2)} + 4\boldsymbol{B}\boldsymbol{a}_{(2)}:\boldsymbol{a}_{(2)} \\
& -51(\boldsymbol{a}_{(2)}:\boldsymbol{B} + 2\mathrm{Tr}[\boldsymbol{B}\cdot\boldsymbol{a}_{(2)}\cdot\boldsymbol{a}_{(2)}])\}.
\end{aligned} \tag{7.29}
$$

All the above closures (HL1,HL2,KS1) correctly reduce to the expected $2\boldsymbol{B}/15$ and $\boldsymbol{B}:\boldsymbol{n}_{(4)}$ for isotropic symmetry ($\boldsymbol{a}_{(2)} = 1/3$) and perfect uniaxial alignment ($\boldsymbol{a}_{(2)} = \boldsymbol{n}_{(2)}$), respectively. In order to compare these closures one can plot the non-vanishing components of the quantity $\boldsymbol{B}:\boldsymbol{a}_{(4)}$ vs the amplitude A of $\boldsymbol{B}$, where $\boldsymbol{B}$ has the following form $\boldsymbol{B} = A((2-a-2b,b,0),(b,b-1,0),(0,0,a+b-1))$. For the (relevant) case that $\boldsymbol{B}$ represents a traceless velocity gradient, and the prefactor a flow rate, the choices $a=0, b=1$ and $a=b=0$ characterize shear (A: shear rate) and uniaxial elongational (A: elongation rate) flow fields, respectively. As for the HLx closures, $\mathrm{Tr}(\boldsymbol{B}:\boldsymbol{a}_{(4)}) = \boldsymbol{B}:\boldsymbol{a}_{(2)}$ holds for (7.29). Any reasonable closure specified by S_4 in terms of S_2 (for 'conventional fluids' with positive order parameters) should at least satisfy $0 < S_4 < S_2$. For example, the ansatz $S_4 = S_2 - S_2(1-S_2)^{\nu}$ parameterized by $0<\nu<1$ has been proposed in [322], the corresponding closures are called KSν-closures, and contain the KS1 closure as a special case. The HLx closures, however, allow to produce pairs S_2, S_4 which fall outside this regime. The closure (7.25), which is immediately extended to higher order tensors, may be preferred if one wants to keep the exact tensorial symmetry while performing a closure relationship between (only) two scalar quantities for a closure involving $\boldsymbol{a}_{(4)}$ and, in general, n scalar functions for a closure involving $\boldsymbol{a}_{(2n)}$.

7.5 Ferrofluids: Dynamics and Rheology

Ferrofluids containing spherical colloidal particles with a permanent ferromagnetic core have been modeled by a system composed of ellipsoidal aggregates (transient chains) along the lines indicated in the previous sections [97, 326, 327]. The stress tensor of this model equals expression (7.8). The Fokker–Planck equation for the orientation distribution function is given by (6.1) with orienting torque (7.11). The potential V_μ for a magnetic moment $\boldsymbol{\mu} = \mu\boldsymbol{u}$ in the local magnetic field $\boldsymbol{H}$ is given by $-\beta V_\mu = \beta\mu\boldsymbol{H}\cdot\boldsymbol{u} = \boldsymbol{h}\cdot\boldsymbol{u}$, with $\beta = 1/(k_B T)$. Hereby the dimensionless magnetic field $\boldsymbol{h} = \mu\boldsymbol{H}/k_B T$ and its amplitude h (Langevin parameter) are introduced. For spheres, $B=0$, the Fokker–Planck equation reduces to the kinetic equation for dilute ferrofluids developed in [328]. Not only the equilibrium magnetization but all equilibrium order parameters are calculated explicitly as function of the magnetic field. From the equilibrium distribution of the Fokker–Planck equation (6.1) upon inserting the above potential V_μ, cf. Chap. 11,

$$
f_{\mathrm{eq}}(\boldsymbol{u}) \propto \mathrm{e}^{-\beta V_\mu} = \frac{h}{4\pi\sinh(h)}\mathrm{e}^{\boldsymbol{h}\cdot\boldsymbol{u}}, \tag{7.30}
$$

one obtains $\langle \boldsymbol{u} \rangle_{eq} = L(h)\boldsymbol{h}/h$, where $L(x) \equiv \coth(x) - 1/x$ is known as 'Langevin function', and the order parameters coincide with ratios of the modified spherical Bessel functions, $S_j^{eq}(h) \equiv \langle P_j(\boldsymbol{u}) \rangle_{\text{eq}} = I_{j+1/2}(h)/I_{1/2}(h)$. Therefore, the following recursion formula is obtained:

$$\begin{aligned} S_{j+1}^{\text{eq}}(h) &= -\frac{2j+1}{h} S_j^{\text{eq}}(h) + S_{j-1}^{eq}(h) , \\ S_0^{\text{eq}}(h) &= 1 , \\ S_1^{\text{eq}}(h) &= L(h) = \coth(h) - 1/h . \end{aligned} \tag{7.31}$$

The equilibrium magnetization $\boldsymbol{M}_{eq} = n\mu \langle \boldsymbol{u} \rangle_{\text{eq}} = n\mu L(h)\boldsymbol{h}/h$ is the classical result for a system of non–interacting magnetic dipoles. The equation for the first moment, i.e., the magnetization, obtained via integration from the Fokker–Planck equation (6.1 with $V = V_\mu$) reads

$$\partial_t \langle \boldsymbol{u} \rangle = \boldsymbol{\omega} \times \langle \boldsymbol{u} \rangle + B \langle (\mathbf{1} - \boldsymbol{uu})\boldsymbol{u} \rangle : \boldsymbol{\gamma} - \frac{1}{\tau} \langle \boldsymbol{u} \rangle + \frac{1}{2\tau} (\mathbf{1} - \langle \boldsymbol{uu} \rangle) \cdot \boldsymbol{h} . \tag{7.32}$$

The one for the second is given in [97] and can be also immediatly derived from the more general equation of change for moments (11.26) in Chap. 11. Using these equations of change, the explicit contribution of the potential V_μ to the full stress tensor can be eliminated. In particular, we obtain for the antisymmetric part of the stress tensor $\boldsymbol{\sigma}^a$, upon inserting the following result

$$\boldsymbol{h} = \tau \boldsymbol{\Pi}^{-1} \cdot (\partial_t \langle \boldsymbol{u} \rangle - \boldsymbol{\omega} \times \langle \boldsymbol{u} \rangle - B[\boldsymbol{\gamma} \cdot \langle \boldsymbol{u} \rangle - \langle \boldsymbol{uuu} \rangle : \boldsymbol{\gamma}] + \tau^{-1} \langle \boldsymbol{u} \rangle) , \tag{7.33}$$

where $\boldsymbol{\Pi}^{-1}$ denotes the inverse of the matrix $\boldsymbol{\Pi} \equiv (\mathbf{1} - \langle \boldsymbol{uu} \rangle)$, an expression in terms of the moments alone: $\boldsymbol{\sigma}^{\text{a}} = -\gamma_1 (\boldsymbol{N} \times \boldsymbol{n}) - \gamma_2 (\boldsymbol{\gamma} \cdot \boldsymbol{n}) \times \boldsymbol{n}$ with the viscosity coefficients $\gamma_1 \propto (3S_1^2)/(2+S_2)$, $\gamma_2 \propto -B\{3S_1(3S_1+2S_3)\}/\{5(2+S_2)\}$, and a shape-dependent proportionality coefficient [327]. By performing NEBD simulation [97] for this system it had been observed that the assumption of uniaxial symmetry can be successfully applied in a wide regime of shear rates and magnetic fields, see Fig. 7.5 for a schematic overview. This figures also summarizes (closure) relationships between the order parameters for different regimes. In [329] the stationary and oscillatory properties of dilute ferromagnetic colloidal suspensions in plane Couette flow were studied. Analytical expressions for the off-equilibrium magnetization and the shear viscosity are obtained within the so-called effective field approximation (EFA), and the predictions of a different approximation based on the linearized moment expansion (LME) were obtained. Direct NEBD simulation of the Fokker–Planck equation were performed in order to test the range of validity of these approximations. It turns out that both EFA and LME provide very good approximations to the stationary off-equilibrium magnetization as well as the stationary shear viscosity in case of weak Couette flow if the magnetic field is oriented in gradient direction. If the magnetic field is oriented in flow direction, and for small amplitude oscillatory Couette flow, the LME should be favored. A sample result which estimates the quality of the approximations is given in Fig. 7.6. Figure 7.7 provides a sample time series

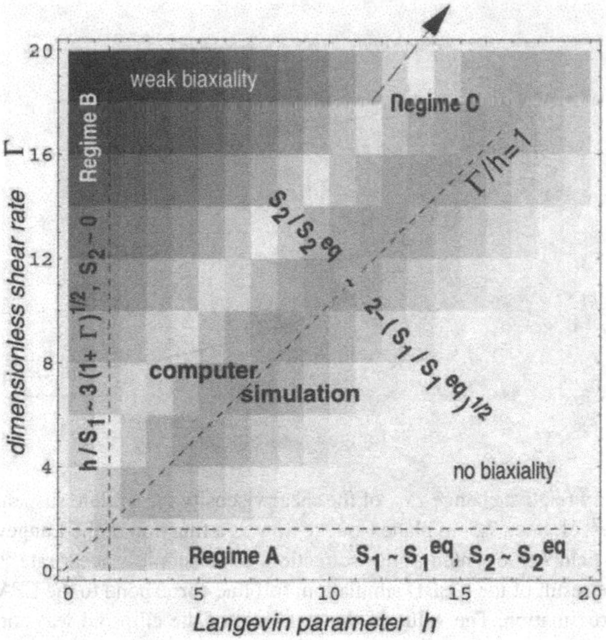

Fig. 7.5. The shaded background represents a measure for the (minor) relevance of biaxiality – obtained via NEBD – on the prediction of the rotational viscosity γ_1 as function of dimensionless magnetic field h and vorticity $\tau\dot{\gamma}$ [97]. Shading ranges from white (uniaxial) to black. In the *top left* corner (data for $\tau\dot{\gamma} = 10$, $h = 1$) we have a 1.2% relative deviation between uniaxial. and biaxial formulas for γ_1. The depicted regimes refer to analytical solutions of the FP equation. A: weak magnetic field, B: weak flow field, C: deterministic limit. The figure summarizes analytical as well as approximative results for these regimes. Adapted from [326]

for a ferrofluid we obtained via MD for a collection of (LJ) repulsive freely rotating permanent magnetic dipoles. Here, it is illustrated why ferrofluids exhibit anisotropic viscosities even in the absence of a magnetic field: often due to chain formation [330]. Not just chains, but other types of agglomerates have been observed via MD as well. Also antiferromagnetic phases belong to this class. This phase can be stabilized if attractive (LJ) interactions – beside dipolar interactions – are present.

7.6 Liquid Crystals: Periodic and Irregular Dynamics

Detailed theoretical studies [332, 333], based on solutions of a generalized Fokker–Planck equation [74, 313], revealed that in addition to the tumbling motion, wagging and kayaking types of motions, as well as combinations thereof occur. Recently, also chaotic motions were inferred from a moment approximation to the same Fokker–Planck equation leading to a 65 dimensional dynamical system [334] for uniaxial particles. While we are going to consider uniaxial particles (following [331]) one

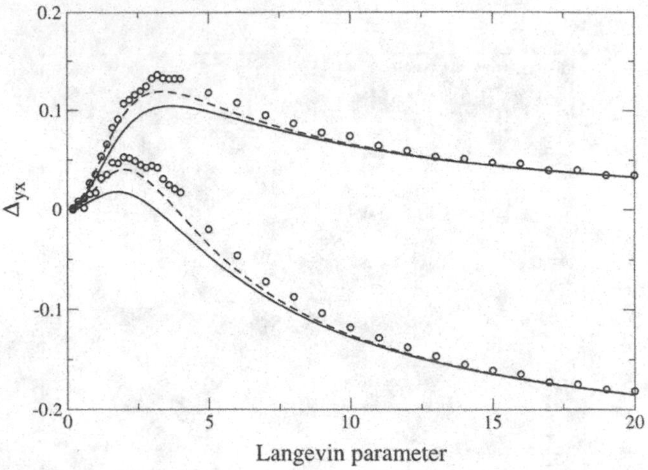

Fig. 7.6. Stationary relative change Δ_{yx} of the shear viscosity for a dilute suspension of ferromagnetic particles, cf. Sect. 7.5, in plane Couette flow as a function of the Langevin parameter h. The magnetic field was oriented in flow direction, dimensionless shear rate $\dot{\gamma} = 0.1$. Symbols represent the result of the NEBD simulation, full line correspond to the EFA, *dashed line* to the LME approximation. The value of the axis ratio of the ellipsoid was chosen as $r = 2$ ($B = 3/5$) for the lower and $r = 5$ ($B = 12/13$) for the *upper curves*. Adapted from [329]

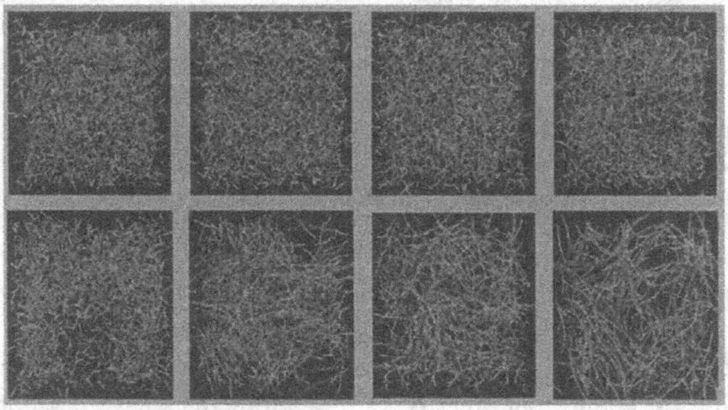

Fig. 7.7. Sample MD shapshot for a simple ferrofluid with increasing (*top left* to *bottom right*) permanent magnetic moment, where external orienting (flow, magnetic) fields are absent. The figure serves to demonstrate, that ferrofluids exhibit anisotropic viscosities even in the absence of a magnetic field (due to chain formation), and that they can be modeled with a combination of the methods presented for colloidal suspensions and FENE-C wormilke micelles

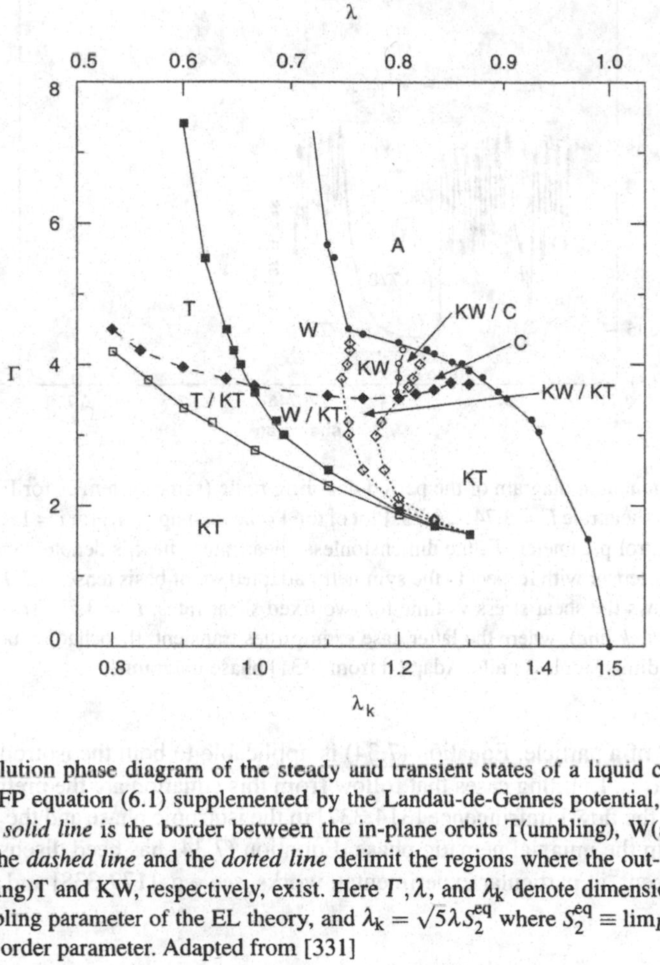

Fig. 7.8. Solution phase diagram of the steady and transient states of a liquid crystal modeled by the FP equation (6.1) supplemented by the Landau-de-Gennes potential, (7.36) with $\vartheta = 0$. The *solid line* is the border between the in-plane orbits T(umbling), W(agging) and A(ligned); the *dashed line* and the *dotted line* delimit the regions where the out-of-plane orbits K(ayaking)T and KW, respectively, exist. Here Γ, λ, and λ_k denote dimensionless shear rate Γ, tumbling parameter of the EL theory, and $\lambda_k = \sqrt{5}\lambda S_2^{eq}$ where $S_2^{eq} \equiv \lim_{\Gamma\to 0} S_2$ is an equilibrium order parameter. Adapted from [331]

may notice that for long triaxial ellipsoidal non-brownian particles chaotic behavior had been also predicted in [335]. Point of departure is the following equation of change for the alignment tensor (notice the similarity with (6.10))

$$\tau_a\left(\partial \boldsymbol{a}_{[2]}/\partial t - 2\,\overline{\boldsymbol{\omega}\times\boldsymbol{a}_{[2]}}\,\right) + \boldsymbol{\Phi}(\boldsymbol{a}_{[2]}) = -\sqrt{2}\,\tau_{ap}\,\boldsymbol{\gamma}\,. \tag{7.34}$$

The quantity $\boldsymbol{\Phi}$ is the derivative of a Landau-de Gennes free energy Φ, (7.36) below, with respect to the alignment tensor. It contains terms of first, second, and third order in $\boldsymbol{a}_{[2]}$. The equation stated here was first derived within the framework of irreversible thermodynamics [301,302], where the relaxation time coefficients $\tau_a > 0$ and τ_{ap} are considered as phenomenological parameters, for their microscopic interpretation see Chap. 7. Equation (7.34) can also be derived, within certain approximations [336], from the Fokker–Planck used there. Then τ_a and the ratio $-\tau_{ap}/\tau_a$ are related to the rotational diffusion coefficent D_r and to a non-sphericity parameter associated with

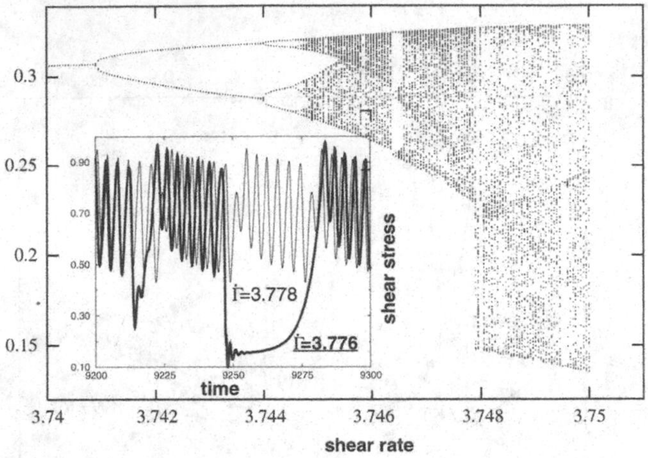

Fig. 7.9. Feigenbaum diagram of the period doubling route (same system as for Fig. 7.8). for the particular shear rate $\Gamma = 3.74 \ldots 3.75$. Plot of the Poincaré map $a_4(t_i)$ for $i = 1 \ldots 82$ at $a_3 = 0$ vs. the 'control parameter' Γ, the dimensionless shear rate. The a_i's denote components of the alignment tensor with respect to the symmetry adapted set of basis tensors (2.9,2.11,2.12). The inset shows the shear stress vs time for two fixed shear rates, $\Gamma = 3.778$ (*thin line*), and $\Gamma = 3.776$ (*thick line*), where the latter case exemplifies transient, rheochaotic, behavior. All quantities in dimensionless units. Adapted from [331] phase diagram

the shape B of a particle. Equation (7.34) is applicable to both the isotropic and the nematic phases. Limiting cases that follow from this equation are the pretransitional behavior of the flow birefringence [314, 337] in the isotropic phase and the EL theory (Sect. 7.1) in the uniaxial nematic phase. Equation (7.34) has beed discussed intensivley in recent, in particular experimental, works, see e.g. [179, 338] and refs. cited herein.

7.6.1 Landau – de Gennes Potential

The five components a_i of $\boldsymbol{a}_{[2]}$ – relative to the symmetry-adapted basis system (2.9, 2.11, 2.12) – are expressed in units of the magnitude of the equilibrium alignment at the temperature (or concentration) where the nematic phase of a lyotropic LC coexists with its isotropic phase. In its dimensionless form the Landau–de Gennes free energy invokes a single model parameter ϑ, viz.,

$$2\Phi = \vartheta a^2 - 2I^{(3)} + a^4 , \tag{7.35}$$

with

$$a^2 \equiv \boldsymbol{a}_{[2]} : \boldsymbol{a}_{[2]}, \quad I^{(3)} \equiv \sqrt{6}\,\mathrm{Tr}(\boldsymbol{a}_{[2]}{}^3), \tag{7.36}$$

with $\boldsymbol{a}_{[2]}{}^3 = \boldsymbol{a}_{[2]} \cdot \boldsymbol{a}_{[2]} \cdot \boldsymbol{a}_{[2]}$. Here $I^{(3)}$ is sometimes[4] called 'third order scalar invariant', a^2 is related to the Maier-Saupe order parameter S_2 as $a^2 = 5\,S_2^2$, and for the derivative with respect to $\boldsymbol{a}_{[2]}$ entering the equation of change of the alignment tensor, we have, using (7.36),

$$\boldsymbol{\Phi}(\boldsymbol{a}_{[2]}) = \frac{\partial \Phi}{\partial \boldsymbol{a}_{[2]}} = \vartheta \boldsymbol{a}_{[2]} - 3\sqrt{6}\,\overline{\boldsymbol{a}_{[2]}{}^2} + 2a^2\,\boldsymbol{a}_{[2]}\,. \tag{7.38}$$

The dynamical system (7.34) has been rewritten in terms of the a_i's in [336] and contains three control parameters two of which are determined by the state point and the material chosen, the third control parameter is a dimensionless shear rate $\Gamma \propto \dot{\gamma}$ [339]. Equation (7.34) with $a_{3,4} = 0$ decribes correctly the flow aligned state as well as the tumbling and wagging behavior of the full system for certain ranges of control parameters, see [336] for a detailed analysis. Here we wish focus on the symmetry breaking solutions with $a_{3,4} \neq 0$. These solutions are associated with kayaking types of motions, but also rather complex and chaotic orbits are found. We use a fourth-order Runge-Kutta method with fixed time step to solve the dynamic system.

7.6.2 In-Plane and Out-of-Plane States

A solution phase diagram of the various in-plane and out-of-plane states is drawn for $\vartheta = 0$ in Fig. 7.9, in its caption we introduce abbreviations for the types of orbits. The three orbits, T, W, A were identified in [336]. The kayaking orbits [332,333], KT and KW, are distinguished from each other according to [340]. Because the physical situation is invariant under the transformation $a_{3,4} \to -a_{3,4}$, two equivalent kayaking states exist. The system shows rather complicated dynamical behavior in region C of the solution diagram where neither one of the simple periodic states nor an aligning state is stable. The specific orbits had been classified in [331] as Periodic KT/KW composite states where the KW sequences are damped with increasing shear rate; Irregular KT or KT/KW states for which the largest Lyapunov exponent is of order 0.01...0.05; Intermittent KT, and iv) Period doubling KT states. The route to chaos for increasing shear rates had been found to depend on the tumbling parameter. When the flow-aligned (A) phase is approached from the complex (C) regime, the oscillation period grows infinitely high, in contrast to the behavior at the KW→A transition, where the amplitude of the oscillation gets damped. The resulting bifurcation plot has a striking similarity to the Feigenbaum diagram of the logistic map,

[4] In view of (10.59) and Tab. 10.1, we can express $I^{(3)}$ in terms of the tensor invariant I_3, $I^{(3)} = 3\sqrt{6}I_3$, further $a^2 = \mathrm{Tr}\,\boldsymbol{a}_{[2]}{}^2 = 2I_2$, such that (7.36) can be also rewritten as

$$\Phi = \vartheta I_2 - \frac{3}{2}\sqrt{6}I_3 + 2I_2^2\,. \tag{7.37}$$

Notice, that due to the Caley-Hamilton theorem even the most general ansatz just contains exactly two independent matrices, $\boldsymbol{a}_{[2]}$ and $\boldsymbol{a}_{[2]}{}^2$. Any higher powers could be rewritten as a linear combination of these two matrices.

$x_{n+1} = rx_n(1-x_n)$. The distance between successive period doubling steps in Fig. 7.9 shrinks rapidly with the order of the period as in the Feigenbaum diagram. Even the chaotic region exhibits the same type of banded structure and has windows of periodic behavior. However, at $\Gamma \approx 3.748$, the chaotic band enlarges abruptly. The reason for this behavior is the equivalence of the states $a_{3,4}$ and $-a_{3,4}$. To test the similarity of the period doubling routes, the values Γ_n where a period of order 2^n emerges and the value Γ_∞ for the beginning of chaos were calculated in [331] for $n = 1 \dots 5$ Like for the logistic map, the Γ_n scale according to a law $\Gamma_n = \Gamma_\infty - C\delta^{-n}$ for $n \gg 1$, with the Feigenbaum constant δ. For our problem, a nonlinear fit yields $\delta = 4.83 \pm 0.02$. The value agrees qualitatively with that for the logistic map. $\delta = 4.669\dots$, and a similar value had been reported in [334]. Irregular behavior of the alignment tensor $\boldsymbol{a}_{[2]}$ immediately converts into irregular behavior for rheological properties, cf. Fig. 7.9 for an example. Based on the findings reported here, the inhomogeneous extension [304, 341, 342] of the present model can be expected to be of relevance in describing experimentally observed instabilities, irregular banded and striped textures [343–346].

8

Connection between Different Levels of Description

8.1 Boltzmann Equation

One of the major issues raised by the Boltzmann equation is the problem of the reduced description. Equations of hydrodynamics constitute a closet set of equations for the hydrodynamic field (local density, local momentum, and local temperature). From the standpoint of the Boltzmann equation, these quantities are low-order moments of the one-body distribution function, or, in other words, the macroscopic variables. The problem of the reduced description consists in deriving equations for the macroscopic variables from kinetic equations, and predicting conditions under which the macroscopic description sets in. The classical methods of reduced description for the Boltzmann equation are: the Hilbert method, the Chapman–Enskog method, and the Grad moment method, reviewed in [98]. The general approach to the problem of reduced description for dissipative system was recognized as the problem of finding stable invariant manifolds in the space of distribution function. The notion of invariant manifold generalizes the normal solution in the Hilbert and in the Chapman–Enskog method, and the finite-moment sets of distribution function in the Grad method. A generalization of the Grad moment method is the concept of the quasiequilibrium approximation, cf. Sect. 2.6 and [94, 98]. Boltzmann's kinetic equation has been expressed in GENERIC form [347], cf. Sect. 8.3, demonstrating that no dissipative potential is required for representing these equations.

8.2 Generalized Poisson Structures

A similar formal structure, namely a symplectic structure, for thermodynamics and classical mechanics has been noted early, e.g., by Peterson in his work about the analogy between thermodynamics and mechanics [51]. Peterson notes the equations of state – by which he means identical relations among the thermodynamic variables characterizing a system – are actually first-order partial differential equations for a function which defines the thermodynamics of the system. Like the Hamilton-Jacobi equation, such equations can be solved along trajectories given by Hamilton's equations, the trajectories being quasistatic processes which obey the given equation of state. This gave rise to the notion of thermodynamic functions as infinitesimal generators of quasistatic processes, with a natural Poisson bracket formulation. This formulation of thermodynamic transformations is invariant under canonical coordinate transformations, just as classical mechanics is. The time-structure invariance of the Poisson bracket as manifested through the Jacobi identity has been used to derive constraint relationships on closure approximations [57]. Next we turn to the modern GENERIC framework [54] which offers a particular useful generalized Poisson structure (GPS). The Poisson structure, together with a Jacobi identity had been recognized recently in two-fluid electrodynamics, in the generalized Heisenberg picture quantum mechanics, fluid models of plasma physics, and other branches of physics, cf. [38,54]. There is a variety of directions, which have not yet been worked out in detail, but extensively discussed. Upon these are nonholonomic constraints [348], boundary conditions [349,350], and extensions to so called super-Poisson structures [351], Nambu–Jacobi brackets [352,353]. For these structures a number of different representations is known such that knowledge can be directly passed over to GENERIC concerning the development of efficient algorithms solving the GENERIC equations.

8.3 GENERIC Equations

The GENERIC equations [38, 54] preserve their structure across different levels (mirco-macro) of description for beyond-equilibrium systems. For a given set of system variables $\boldsymbol{x}$ (defining the actual state space) the following (reversible and dissipative) brackets

$$\{A,B\} \equiv \frac{\delta A}{\delta \boldsymbol{x}} \cdot \boldsymbol{L} \cdot \frac{\delta B}{\delta \boldsymbol{x}}, \quad [A,B] \equiv \frac{\delta A}{\delta \boldsymbol{x}} \cdot \boldsymbol{M} \cdot \frac{\delta B}{\delta \boldsymbol{x}} \tag{8.1}$$

for arbitrary functionals A,B on state space, the time evolution equation for A,

$$\frac{\mathrm{d}A}{\mathrm{d}t} = \{A,H\} + [A,S]\,, \tag{8.2}$$

the degeneracy conditions

$$\boldsymbol{M} \cdot \frac{\delta H}{\delta \boldsymbol{x}} = 0, \quad \boldsymbol{L} \cdot \frac{\delta S}{\delta \boldsymbol{x}} = 0\,, \tag{8.3}$$

the antisymmetry of $\boldsymbol{L}$, the Casimir symmetry of $\boldsymbol{M}$, together with the positive definiteness of $\boldsymbol{M}$ and the following Jacobi identity (for arbitrary functionals A, B, C)

$$0 = \{\{A,B\},C\} + \{\{B,C\},A\} + \{\{C,A\},B\}\,, \tag{8.4}$$

constitute the GENERIC framework [38, 54]. The Hamiltonian H and entropy S essentially model the system under consideration, whereas $\boldsymbol{L}$ can be motivated by analyzing the transformation behavior of variables, and $\boldsymbol{M}$ models the dissipative motion of variables. The requirement for energy conservation and increasing entropy, respectively, implies the antisymmetry of $\boldsymbol{L}$ and a degeneracy condition and positive semidefinite block $\boldsymbol{M}$. The Jacobi identity must hold in order to ensure a self-consistent time-invariant description. A large number of thermodynamically admissible (generalized and extended) physical models has been collected at www.polyphys.mat.ethz.ch.

For a GENERIC bracket one can deduce the following evolution equation

$$\frac{d\{A,B\}}{dt} = \left\{\frac{dA}{dt},B\right\} + \left\{A,\frac{dB}{dt}\right\} = \{\{A,B\},H\}\,. \tag{8.5}$$

This expression reflects the time structure invariance of a Poisson bracket, i.e., the operator $\boldsymbol{L}$ behaves as a 'conserved' quantity. If the subscript t denotes the time-dependent solution F_t of the evolution equation $dF_t/dt = \{F_t, H\}$, then the Jacobi identity implies time structure invariance in the sense that $\{A,B\}_t = \{A_t, B_t\}$ for arbitrary functions A, B on state space. The definition (8.1) implies that when evaluating the Jacobi identity (8.4) second derivatives of the functions A, B, C would appear in principle. However, these second derivatives cancel pairwise, simplifying the Jacobi identity. The bracket of classical point mechanics fulfills the Jacobi identity since all elements of the matrix $\boldsymbol{L}$ are constant. It is sufficient to test the Jacobi identity against three *linear* functions [354] (this reference also provides a code for evaluating Jacobi identities). Worked out examples are given in [38, 354].

Dynamic GENERIC equations for a single-segment reptation model without independent alignment, incorporating ideas of convective constraint release and anisotropic tube cross section in flow [355] have been developed by Öttinger [356], and investigated in [175], see also Sect. 6.1.

8.3.1 Building Block *L*

A large number of explicit expressions, and procedures for the construction of GENERIC building blocks can be found in the literature, cf. [54]. In order to just write down an expression for a simple example, we choose the rarefied gas, since its building blocks will be explicitly needed to perform the simulations of Sect. 8.9. Further simple examples, including classical hydrodynamics, piston and cylinder, reptation model for polymer melts, where, at the same time also the Jacobi identity is explicitly tested, can be found in [354].

In the operator notation for Poisson matrices [38] the transformation formula $\boldsymbol{L}_0 \to \boldsymbol{L}$ leads from the Poisson operator for the Boltzmann equation [357] to the following element of the Poisson matrix associated with the observables A and B,

$$L_{AB}(\boldsymbol{r}) = -\int \hat{A}(\boldsymbol{r},\boldsymbol{p})\frac{\partial}{\partial \boldsymbol{r}}\cdot f(\boldsymbol{r},\boldsymbol{p})\frac{\partial \hat{B}(\boldsymbol{r},\boldsymbol{p})}{\partial \boldsymbol{p}}\mathrm{d}^3 p$$

$$-\int f(\boldsymbol{r},\boldsymbol{p})\frac{\partial \hat{A}(\boldsymbol{r},\boldsymbol{p})}{\partial \boldsymbol{p}}\cdot\frac{\partial}{\partial \boldsymbol{r}}\hat{B}(\boldsymbol{r},\boldsymbol{p})\mathrm{d}^3 p\,. \tag{8.6}$$

Rarefied Gas

Specializing this result to the pairs of components of $\boldsymbol{x} \equiv (\rho, \boldsymbol{v}, T, \boldsymbol{x}_4)$ with

$$\boldsymbol{x}_4 \equiv \frac{m}{k_{\mathrm{B}}TN}\boldsymbol{cc}, \quad \boldsymbol{c} \equiv \frac{\boldsymbol{p}}{m} - \boldsymbol{\kappa}\cdot\boldsymbol{r}\,, \tag{8.7}$$

cf. Sect. 8.9, we obtain for the homogeneous case (no spatial temperature or density gradients)

$$L(\boldsymbol{r}) = -\begin{pmatrix} 0 & 0 & 0 & 0 \\ 0 & -\rho^{-1}\boldsymbol{\omega} & 0 & \boldsymbol{L}_{24} \\ 0 & 0 & 0 & \boldsymbol{L}_{34} \\ 0 & \boldsymbol{L}_{42} & \boldsymbol{L}_{43} & \boldsymbol{L}_{44} \end{pmatrix}, \tag{8.8}$$

where $\boldsymbol{\Omega} \equiv (\boldsymbol{\kappa} - \boldsymbol{\kappa}^T)/2$ is the vorticity tensor, and the beyond-hydrodynamic entries in this Poisson matrix are

$$(L_{24})_{ijk} = \frac{1}{\rho}\left\{x_{4ik}\frac{\partial}{\partial r_j} + x_{4ji}\frac{\partial}{\partial r_k} - \frac{2}{3}x_{4jk}\left[\boldsymbol{x}_4\cdot\frac{\partial}{\partial \boldsymbol{r}}\right]_i\right\}, \tag{8.9}$$

$$(L_{42})_{ijk} = \frac{1}{\rho}\left\{x_{4kj}\frac{\partial}{\partial r_i} + x_{4ik}\frac{\partial}{\partial r_j} - \frac{2}{3}x_{4ij}\left[\boldsymbol{x}_4\cdot\frac{\partial}{\partial \boldsymbol{r}}\right]_k\right\}, \tag{8.10}$$

$$\boldsymbol{L}_{34} = -\boldsymbol{L}_{43} = \frac{4m}{3\rho k_{\mathrm{B}}}(\boldsymbol{\Omega}\cdot\boldsymbol{x}_4 - \boldsymbol{x}_4\cdot\boldsymbol{\Omega})\,, \tag{8.11}$$

$$(L_{44})_{ijkl} = -\frac{2m}{\rho k_{\mathrm{B}}T}\left[\frac{2}{3}x_{4ij}(\boldsymbol{\Omega}\cdot\boldsymbol{x}_4 - \boldsymbol{x}_4\cdot\boldsymbol{\Omega})_{kl} + \frac{2}{3}(\boldsymbol{\Omega}\cdot\boldsymbol{x}_4 - \boldsymbol{x}_4\cdot\boldsymbol{\Omega})_{ij}x_{4kl}\right.$$
$$\left. + x_{4ik}\Omega_{jl} + x_{4il}\Omega_{jk} + x_{4jl}\Omega_{ik} + x_{4jk}\Omega_{il}\right]. \tag{8.12}$$

In the above, x_{4ij} denote components of $\boldsymbol{x}_4$. The components $\boldsymbol{L}_{24}$ and $\boldsymbol{L}_{42}$ vanish for homogeneous systems. In determining the explicit Poisson matrix (8.8), all third moments of the peculiar velocity $\boldsymbol{c}$ vanish. The precise functional form of the distribution $f(\boldsymbol{r},\boldsymbol{p})$ has actually been irrelevant and, in particular, the same Poisson matrix would have been obtained with Grad's ten moment approximation [358]. As a consequence of the rigorous nature of the procedure, the Poisson bracket implied by (8.8) satisfies the Jacobi identity [354].

Energy for the Rarefied Gas, Evolution Equation

The energy E for the rarefied gas reads

$$E(\boldsymbol{x}) = \int \left(\frac{1}{2} \rho v^2 + \frac{3}{2} \frac{\rho}{m} k_B T \right) d^3r, \tag{8.13}$$

in which the kinetic energy and the internal energy of an ideal gas can be recognized. Here, $\boldsymbol{x} = (\rho, \boldsymbol{v}, T, \boldsymbol{x}_4)$ denotes the full list of state variables, i.e., hydrodynamic ones supplemented by Grad's moment $\boldsymbol{x}_4 \equiv \langle \boldsymbol{\Pi}_4 \rangle_{\boldsymbol{x}}$ from (8.50). A contribution due to the interaction potential, (4.1), can be neglected in (8.13) because for a rarefied gas it is small compared to the kinetic counterpart. It is also consistent with the fact that in the limit of vanishing volume of particles a finite mean free path remains, known as Grad's limit. With E from expression (8.13) we have

$$\delta E / \delta \boldsymbol{x} = \begin{pmatrix} \frac{1}{2m}(mv^2 + 3k_B T) \\ \rho \boldsymbol{v} \\ 3\rho k_B/(2m) \\ \boldsymbol{0} \end{pmatrix}. \tag{8.14}$$

The summation of products $\sum_{k=1}^{4} L_{4k} \cdot \delta E / \delta x_k$ then yields the reversible part of the evolution equation for $\boldsymbol{x}_4$, which appears on the rhs of (8.54). For the lhs of this equation, see Sect. 8.3.2.

8.3.2 Building Block *M*

Friction matrices can be calculated from the Green-Kubo-type formula,

$$\boldsymbol{M} = \frac{1}{k_B} \int_0^{\tau} \langle \dot{\boldsymbol{x}}^f(t) \dot{\boldsymbol{x}}^f(0) \rangle_{\boldsymbol{x}} \, dt, \tag{8.15}$$

while using the symbolic notation of [54]. Here τ is an intermediate time scale separating the slow degrees of freedom from the fast ones, $\dot{\boldsymbol{x}}^f$ is the rapidly fluctuating part of the time derivative of the atomistic expressions for the slow variables $\boldsymbol{x}$, and the average indicated by the pointed brackets is over an ensemble of atomistic trajectories consistent with the coarse-grained state $\boldsymbol{x}$ at $t = 0$ and evolved according to the atomistic dynamics to the time t. We assume that the correlation function (8.15) decays sufficiently fast with increasing time difference t so that the integral is finite. In order to get a converged expression for the friction matrix we would like to take τ large, but then we are faced with two problems: (i) the separation of time scales may not be sufficiently pronounced, and (ii) the evaluation of (8.15) in simulations leads to large error. As a consequence of problem (i), the slow variables begin to change in the time range considered in the integral (8.15), and one obtains corrections depending on the ratio of fast to slow time scales [359]. If one would like to to consider large τ then one should modify the slow time evolution by a control mechanism (cf. our multiplostatted equations of Sect. 8.9.1) that artificially keeps $\boldsymbol{x}$ constant, where the effect of the control mechanism should be kept of the same order as the unavoidable corrections depending on the ratio of fast to slow time scales. The present discussion follows strictly [54]. By assuming that the system is invariant under time translations and that it possesses a usual bare Onsager–Casimir symmetry (which leads to the factor 2 in the denominator), we obtain

$$\begin{aligned} \boldsymbol{M}(\boldsymbol{x}) &= \frac{1}{2k_{\mathrm{B}}} \int_{-\tau}^{\tau} \langle \dot{\boldsymbol{x}}^f(t) \dot{\boldsymbol{x}}^f(0) \rangle_{\boldsymbol{x}} \, dt \\ &= \frac{1}{2k_{\mathrm{B}}\tau} \int_0^{\tau} \int_0^{\tau} \langle \dot{\boldsymbol{x}}^f(t'') \dot{\boldsymbol{x}}^f(t') \rangle_{\boldsymbol{x}} \, dt'' dt' , \end{aligned} \tag{8.16}$$

For the last step in (8.16), which is valid for large τ only, we have exploited the fact that only small time-differences $t'' - t'$ contribute to the integral. Equation (8.16) can be rewritten as

$$\boldsymbol{M}(\boldsymbol{x}) = \frac{\tau}{2k_{\mathrm{B}}} \left\langle \left\{ \frac{1}{\tau} \int_0^{\tau} \dot{\boldsymbol{x}}^f(t'') dt'' \right\} \left\{ \frac{1}{\tau} \int_0^{\tau} \dot{\boldsymbol{x}}^f(t') dt' \right\} \right\rangle . \tag{8.17}$$

Using the exact time evolution operator $\hat{L}$ for the microscopic dynamics, we have

$$\Delta_\tau \boldsymbol{\Pi}(z) = \int_0^{\tau} \dot{\boldsymbol{x}}^f(t') dt' = \boldsymbol{\Pi}(z(\tau)) - \boldsymbol{\Pi}(z(0)) - \int_0^{\tau} \langle i\hat{L}\boldsymbol{\Pi} \rangle_{\boldsymbol{\Pi}(z(t'))} dt' , \tag{8.18}$$

where

$$\int_0^{\tau} \langle i\hat{L}\boldsymbol{\Pi} \rangle_{\boldsymbol{\Pi}(z(t'))} dt' \approx \langle \boldsymbol{\Pi}(z(\tau)) \rangle_{\boldsymbol{x}} - \langle \boldsymbol{\Pi}(z(0)) \rangle_{\boldsymbol{x}} \tag{8.19}$$

can usually be neglected in the average (8.17) as a small deterministic effect (corresponding to the slow dynamics) on top of the dominating fluctuations (resulting from fast variables). We thus have the integrated version of (8.17),

$$\boldsymbol{M}(\boldsymbol{x}) = \frac{1}{2k_{\mathrm{B}}\tau} \langle \Delta_\tau \boldsymbol{\Pi}(z) \Delta_\tau \boldsymbol{\Pi}(z) \rangle_{\boldsymbol{x}} . \tag{8.20}$$

This equation is a generalization of a standard result (see, e.g., (2.107), (2.108) of [360]). In the context of diffusion, (8.15) corresponds to an expression for the diffusion coefficient in terms of the velocity autocorrelation function associated with the names of Green and Kubo, whereas (8.20) corresponds to a diffusion coefficient in terms of the mean square displacement often associated with the name of Einstein. Alternate formulas have been discussed in [361, 362] and Sect. 6.5.5 of [360]. The friction matrix $\boldsymbol{M}(\boldsymbol{x})$ (8.20) is used in (8.49).

Rarefied Gas

In the present setting, the components M_{ij} with $i, j \in \{1,2,3\}$ related to density, velocity, and temperature, naturally vanish for an isothermal, homogeneous, rarefied gas. Analytic considerations in the context of the structured version of Grad's moment method result in $\boldsymbol{M}_{i4} = \boldsymbol{M}_{4i} = 0$ for $i \in \{1,2,3\}$ and we are left with the coefficient $\boldsymbol{M}_{44}$ which is obtained from the fluctuations of the nonequilibrium variable $\boldsymbol{\Pi}_4$ and enters (8.54) via $\sum_{k=1}^{4} \boldsymbol{M}_{4k} \cdot \delta S/\delta x_k = \boldsymbol{M}_{44} \cdot \delta S/\delta x_4 = k_{\mathrm{B}} \boldsymbol{M}_{44} \cdot \boldsymbol{\lambda}_4$.

8.4 Dissipative Particles

Because large-scale NEMD simulation can bridge time scales dictated by fast modes of motion together with slow modes, which determine viscosity, it can capture the effects of varying molecular topology on fluid rheology resulting, e.g., from chemical reactions. Mesoscopic regimes involving scales exceeding several nsec and/or micrometers require the 'fast' molecular modes of motion to be eliminated in favor of a more coarse grained representation, where the internal degrees of freedom of the fluid are ignored and only their center of mass motion is resolved. On this level, the particles will represent clusters of atoms or molecules, so called, dissipative particles (DPD). It is possible to link and pass the averaged properties of molecular ensembles onto dissipative particles by using bottom up approach from NEMD/NEBD by means of the somewhat systematic coarse-graining procedure [363]. GENERIC had been used to construct modifications of Smoothed Particle Hydrodynamics (SPH) including thermal fluctuations and DPD in [364]. A method suited for the efficient treatment of polymer solution dynamics is the Lattice Boltzmann (LB) method and its improved versions [365]. A GENERIC formulation of LB has been discussed in [366]. In its application to polymer solution dynamics, the polymer itself is still treated on a simple molecular level using a bead-spring lattice model, but the solvent molecules are treated on the level of a discretized Boltzmann equation. In this way the hydrodynamics of the solvent is correctly captured, and the hydrodynamic interaction between different units on the polymer chain, which is mediated by the hydrodynamic flow generated within the solvent through the motion of the polymer, is present in the simulation without explicit treatment of all solvent molecules. It is expected, that NEMD, DPD and LB together can capture both microscopic and macroscopic scales [367].

8.5 Langevin and Fokker–Planck Equation, Brownian Dynamics

8.5.1 Motivation

In order to 'derive' a diffusion, or Fokker–Planck equation, we consider random variables $\boldsymbol{X}(t)$ related to a stochastic process, characterized by realizations $(\boldsymbol{x}_1,t_1)$, $(\boldsymbol{x}_1,t_1)$ etc. with $t_1 \geq t_2$. For a Markov process, the conditional probability for a realization becomes

$$\begin{aligned} p(\boldsymbol{x}_1,t_1;\boldsymbol{x}_2,t_2;..|\boldsymbol{y}_1,\tau_1;\boldsymbol{y}_2,\tau_2;..) &= \frac{p(\boldsymbol{x}_1,t_1;\boldsymbol{x}_2,t_2;..;\boldsymbol{y}_1,\tau_1;\boldsymbol{y}_2,\tau_2;..)}{p(\boldsymbol{y}_1,\tau_1;\boldsymbol{y}_2,\tau_2;..)} \\ &= p(\boldsymbol{x}_1,t_1;\boldsymbol{x}_2,t_2;..|\boldsymbol{y}_1,\tau_1)\,, \end{aligned} \tag{8.21}$$

i.e., the 'future' depends on the present state only, but not on information about past states. Now making use of the so called Chapman-Kolmogorov equation for the jump probability

$$p(\boldsymbol{x}_1,t_1|\boldsymbol{x}_3,t_3) = \int p(\boldsymbol{x}_1,t_1|\boldsymbol{x}_2,t_2)p(\boldsymbol{x}_2,t_2|\boldsymbol{x}_3,t_3)\,, \tag{8.22}$$

one obtains,[1] up to terms of order τ^2:

$$p(\boldsymbol{x},t+\tau|\boldsymbol{x}',t) = (1-\tau a(\boldsymbol{x},t))\delta(\boldsymbol{x}-\boldsymbol{x}') + \tau w(\boldsymbol{x},\boldsymbol{x}',t) + o(\tau^2)\,, \tag{8.23}$$

with jump rate $a(\boldsymbol{x}',t) = \int w(\boldsymbol{x},\boldsymbol{x}',t)\mathrm{d}\boldsymbol{x}$, and w is a transition rate $w(\boldsymbol{x}\to\boldsymbol{x}')$ at time t. For

$$p(\boldsymbol{x},t) = \int p(\boldsymbol{x},t|\boldsymbol{y},t_0)p(\boldsymbol{y},t_0)\mathrm{d}\boldsymbol{y}\,, \tag{8.24}$$

in the limiting case $\tau\to 0$, the Chapman-Kolmogorov equation (8.22) is a master equation for the probability function

$$\frac{\partial}{\partial t}p(\boldsymbol{x},t) = \int \mathrm{d}\boldsymbol{x}'\, w(\boldsymbol{x},\boldsymbol{x}',t)p(\boldsymbol{x}',t) - w(\boldsymbol{x}',\boldsymbol{x},t)p(\boldsymbol{x},t)\,, \tag{8.25}$$

expressing the change of a 'long-time' quantity in terms of an integral over short time information contained in the transition probability. If we further restrict ourself to a diffusion processes, where the random variable is allowed to perform 'small jumps', and where p is a smooth function of $\boldsymbol{x}$, the diffusion process is characterized by its first and second moments [368]

$$\int (\boldsymbol{y}-\boldsymbol{x})p(\boldsymbol{y},t+\tau|\boldsymbol{x},t)\mathrm{d}\boldsymbol{y} \equiv \boldsymbol{A}(\boldsymbol{x},t)\tau + o(\tau^2)\,, \tag{8.26}$$

$$\int (\boldsymbol{y}-\boldsymbol{x})(\boldsymbol{y}-\boldsymbol{x})p(\boldsymbol{y},t+\tau|\boldsymbol{x},t)\mathrm{d}\boldsymbol{y} \equiv \boldsymbol{D}(\boldsymbol{x},t)\tau + o(\tau^2)\,, \tag{8.27}$$

i.e., the transition probability for small times is Gaussian distributed with mean value $\boldsymbol{A}(\boldsymbol{x},t)$ and variance (matrix) $\boldsymbol{D}(\boldsymbol{x},t)$. We have $\boldsymbol{A}(\boldsymbol{x},t) = \int(\boldsymbol{y}-\boldsymbol{x})w(\boldsymbol{y},\boldsymbol{x},t)\mathrm{d}\boldsymbol{y}$. Combing (8.24), (8.25), and (8.27) directly yields the Fokker–Planck equation (8.29). For a more rigorous introduction to Fokker–Planck equations see [368, 369], for a large range of applications for anisotropic and polymeric fluids see Chap. 11.

8.5.2 Interpretation, and Langevin Equation

In order to apply the GENERIC framework it is important to identity the relevant (state) variables which may sufficiently describe the given physical system. In Chap. 6 we dealt with primitive path models which certainly are more abstract and less dimensional objects than FENE chains discussed in the foregoing sections. With the treatment of elongated particles (Chap. 7) we continued the way through models possessing a decreasing number of molecular details. We therefore provide some general comments on how to reduce the number of variables in those dynamical model systems, which are described in terms of stochastic differential equations, such as Langevin equations for a set of stochastic variables $\boldsymbol{x}$, whose typical structure is to split the equation of motion for a variable into a deterministic (drift) plus a stochastic (diffusion) part

$$\frac{d}{dt}\boldsymbol{X} = \boldsymbol{A}(\boldsymbol{X}) + \boldsymbol{B}\cdot\tilde{\boldsymbol{\eta}} \tag{8.28}$$

[1] Just use $\delta(\boldsymbol{x}_1-\boldsymbol{x}_2)\delta(\boldsymbol{x}_2-\boldsymbol{x}_3) = \delta(\boldsymbol{x}_1-\boldsymbol{x}_3)$ below the integral.

with time t and 'noise' $\tilde{\boldsymbol{\eta}}$ or equivalent Fokker–Planck equations (used at several places throughout this monograph) for the corresponding distribution function $f(\boldsymbol{x},t)$

$$\frac{\partial f}{\partial t} = \mathcal{L}_{\rm FP}\, f, \quad \mathcal{L}_{\rm FP} = -\frac{\partial}{\partial \boldsymbol{x}} \cdot \boldsymbol{A}(\boldsymbol{x},t) + \frac{\partial}{\partial \boldsymbol{x}\partial \boldsymbol{x}} : \boldsymbol{D}(\boldsymbol{x},t) \tag{8.29}$$

with diffusion tensor $\boldsymbol{D} = \boldsymbol{B}^\dagger \cdot \boldsymbol{B}$ using Ito's interpretation. The difficulty of solving the Fokker–Planck equation like any other partial differential equation increases with increasing number of independent variables. It is therefore advisable to eliminate as many variables as possible. For an introduction to stochastic modeling, including an introduction to brownian dynamics (NEBD) computer simulation which rigorously solves (8.28), see [58,368,369]. A brief introduction to Fokker–Planck and Langevin equations as well as brownian dynamics, however, can be found in Chap. 11, and sample codes are availoable in Chap. 12 of this monograph.

8.6 Projection Operator Methods

If the drift and diffusion coefficients do not depend on some variables, the Fourier transform of the probability density for these variables can then be obtained by an equation where the variables no longer appear. To be more specific, if the drift and diffusion coefficients do not depend on $x_1,..,x_n$ with $N > n$ being the total number of variables, making a Fourier transform of p with respect to the first n variables, by using the Fokker–Planck equation (8.29) and performing partial integrations the following equation for $\hat{f} = \hat{f}(x_{n+1},...,x_N)$ must be solved: $\partial \hat{f}/\partial t = \hat{\mathcal{L}}_{\rm FP}\hat{f} = \hat{\mathcal{L}}_{\rm FP}(x_{n+1},...,x_N)$ with

$$\begin{aligned}\hat{\mathcal{L}}_{\rm FP} = & -i\sum_{i=1}^{n} k_i A_i - \sum_{i=n+1}^{N} \frac{\partial A_i}{\partial x_i} - \sum_{i,j=1}^{n} k_i k_j D_{ij} \\ & + 2i\sum_{i=1}^{n}\sum_{j=n+1}^{N} k_i \frac{\partial D_{ij}}{\partial x_j} + \sum_{i,j=n+1}^{N} \frac{\partial^2 D_{ij}}{\partial x_i \partial x_j} \,.\end{aligned} \tag{8.30}$$

Generally, (8.30) must be resolved for every $\boldsymbol{k}$. If one is looking only for periodic solutions in the variables x_i $(i \le n)$, the wave numbers k_i must be integers and the integral (for the Fourier transform) must be replaced by a sum over these integer numbers. Furthermore, if one is interested only in some expectation values of the form $\langle \exp imx_i(t)\rangle$ (for a specific $i \le n$), only the solution of (8.30) with $k_i = -m$ needs to be calculated. A class of Fokker–Planck equations with two variables where the drift and diffusion coefficients do not depend on one variable and where solutions are given in terms of hypergeometric functions, see [370] and App. A6 of [369]. If the decay constants for some variables are much larger than those for other ones, the 'fast' variables can then approximately be eliminated. This is achieved by adiabatic elimination of the fast variables. Starting from the Langevin equation (8.28) for the slow ($\equiv x_1$) and fast ($\equiv x_2$) variables, the Fokker–Planck equation for the distribution function $f(\boldsymbol{x})$ is rewritten as $\partial f/\partial t = [\mathcal{L}_1 + \mathcal{L}_2]f$, with – $i = 1$ (slow) and $i = 2$ (fast) –

$$\mathcal{L}_i = \frac{\partial \tilde{A}_i(\boldsymbol{x})}{\partial x_i} + \frac{\partial^2}{\partial x_i^2} D_{ii}(\boldsymbol{x}), \quad \tilde{A}_i(\boldsymbol{x}) = A_i(\boldsymbol{x}) + B_{ii}\frac{\partial}{\partial x_i} B_{ii} \,. \tag{8.31}$$

In the spirit of the Born-Oppenheimer approximation in quantum mechanics one first looks for eigenfunctions of the operator $\mathcal{L}_2$. Here the variable x_1 appears as a parameter. We assume that for every parameter a stationary solution and discrete eigenvalues λ_n and eigenfunctions ϕ_n exist ($n \geq 0$). These generally depend on the parameter x_1: $\mathcal{L}_2(\boldsymbol{x})\phi_n(\boldsymbol{x}) = \lambda_n(x_1)\phi_n(\boldsymbol{x})$. For $n = 0$, $\lambda_0 = 0$ we have the stationary solution $f_{stat} = \phi_0(\boldsymbol{x})$. By expanding the distribution function f into the complete set ϕ_n of the operator $\mathcal{L}_2$ $f(\boldsymbol{x}) = \sum_m c_m(x_1,t)\phi_m(\boldsymbol{x})$, and inserting this expansion into the Fokker–Planck equation involving $\mathcal{L}_{1,2}$ one obtains $[\frac{\partial}{\partial t} + \lambda_n(x_1)]c_n = \sum_{m=0}^{\infty} L_{n,m}c_m$, with $L_{n,m} \equiv \int \phi_n^+ \mathcal{L}_1(\boldsymbol{x})\phi_m(\boldsymbol{x})\mathrm{d}x_2$, and the functions ϕ^+ denote the eigenfunctions of the adjoint operator $\mathcal{L}_1^\dagger$. The orthonormalization and completeness relations read $\int \phi_n^\dagger \phi_m dx_2 = \delta_{nm}$ and $\int \phi_n^\dagger(x_1,x_2)\phi_n(x_1,x_2') = \delta(x_2 - x_2')$, respectively. The $L_{n,m}$ are operators with respect to the slow variable x_1. Because we are interested only in the time scale large compared to the decay coefficient of the fast variable, we may neglect the time derivative in the equation with $n \geq 1$. Finally, the equation of motion for the distribution function $f(x_1,t) = c_0(x_1,t)$ of the relevant variable x_1 reads

$$\frac{\partial f(x_1,t)}{\partial t} = \mathcal{L}_0 f(x_1,t), \quad \mathcal{L}_0 = L_{0,0} + \sum_{n=1}^{\infty} L_{0,n}\lambda_n(x_1)^{-1}L_{n,0} + \cdots \tag{8.32}$$

where the dots denote higher order terms and, in particular,

$$L_{0,0} = -\frac{\partial}{\partial x_1}\int \tilde{A}_1(\boldsymbol{x})\phi_0(\boldsymbol{x})dx_2 + \frac{\partial^2}{\partial x_1^2}\int D_{ii}(\boldsymbol{x})\phi_0(\boldsymbol{x})dx_2 \,. \tag{8.33}$$

To solve (8.32) explicitely for the distribution function $f(x_1,t)$ for the slow variables, the operator $\mathcal{L}_0$ should be given analytically. This is the case only if the eigenvalues and eigenfunctions of $\mathcal{L}_2$ are known analytically and if the matrix elements occurring in (8.32) can be calculated analytically. An application of this procedure is given on page 192 of [369].

Quite often the elimination of one or more variables is done with the Nakajima-Zwanzig projector operator formalism [371–374]. This formalism can be alternatively applied, whereby a projection operator $\mathcal{P}$ is defined by $\mathcal{P}f = (\int \phi_0^\dagger f dx_2)\phi_0$, where ϕ_0 is the (above) stationary solution. In view of the orthogonality relations given above, $\mathcal{P}^2 = \mathcal{P}$ for a projection operator holds. Because the system $\phi_n, \phi_n^\dagger$ is complete, the operator $1 - \mathcal{P}$ may be cast in the form $\mathcal{Q}f \equiv (1-\mathcal{P})f = \sum_{n=1}^{\infty}(\int \phi_n^\dagger f dx_2)\phi_n$. In the projection operator formalism, the equation of motion is split up into two coupled equations for $\mathcal{P}f$ and $(1-\mathcal{P})f$, i.e., into

$$\frac{\partial}{\partial t} f = \mathcal{L}_{\mathrm{FP}}\, f = \mathcal{P}\mathcal{L}_{\mathrm{FP}}\, f + \mathcal{Q}\mathcal{L}_{\mathrm{FP}}\, f \,, \tag{8.34}$$

with $\mathcal{P}\mathcal{L}_{\mathrm{FP}}\, f = \mathcal{P}\mathcal{L}\mathcal{P}f + \mathcal{P}\mathcal{L}\mathcal{Q}f$, and $\mathcal{Q}\mathcal{L}_{\mathrm{FP}}\, f = \mathcal{Q}\mathcal{L}\mathcal{P}f + \mathcal{Q}\mathcal{L}\mathcal{Q}f$. The usual Markov approximation to the formal solution of this problem consists in neglecting the time derivative, as used here in order to derive (8.32).

An appropriate way of systematic coarse-graining is provided by GENERIC [38] and its statistical foundation based on projection operator techniques for separating time scales [375]. For Monte Carlo simulations, nonequilibrium ensembles corresponding to the deformations of polymer molecules in flows can be introduced and used in order to determine deformation-dependent energies and entropies [376], which are the generators of reversible and irreversible time-evolution, (8.1), respectively. For MD simulations, the projection-operator formalism shows that all dynamic material information can and actually should be evaluated in a systematic way from simulations over time spans much shorter than the final relaxation time [377].

8.7 Stress Tensors: Giesekus – Kramers – GENERIC

Within so called GENERIC Canonical Monte Carlo (GCMC) [376] and the 'reduced description' mentioned in Sect. 2.6 the relevant distribution function is approximated using a reduced set of (slow) variables. These may be particular moments of the distribution function itself. Using the underlying Fokker–Planck equation from this representation one can derive equations of change for the slow variables, and sometimes solve the set of equations for the 'conjugate' or 'dual' variables efficiently. Within GCMC the distribution function (based on all 'atomistic' phase space coordinates abbreviated as z) involves unknown Lagrange parameters $\boldsymbol{\Lambda}$ and a 'phase space function' $\boldsymbol{\Pi}(z)$:

$$f(z)_{\boldsymbol{\Lambda}} = f_{\mathrm{eq}}\frac{1}{Z}e^{-\boldsymbol{\Lambda}:\boldsymbol{\Pi}}, \quad f_{\mathrm{eq}} \propto e^{-\beta E_0} \tag{8.35}$$

normalized by Z. Here, For the case of the homogeneous Hookean bead-spring model (Rouse model) with bond energy $E_0 \equiv \frac{H}{2}\sum_k \boldsymbol{Q}_k \cdot \boldsymbol{Q}_k$ we wish to see under which conditions the three different representations for the stress tensor (Gieskus, Kramers, GENERIC) are equal to each other, and we want to provide an expression of the Lagrange parameter in terms of flow parameters. See [378] for a discussion about material objectivity and thermodynamical consistency of stress tensor expressions.

Let us consider a single (arbitrary) normal mode $\boldsymbol{\Pi} \equiv \boldsymbol{X}_P\boldsymbol{X}_P$ $(P \in 1,\ldots,N-1)$ as slow variable. The first mode, for example, is given by $\boldsymbol{X}_1 \equiv \sum_i (2/N)^{1/2}\sin(\frac{i\pi}{N})\boldsymbol{Q}_i$ [58]. The Gieskus expression for the stress tensor is known as

$$\boldsymbol{\sigma}^{\mathrm{GIE}} = -\frac{1}{2}n\zeta \sum_{i,j=1}^{N-1} C_{i,j}\left(\boldsymbol{\kappa}\cdot\langle \boldsymbol{Q}_i\boldsymbol{Q}_j\rangle + \langle \boldsymbol{Q}_i\boldsymbol{Q}_j\rangle\cdot\boldsymbol{\kappa}^T\right) \tag{8.36}$$

with the useful properties $\sum_{ij} C_{ij}\boldsymbol{Q}_i\boldsymbol{Q}_j = \sum_k c_k \boldsymbol{X}_k\boldsymbol{X}_k$, $c_k = 1/a_k$ and $a_k = 4\sin^2(k\pi/(2N))$ and $\sum_{k=1}^{N-1} c_k = (N^2-1)/6$ [4]. The Kramers expression reads [4]

$$\boldsymbol{\sigma}^{\mathrm{KRA}} = n\sum_i^{N-1}\langle \boldsymbol{Q}_i\boldsymbol{F}_i\rangle + (N-1)nk_{\mathrm{B}}T\,\mathbf{1} \tag{8.37}$$

with $\boldsymbol{F}_i = -dE_0/d\boldsymbol{Q}_i$, and the GENERIC expression for the same problem (assuming a symmetric stress tensor) reads [38]

$$\boldsymbol{\sigma}^{\mathrm{GEN}} = nk_{\mathrm{B}}T(\boldsymbol{\Lambda}\cdot\boldsymbol{X}+\boldsymbol{X}^T\cdot\boldsymbol{\Lambda}^T)), \quad \boldsymbol{X} = \langle\boldsymbol{\Pi}\rangle \tag{8.38}$$

In the above equations the average is defined via $\langle F\rangle = \int\int F f_{\boldsymbol{\Lambda}}\, d\boldsymbol{z}$ where $\boldsymbol{z} = \{\boldsymbol{Q}_1,\boldsymbol{Q}_2,...,\boldsymbol{Q}_{N-1}\}$. Inserting the special form $\boldsymbol{\Pi}$ into (8.35) we obtain $f(\boldsymbol{z})_{\boldsymbol{\Lambda}} = Z^{-1}\exp\{-\boldsymbol{\Lambda}:\boldsymbol{X}_1\boldsymbol{X}_1 - \frac{\beta H}{2}\sum_k \boldsymbol{X}_k\cdot\boldsymbol{X}_k\}$. and $\boldsymbol{X} = \langle\boldsymbol{\Pi}\rangle = \frac{1}{2}(\boldsymbol{\Lambda}+\frac{\beta H}{2}\mathbf{1})^{-1}$, or equivalently, an expression of the Lagrange parameter in terms of the averaged normal mode $\boldsymbol{\Lambda} = \frac{1}{2}(\boldsymbol{X}^{-1} - \beta H\mathbf{1})$. The GENERIC stress is thus rewritten as

$$\boldsymbol{\sigma}^{\mathrm{GEN}} = nk_{\mathrm{B}}T(\mathbf{1}-\beta H\boldsymbol{X})\,. \tag{8.39}$$

By using the identity

$$\langle\boldsymbol{X}_k\boldsymbol{X}_k\rangle = \frac{1}{\beta H}\mathbf{1} + \delta_{k,P}\left(\boldsymbol{X} - \frac{1}{\beta H}\mathbf{1}\right) \tag{8.40}$$

we immediately see, that $\boldsymbol{\sigma}^{\mathrm{KRA}} = \boldsymbol{\sigma}^{\mathrm{GEN}}$ rigorously holds. Concerning the correspondence between Gieskus and GENERIC stresses we arrive at the following condition for $\boldsymbol{X}$ in terms of the flow field: $\boldsymbol{\sigma}^{\mathrm{GEN}} = \boldsymbol{\sigma}^{\mathrm{GIE}}$ if and only if

$$-4\lambda_H\boldsymbol{\gamma}\left(\frac{N^2-1}{6} - c_P\right) - 2c_P\lambda_H(\beta H)(\boldsymbol{\kappa}\cdot\boldsymbol{X}+\boldsymbol{X}\cdot\boldsymbol{\kappa}^T) = \mathbf{1} - \beta H\boldsymbol{X} \tag{8.41}$$

with the time constant of Hookean dumbbell $\lambda_H = \zeta/(4H)$. In order to apply these findings, let us consider simple shear flow with dimensionless shear rate $\Gamma = \dot{\gamma}\lambda_H$. For that particular case we obtain the following moment $\boldsymbol{X}$ and Lagrange parameter $\boldsymbol{\Lambda}$ in terms of the shear rate:

$$\boldsymbol{X} = \frac{1}{\beta H}\begin{pmatrix} 1+4c_P\frac{(N^2-1)}{3}\Gamma^2 & \frac{N^2-1}{3}\Gamma & 0 \\ & 1 & 0 \\ & & 1 \end{pmatrix}, \tag{8.42}$$

$$\boldsymbol{\Lambda} = \beta H\begin{pmatrix} \frac{(N^2-1)(N^2-1-12c_P)}{2(9-\Gamma^2(N^2-1)(N^2-1-12c_P))}\Gamma^2 & -\frac{3(N^2-1)}{2(9-\Gamma^2(N^2-1)(N^2-1-12c_P))}\Gamma & 0 \\ & \frac{(N^2-1)}{2(9-\Gamma^2(N^2-1)(N^2-1-12c_P))}\Gamma^2 & 0 \\ & & 0 \end{pmatrix}. \tag{8.43}$$

Note that $N^2-1-12c_P < 0$ for $P=1$, $N^2-1-12c_P > 0$ for $P=2,3$, both signs (dependent on N) otherwise. The first mode should always be taken into account within the set of slow variables. $\boldsymbol{\Lambda}$ is nontrivial and singular. When considering a single mode P we therefore recover the expected form of the stress tensor and the exact Rouse viscosity by matching the stresses, but we have disagreement for the first normal stress. To be more specific,

$$\eta = nk_{\mathrm{B}}T\lambda_H\frac{(N^2-1)}{3} = \eta^{\mathrm{Rouse}},$$

$$\Psi_1 = nk_{\mathrm{B}}T\lambda_H^2\frac{4(N^2-1)}{3}c_P \neq \Psi_1^{\mathrm{Rouse}}. \tag{8.44}$$

This example can be generalized to other types of flow and other (more suitable) choices for the phase space function $\boldsymbol{\Pi}$ in terms of 'atomistic coordinates'. Several examples are discussed in [376]. The goal is to approximate the correct distribution function in a most efficient way by considering a small number of relevant variables. These must not necessarily be the normal coordinates we had just chosen for illustrative purpose.

8.8 Generalized Canonical Ensemble and Friction Matrix

Let $\boldsymbol{\Pi}(z)$ denote a yet unspecified n-dimensional list of microscopic observables in terms of the atomistic phase space coordinates $\boldsymbol{z}(t) = \{\boldsymbol{r}_j(t), \boldsymbol{p}_j(t)\}$. The generalized canonical ensemble in terms of $\boldsymbol{\Pi}(\boldsymbol{z})$ is characterized by a distribution function $\rho_{\boldsymbol{x}}(\boldsymbol{z})$ with

$$\rho_{\boldsymbol{x}}(\boldsymbol{z})/\rho_0(\boldsymbol{z}) = \frac{1}{Z(\boldsymbol{x})}\, \mathrm{e}^{-\boldsymbol{\lambda}(\boldsymbol{x})\cdot\boldsymbol{\Pi}(\boldsymbol{z})}, \tag{8.45}$$

which contains an n-dimensional list of Lagrange parameters $\boldsymbol{\lambda}$. The dot between $\boldsymbol{\lambda}$ and $\boldsymbol{\Pi}$ stands for the n-fold contraction between both quantities resulting in a scalar argument for the exponential. The average $\boldsymbol{x} \equiv \langle\boldsymbol{\Pi}\rangle_{\boldsymbol{x}} \equiv \int \rho_{\boldsymbol{x}}(\boldsymbol{z})\boldsymbol{\Pi}(\boldsymbol{z})\, d\boldsymbol{z}$ depends just on $\boldsymbol{\lambda}$, and vice versa, motivating the notation $\boldsymbol{\lambda}(\boldsymbol{x})$ and $\rho_{\boldsymbol{x}}$ used in (8.45). The prefactor $Z(\boldsymbol{x}) \equiv \int \exp\{-\boldsymbol{\lambda}(\boldsymbol{x})\cdot\boldsymbol{\Pi}\}\, d\boldsymbol{z}$ ensures suitable normalization. For convenience, we include the 'equilibrium' distribution function $\rho_0(\boldsymbol{z})$ in (8.45) since the relevant variables and Lagrange parameters to be discussed in the following handle nonequilibrium situations. To be specific, in Sect. 8.9 the quantity $\boldsymbol{x}$ becomes $\boldsymbol{x} = (x_1, \boldsymbol{x}_2, x_3, \boldsymbol{x}_4)$ where $x_1 = \rho$, $\boldsymbol{x}_2 = \boldsymbol{v}$, and $x_3 = T$ stand for the hydrodynamic field variables mass density, velocity, and temperature, respectively. The quantity $\boldsymbol{x}_4$ will denote the additional slow structural nonequilibrium variable. The stationary ensemble (8.45) determines averages for any phase space function in terms of $\boldsymbol{\lambda}$. In practice, these averages can be always obtained by using Monte Carlo methods. The distribution (8.45) implies a 'closure relationship', which couples the moments of the distribution function, and results in restrictions for the applicability of BEMD to be discussed in Sect. 8.9.2. Within the generalized canonical ensemble, the entropy $S(\boldsymbol{x}) \equiv -k_{\mathrm{B}}\langle\ln\rho_{\boldsymbol{x}}\rangle_{\boldsymbol{x}}$ is available in terms of the macroscopic observable $\boldsymbol{x}$ (or alternatively $\boldsymbol{\lambda}$), most conveniently as

$$S(\boldsymbol{x}) = k_{\mathrm{B}}(\ln Z(\boldsymbol{x}) + \boldsymbol{\lambda}(\boldsymbol{x})\cdot\boldsymbol{x})\,. \tag{8.46}$$

The Lagrange parameter thus is interpreted as the variable conjugate to $\boldsymbol{x}$, while $k_{\mathrm{B}}\boldsymbol{\lambda} = \delta S/\delta\boldsymbol{x}$.

The GENERIC equation for the time evolution of the macroscopic variable $\boldsymbol{x}$ (to be considered as relevant, and 'slow') reads, according to (8.1),

$$\frac{d\boldsymbol{x}}{dt} = \boldsymbol{L}(\boldsymbol{x})\cdot\frac{\delta E(\boldsymbol{x})}{\delta\boldsymbol{x}} + \boldsymbol{M}(\boldsymbol{x})\cdot\frac{\delta S(\boldsymbol{x})}{\delta\boldsymbol{x}}, \tag{8.47}$$

where $\boldsymbol{L}$ is an antisymmetric linear operator that obeys the Jacobi identity, and $\boldsymbol{M}$ is Onsager/Casimir symmetric and positive-semidefinite. The following degeneracy conditions $\boldsymbol{L} \cdot \delta S(\boldsymbol{x})/\delta \boldsymbol{x} = 0$ and $\boldsymbol{M} \cdot \delta E(\boldsymbol{x})/\delta \boldsymbol{x} = 0$ supplement (8.47), which contains four 'building blocks' $\boldsymbol{L}, E, \boldsymbol{M}, S$. The quantities E and S represent energy and entropy, respectively, as described in more detail elsewhere [38, 379]. See also App. 8.3.1 for particular representations to be used in the next section. Within the generalized canonical ensemble we essentially can extract all building blocks in terms of $\boldsymbol{x}$ by varying $\boldsymbol{\lambda}$ for a given atomistic model system through Monte Carlo simulation. More specifically, energy is obtained via $E(\boldsymbol{x}) = \langle E_0 \rangle_{\boldsymbol{x}}$ (E_0 denotes the atomistic energy function). Entropy $S(\boldsymbol{x})$ has been already expressed through $\boldsymbol{x}$ above. Further, $\boldsymbol{L}(\boldsymbol{x}) = \langle \{\boldsymbol{\Pi}, \boldsymbol{\Pi}\}_0 \rangle_{\boldsymbol{x}}$, where the classical Poisson bracket reads $\{A, B\}_0 \equiv (\partial A/\partial z) \cdot \boldsymbol{L}_0 \cdot (\partial B/\partial z)$ with the symplectic matrix $\boldsymbol{L}_0$; it represents classical Hamilton's equations for the atomistic system. The matrix $\boldsymbol{M}$, in the stationary regime, obeys

$$k_{\mathrm{B}} \boldsymbol{M} \cdot \boldsymbol{\lambda}(\boldsymbol{x}) = -\boldsymbol{L} \cdot \delta E/\delta \boldsymbol{x} \,. \tag{8.48}$$

A key point to the understanding of the BEMD method described in Sect. 8.9 is the following. The components of the Lagrange parameter cannot be varied arbitrarily to obtain the physical realizations of $\boldsymbol{x}$. We have to respect interrelations between its components. In this work we are interested in determining the Lagrange parameter for the structural variable ($\boldsymbol{x}_4$) for given hydrodynamic variables ($\rho, \boldsymbol{v}, T$). Dynamical information is needed to obtain the friction matrix from (8.48) self-consistently. As discussed in App. 8.3.2, the friction matrix $\boldsymbol{M}$ entering the time evolution equation for the slow variables $\boldsymbol{x}$ is obtained via brownian or molecular dynamics simulation as

$$\boldsymbol{M} = \frac{1}{2 k_{\mathrm{B}} \tau_s} \langle \triangle_{\tau_s} \boldsymbol{\Pi}(z) \triangle_{\tau_s} \boldsymbol{\Pi}(z) \rangle_{\boldsymbol{x}} \,, \tag{8.49}$$

with $\triangle_{\tau_s} \boldsymbol{\Pi}(z) \equiv \boldsymbol{\Pi}(z(\tau_s)) - \boldsymbol{\Pi}(z(0))$ and a time scale τ_s that is large compared with the one characterizing the rapid fluctuations of $\boldsymbol{\Pi}$, but small compared with the time scale on which its average $\boldsymbol{x}$ varies.

8.9 Beyond-Equilibrium Molecular Dynamics (BEMD)

Beyond-equilibrium molecular dynamics (BEMD) demonstrates the application of GENERIC (Sect. 8.3) and the calculation of the friction matrix. The method is based on – and restricted to – the regime, where a generalized canonical ensemble provides a sufficiently rigorous description in terms of microscopic expressions for non-equilibrium variables. Multiplostatted equations of motion (Nosé-Hoover variants of Hamilton's classical equations) are employed in Sect. 8.9.1 to maintain this ensemble. BEMD makes use of the generalized canonical ensemble elaborated in Sect. 8.8, where the connection with the GENERIC equations is clarified. The friction matrix appearing in the dynamical equation is iteratively obtained employing a Green-Kubo type expression. Since the remaining 'building blocks' for the GENERIC equation are readily accessible via static Monte Carlo simulation (for the application of

Sect. 8.9 they are available analytically), BEMD provides the desired information to perform multiscale simulations.

Application: Rarefied Gas

The rarefied gas case considered here is exemplary in the sense that the behavior of such a simple system is known, and that we can focus on the procedural aspects. It is atypical in the sense that the pressure tensor is dominated by its kinetic contribution.

Without doubt the nonlinear flow behavior of gases can be considered as well understood. It was known 70 years ago that a gas can display nonlinear effects [380] and several authors have discussed the problem since then [381–383]. For a rarefied gas, experimental data is lacking to check the theories. Nonlinear effects occur at strain rates large compared with the ones accessible in the laboratory. For analytic approximate solutions for material functions of the rarefied gas we refer the reader to [54, 383]. For the present model system BEMD operates in the relevant (Newtonian) domain. For more complex fluids, nonlinear effects occur at accessible strain rates, and conventional NEMD may be applied to an appropriately coarse-grained level of description to investigate non-Newtonian effects. But for complex fluids, the zero-strain-rate viscosities, which are difficult to access via NEMD, are also of major importance. Here, BEMD may be used to extend efficiently the simulation window to lower strain rates. Alternate methods particularly useful at low rates have been proposed by Ciccotti [384] (evaluating differences between equilibrium and nonequilibrium trajectories) and Morriss and Evans [385] (employing a transient time correlation function).

For this simple and exemplary case we are able to obtain $\boldsymbol{\lambda}$ in terms of the slow nonequilibrium variable as well as $\boldsymbol{L} \cdot \delta E/\delta \boldsymbol{x}$ analytically. 'Multiplostatted' equations of motion are employed to perform $\boldsymbol{\lambda}$-biased nonequilibrium molecular dynamics simulations in such a way that $\boldsymbol{\lambda}$ (respectively $\boldsymbol{x}$) is iteratively and self-consistently obtained by evaluating, on the one hand, $\boldsymbol{M}$ from (8.49), and, on the other hand, $\boldsymbol{L} \cdot \delta E/\delta \boldsymbol{x}$. The latter can be directly expressed in terms of $\boldsymbol{x}$; the former is evaluated for given $\boldsymbol{x}$, which allows us to iteratively obtain the correct Lagrange parameter. Having determined all building blocks in terms of the 'slow' variables $\boldsymbol{x}$, we can also solve (8.47) to study the transient behavior on 'large' time scales.

Target Level

Besides using mass density $x_1 = \rho$, velocity field $\boldsymbol{x}_2 = \boldsymbol{v}$, and temperature $x_3 = T$ as hydrodynamic state variables for the description of (ideal, homogeneous, isotropic) gases, a coarse target level for an anisotropic, rarefied gas is provided by the structured version of Grad's moment method [386], which suggests using the (dimensionless) kinetic pressure tensor as an additional nonequilibrium variable $\boldsymbol{x}_4 = \langle \boldsymbol{\Pi}_4 \rangle_{\boldsymbol{x}}$. We wish to study a nonequilibrium ensemble for which density, temperature, and flow field are constant. As discussed in Sect. 8.3.2, all entries in the friction matrix $\boldsymbol{M}$ (except $\boldsymbol{M}_{44}$) vanish and therefore just the Lagrange parameters $\boldsymbol{\lambda}_4$ needs to be considered.

Nonequilibrium Variable for the Rarefied Gas

Consider a macroscopically homogeneous rarefied gas composed of a fixed number N of particles with equal masses m at constant temperature and volume. The system at rest is described by a canonical equilibrium ensemble. According to the rules of statistical physics [387], the kinetic part of the pressure tensor $\boldsymbol{\pi}_{\text{kin}}$ for a particle system subjected to a macroscopic flow field $\boldsymbol{v}(\boldsymbol{r})$ reads $\boldsymbol{\pi}_{\text{kin}} = p\boldsymbol{\Pi}_4(\boldsymbol{z})$ with the dimensionless instantaneous kinetic pressure tensor

$$\boldsymbol{\Pi}_4(\boldsymbol{z}) \equiv \frac{m}{k_{\text{B}}T}\frac{1}{N}\sum_{j=1}^{N}\boldsymbol{c}_j\boldsymbol{c}_j\,, \tag{8.50}$$

where $p = \rho k_{\text{B}}T/m$ is the ideal gas pressure, $\rho = Nm/V$ the particle mass density, $\boldsymbol{c}_j = \boldsymbol{p}_j/m - \boldsymbol{v}(\boldsymbol{r}_j)$ denotes the peculiar velocity of particle j, and $\boldsymbol{p}_j$ its canonical momentum. Homogeneous steady flows are characterized by a position-independent transposed velocity gradient $\boldsymbol{\kappa}$, i.e.,

$$\boldsymbol{v}(\boldsymbol{r}) = \boldsymbol{\kappa}\cdot\boldsymbol{r}, \quad \boldsymbol{\kappa} = (\nabla\boldsymbol{v})^T\,. \tag{8.51}$$

More specifically, we consider shear flows with $\kappa_{ij} = \dot{\gamma}\delta_{i1}\delta_{j2}$ and shear rate $\dot{\gamma}$ in the examples below. The additional potential part of the pressure tensor stems from collisions through the interaction potential between particles. For the present investigation we choose the spherically symmetric two-body Lennard–Jones (LJ) potential, defined in (4.1). For the pressure tensor of rarefied gases the potential contribution is of minor interest, i.e., negligible compared to its kinetic counterpart [383]. The potential part dominates in dense fluids [233, 360, 388]. The interaction potential itself, however, produces momentum transfer and generates dissipation to be quantified through the friction matrix $\boldsymbol{M}$.

The isotropic pressure is considered constant, i.e., $\text{Tr}(\boldsymbol{\Pi}_4) = 3$. We might have introduced the traceless part of (8.50) rather than using the present definition, (8.50); the former choice allows us to handle scalar pressure variations, cf. [54], where a slightly different notation is used.

Implications for Lagrange Parameter and Reversible Motion

For the chosen system and structural variable $\boldsymbol{\Pi}_4$, the corresponding Lagrange parameter is identified analytically by multiplying $\rho_{\boldsymbol{x}}$ with $\boldsymbol{\Pi}_4(\boldsymbol{z})$ and subsequent (standard Gaussian, cf. Page 14.8) integration over $\boldsymbol{z}$ to obtain

$$\boldsymbol{x}_4 = \frac{\int \rho_{\boldsymbol{x}}\boldsymbol{\Pi}_4\,d\boldsymbol{z}}{\int \rho_{\boldsymbol{x}}\,d\boldsymbol{z}} = \int \mathrm{e}^{-\boldsymbol{\lambda}_4:\boldsymbol{\Pi}_4}\,d\boldsymbol{p}^{(N)} = \frac{N}{2}\boldsymbol{\lambda}_4^{-1}\,. \tag{8.52}$$

This is an analytic expression for the Lagrange parameter in terms of the slow variable $\boldsymbol{x}_4 \equiv \langle\boldsymbol{\Pi}_4\rangle_{\boldsymbol{x}}$, and vice versa, $\boldsymbol{\lambda}_4 = N\boldsymbol{x}_4^{-1}/2$. As discussed in detail, these quantities depend on the remaining components of $\boldsymbol{x}$, in particular on the flow field, and it is our goal to obtain this relationship using the stationary GENERIC equation.

The energy E and the operator $\boldsymbol{L}$ characterizing the reversible dynamics of a rarefied gas are collected in the Sect. 8.3.1. Upon inserting $E, \boldsymbol{L}$ from (8.13), (8.8) into the stationary GENERIC equation, (8.48), and by expressing $\boldsymbol{\lambda}_4$ through $\boldsymbol{x}_4$ via (8.52) we obtain

$$\sum_i \boldsymbol{L}_{4i} \cdot \delta E / \delta \boldsymbol{x}_i = \boldsymbol{\kappa} \cdot \boldsymbol{x}_4 + \boldsymbol{x}_4 \cdot \boldsymbol{\kappa}^T - \frac{2}{3} \boldsymbol{x}_4 \boldsymbol{x}_4 : \boldsymbol{\kappa} \,. \tag{8.53}$$

This is an expression for the reversible contribution to the time evolution of the slow component $\boldsymbol{x}_4$ in terms of $\boldsymbol{x} = (\ldots, \boldsymbol{v}, \ldots, \boldsymbol{x}_4)$, while $\boldsymbol{\kappa} = (\nabla \boldsymbol{v})^T$. Evolution equations for the remaining components can be also deduced from Sect. 8.3.1. For the present purpose, however, these equations contain irrelevant information.

Stationary GENERIC Equation for the Rarefied Gas

As discussed in Sect. 8.3.2, for the structured version of Grad's moment method the quantity $\boldsymbol{M}_{44}$ is the only nonvanishing component of the friction matrix for a rarefied gas. Upon inserting (8.53) into (8.48) we therefore arrive at the following stationary GENERIC equation for the structural variable of a rarefied gas:

$$\frac{k_B N}{2} \boldsymbol{M}_{44} : {\boldsymbol{x}_4}^{-1} = \boldsymbol{\kappa} \cdot \boldsymbol{x}_4 + \boldsymbol{x}_4 \cdot \boldsymbol{\kappa}^T - \frac{2}{3} \boldsymbol{x}_4 \boldsymbol{x}_4 : \boldsymbol{\kappa} \,. \tag{8.54}$$

The right hand side (rhs) of (8.54) represents the deterministic transformation behavior of $\boldsymbol{x}_4$ due to the flow field, where the shape of the distribution function, (8.45), has eliminated the appearance of higher moments of $\boldsymbol{\Pi}_4$. The rhs coincides with the change of $\boldsymbol{x}_4$ due to a flow field derived by Grad [386]. Close to equilibrium, where the traceless quantity $\boldsymbol{x}_4 - \boldsymbol{1}$ is small, and ${\boldsymbol{x}_4}^{-1} \approx \boldsymbol{2} - \boldsymbol{x}_4$, the rhs of this equation reduces to the traceless quantity $\boldsymbol{\kappa} + \boldsymbol{\kappa}^T$ (incompressible flow), and the left hand side (lhs) becomes proportional to $\boldsymbol{N} : {\boldsymbol{x}_4}^{-1} = 2(1 - \boldsymbol{x}_4)$ with the 4th-rank tensor $\boldsymbol{N}_{ijkl} = \delta_{ik}\delta_{jl} + \delta_{il}\delta_{jk} - (2/3)\delta_{ij}\delta_{kl}$.

The friction matrix is estimated from short pieces of the trajectory of duration τ_s. The BEMD method uses this value for the friction matrix, at a given value for $\boldsymbol{x}$, to obtain an updated value for this variable from (8.54) (more generally from (8.48)). BEMD proceeds with this iterative process until convergence of the value for $\boldsymbol{x}$ and the Lagrange parameter is reached (see also [389] for further details on how we iteratively determine a consistent nonequilibrium state $\boldsymbol{x}$). For the rarefied gas, once we obtain $\boldsymbol{x}_4(\boldsymbol{x})$ (or $\boldsymbol{\lambda}_4(\boldsymbol{x})$) the effect of flow, density, and temperature on the macroscopic kinetic contribution to the pressure tensor and the corresponding material functions are directly evaluated from $\boldsymbol{\pi}_{\text{kin}} = \rho k_B T \boldsymbol{x}_4 / m$, and all building blocks $\boldsymbol{L}, E, \boldsymbol{M}, S$ are expressed in terms of $\boldsymbol{x} = (\rho, \boldsymbol{\kappa} \cdot \boldsymbol{r}, T, \boldsymbol{x}_4)$.

We remind the reader that (8.54) depends on the choice of nonequilibrium variable. For dense fluids, where the current choice $\boldsymbol{\Pi}_4$ is inappropriate, one has to obtain the analog of (8.54) from (8.48). In [390], where another structural variable has been used to describe polymer melts, $\boldsymbol{M}_{44}$ has been assumed to be the only nonvanishing component that couples to $\boldsymbol{x}_4$. For the rarefied gas, however, this simplification of the equation of change is rooted in the structured version of Grad's moment expansion.

8.9.1 Multiplostatted Equations

For calculating friction matrices such as $\boldsymbol{M}_{44}$ with

$$\boldsymbol{M}_{44} = \frac{1}{2k_{\mathrm{B}}\tau_s} \langle \triangle_{\tau_s} \boldsymbol{\Pi}_4(\boldsymbol{z}) \triangle_{\tau_s} \boldsymbol{\Pi}_4(\boldsymbol{z}) \rangle_{\boldsymbol{x}} \tag{8.55}$$

by molecular dynamics, we need to generate initial configurations according to some ensemble characterized by $\boldsymbol{x}$, and we then need to evolve these initial configurations according to Hamilton's equations of motion. For simulation purposes it is advantageous to modify the time-evolution equations such that the initial ensemble remains invariant. The intrinsic inaccuracy associated with time-scale separation is then shifted from the slow change of the ensemble to a minor modification of the time-evolution equations.

Following the ideas of Nosé [391] and Hoover [392] maintaining the Lagrange multiplier in the course of the time-evolution actually means that we need to preserve a multiplostatted (ms) generalized canonical nonequilibrium ensemble

$$\begin{aligned} \rho_{\boldsymbol{x}}^{\mathrm{ms}}(\boldsymbol{z}) &= \rho_{\boldsymbol{x}}(\boldsymbol{z})\, \mathrm{e}^{-\beta H_{\mathrm{Nose}}} \\ &\propto \rho_0(\boldsymbol{z}) \mathrm{e}^{-\boldsymbol{\lambda}_4 : \boldsymbol{\Pi}_4}\, \mathrm{e}^{-\beta H_{\mathrm{Nose}}} , \end{aligned} \tag{8.56}$$

where $\rho_{\boldsymbol{x}}(\boldsymbol{z})$ has been inserted from (8.45). Due to our choice for $\boldsymbol{\Pi}_4$, which depends on the momenta, the kinetic energy is contained in $\boldsymbol{\lambda}_4 : \boldsymbol{\Pi}_4$. More specifically, $E_{\mathrm{kin}}(\boldsymbol{z}) = -\boldsymbol{\lambda}_4 : \boldsymbol{\Pi}_4$ with an isotropic Lagrange parameter $\boldsymbol{\lambda}_4 = N\boldsymbol{1}/2$. For this reason, here ρ_0 contains only the potential energy $E_{\mathrm{pot}}(\boldsymbol{z})$ due to the interaction potential rather than the total energy, i.e., $\rho_0(\boldsymbol{z}) = \exp\{-\beta E_{\mathrm{pot}}\}$. The term $H_{\mathrm{Nose}} \equiv \frac{1}{2} p_T^2/M_T$ contains an additional variable p_T representing a reservoir. The temperature explicitly enters through $\beta = (k_{\mathrm{B}}T)^{-1}$ and definition (8.50).

While considering fixed volume and fixed number of particles, the strategy for obtaining a dynamics consistent with (8.56) is based on introducing canonical variables $\boldsymbol{p}'_j$ and scalars q_T, p_T, where the canonical particle momenta $\boldsymbol{p}'_j$ are related to the physical ones, $\boldsymbol{p}_j$ by $\boldsymbol{p}_j = \boldsymbol{p}'_j/q_T$. In order to obtain all time-evolution equations that constitute the multiplostat we choose as the Hamiltonian

$$H' = q_T \left(E_{\mathrm{pot}} + H_{\mathrm{Nose}} + \beta^{-1} \boldsymbol{\lambda}_4 : \boldsymbol{\Pi}_4 + f k_{\mathrm{B}} T \ln q_T \right) , \tag{8.57}$$

with f denoting the number of degrees of freedom, $f = 3N$. This Hamiltonian is consistent with (8.56). The logarithmic form in q_T is crucial because it leads to the exponential in the generalized canonical distribution. From (8.57) we obtain the modified Hamiltonian equations $\dot{\boldsymbol{r}}_j = dH'/d\boldsymbol{p}'_j$, $\dot{\boldsymbol{p}}'_j = -dH'/d\boldsymbol{r}_j$, $\dot{q}_T = dH'/dp_T$ and $\dot{p}_T = -dH'/dq_T$. When re-expressing these equations of change in physical variables (simple chain rule for $\dot{\boldsymbol{p}}_j$, mechanical force $\boldsymbol{F}_j \equiv -d\phi/d\boldsymbol{r}_j$) and after eliminating q_T by making use of its time evolution equation, which reads $d\ln q_T/dt = p_T/M_T$, we have

$$\dot{\boldsymbol{r}}_j = k_{\mathrm{B}}T \frac{\partial \boldsymbol{\Pi}_4 : \boldsymbol{\lambda}_4}{\partial \boldsymbol{p}_j} , \tag{8.58}$$

$$\dot{\boldsymbol{p}}_j = \boldsymbol{F}_j - \frac{k_B T}{m} \frac{\partial \boldsymbol{\Pi}_4 : \boldsymbol{\lambda}_4}{\partial \boldsymbol{r}_j} - \frac{p_T}{M_T} \boldsymbol{p}_j , \tag{8.59}$$

$$\dot{p}_T = \sum_j \left(\boldsymbol{p}_j \cdot k_B T \frac{\partial \boldsymbol{\Pi}_4 : \boldsymbol{\lambda}_4}{\partial \boldsymbol{p}_j} \right) - f k_B T . \tag{8.60}$$

In practice, periodic boundary conditions [393] supplement this set of equations when considering bulk properties. For obtaining (8.64), we require that the value of H' is kept at zero at all times; otherwise a term H'/q_T has to be added to (8.64). In view of the conservation of H' as the generator for the Hamiltonian evolution of the canonical variables, this assumption seems to require a proper choice of initial conditions. However, with $H' = 0$ we eliminate q_T from (8.58)–(8.60) and the initial $q_T(0)$ is uniquely determined from $H'(0)$ using (8.57). In Nosé's derivation the term H'/q_T does not appear since $dt = q_T d\tau$ and H' are defined without the prefactor q_T. The parameter M_T (also related to a characteristic response frequency $\nu = f k_B T / M_T$ [394]) describes the inertia of the scale factor p_T and must be chosen carefully. Details are given in [389]. Here we prefer to keep the notation of [54], i.e., use M_T (proportional to the number of degrees of freedom f) rather than ν (independent of f).

Multiplostatted Equations for the Rarefied Gas

Specializing to our structural noncequilibrium variable the following derivatives are readily evaluated using the definition of $\boldsymbol{\Pi}_4$ and the relationship between peculiar velocity and canonical momentum,

$$k_B T \frac{\partial \boldsymbol{\Pi}_4 : \boldsymbol{\lambda}_4}{\partial \boldsymbol{p}_j} = \frac{2\boldsymbol{\lambda}_4}{N} \cdot \boldsymbol{c}_j ,$$

$$k_B T \frac{\partial \boldsymbol{\Pi}_4 : \boldsymbol{\lambda}_4}{\partial \boldsymbol{r}_j} = -m \boldsymbol{\kappa}^T \cdot \frac{2\boldsymbol{\lambda}_4}{N} \cdot \boldsymbol{c}_j , \tag{8.61}$$

when $\boldsymbol{\lambda}_4$ does not depend on phase space variables. Finally, we eliminate $\boldsymbol{\lambda}_4$ using (8.52) to obtain thc multiplostatted equations of change (8.58)–(8.60) for the rarefied gas:

$$\dot{\boldsymbol{r}}_j = \boldsymbol{x}_4{}^{-1} \cdot \boldsymbol{c}_j , \tag{8.62}$$

$$\dot{\boldsymbol{p}}_j = \boldsymbol{F}_j + \boldsymbol{\kappa}^T \cdot \boldsymbol{x}_4{}^{-1} \cdot \boldsymbol{c}_j - \frac{p_T}{M_T} \boldsymbol{p}_j , \tag{8.63}$$

$$\dot{p}_T = \sum_j (\boldsymbol{p}_j \cdot \boldsymbol{x}_4{}^{-1} \cdot \boldsymbol{c}_j) - f k_B T . \tag{8.64}$$

This set reduces to Nosé's equilibrium thermostat for $\boldsymbol{v}(\boldsymbol{r}) = \boldsymbol{0} \rightarrow \boldsymbol{\kappa} = \boldsymbol{0}, \boldsymbol{x}_4 = \boldsymbol{1}, \boldsymbol{c}_j = \boldsymbol{p}_j/m$.

Due to the velocity gradient $\boldsymbol{\kappa}^T$ appearing in (8.63) the time evolution equations (8.62)–(8.64) offer similarities with the DOLLS [360] equations, whereas it is known that the SLLOD equations (with $\boldsymbol{\kappa}$ instead of $\boldsymbol{\kappa}^T$ in (8.63))give an exact description of shear flow (but not of elongational flow) arbitrarily far from equilibrium. We comment on this similarity in Sect. 8.9.3.

Multiplostatted Equations for Arbitrary Π_4

For a general application of multiplostatted equations, when neither the kinetic nor the potential energy are contained in the definition of $\boldsymbol{\Pi}_4$, the quantity ρ_0 is the full equilibrium distribution function. The equations (8.58)–(8.60) remain valid upon adding the contributions (8.61) with $2\boldsymbol{\lambda}_4/N = \mathbf{1}$ resulting from the conventional kinetic energy term.

8.9.2 Applicability of BEMD

In order to appreciate the set of multiplostatted equations and the iteration concept (and before going to apply them to a rarefied gas) some words of caution are in order. One can ask under which conditions the multiplostatted equations (8.62)–(8.64) can be considered as small modifications of the original equations of motion. The term involving the scale factor p_T can be kept small by choosing the parameter M_T sufficiently large. It has to be chosen such that p_T changes on the time scale of the slowest relevant variable $\boldsymbol{x}_4$, thus allowing one to explore the physically achievable values of energy on that time scale. Equation (8.63) requires velocity gradients to be small compared to typical momentum relaxation rates, while (8.62) requires the average flow velocity to be small compared to the individual particle velocities. We hence use a coordinate system in which the initial total momentum vanishes, and the origin coincides with the initial center of mass of the particles. This particular frame of reference is crucial and also convenient (see also [395] and Sect. 8.9.3). For the flow terms in the multiplostat to be small we obtain from (8.62) the further condition that the simulated system can be taken sufficiently small so that the variations of the flow velocity across the system are small compared to thermal velocities. Finally, the smallness of the terms maintaining a constant Lagrange multiplier needs to be checked for every particular choice.

These general restrictions propagate to the case of a rarefied gas as follows. Here and in the remaining section we use dimensionless Lennard–Jones units (in particular, $m = 1, k_B = 1$). The average velocity $v = |\boldsymbol{v}|$ for particles in a rarefied gas at rest is $\langle v \rangle = \sqrt{8k_B T/\pi}$. The mean free path λ is defined as $\lambda = \langle v \rangle \tau$. The so-called Boltzmann viscosity (equal to the zero shear-rate shear viscosity) takes the value $\eta_B \equiv p\tau$. The strength of the friction matrix becomes $M_{\text{eq}} \propto (\rho\tau)^{-1}$, where the coefficient of proportionality is very well approximated by unity, as confirmed by the present simulations. In equilibrium, $\boldsymbol{M}_{44} = M_{\text{eq}}\boldsymbol{N}$ with $\boldsymbol{N}$ already defined in Sect. 8.9. Along with the above considerations we choose a box size $L = \alpha\lambda$ with a parameter $\alpha \gg 1$ to be discussed below. For variations of the flow velocity across the system to be small compared to thermal velocities requires $L\dot{\gamma} \ll \langle v \rangle$, or equivalently, $\dot{\gamma} \ll \langle v \rangle / L$. Since $\alpha \gg 1$, the latter inequality is actually not restrictive since $\langle v \rangle / L < \tau^{-1}$ then poses the stronger restriction for the case of an Lennard–Jones gas. The Maxwell relaxation time τ is related to a Chapman-Cowling collision integral [389] and becomes, for the case of an Lennard–Jones gas, $\tau \approx (6\rho\sqrt{T})^{-1}$, which equals $(\sigma\rho \langle v \rangle)^{-1}$ with an effective particle diameter $d = 0.92$ (for the case of a Lennard–Jones gas), and 'cross section' $\sigma \equiv \sqrt{2}\pi d^2$.

BEMD of a rarefied Lennard–Jones gas subjected to period boundary conditions and weak shear flow had been fully implemented in [389], where all simulation details are explicitely given.

Once we measure $\boldsymbol{M}_{44}$ defined in (8.49) at given $\boldsymbol{x}$, we determine a corrected 'Lagrange parameter' $\boldsymbol{x}_4$ from (8.54), using the current value for $\boldsymbol{x}_4$ on its rhs. The friction matrix $\boldsymbol{M}_{44}$ is traceless in its first and last two components, thus $\boldsymbol{M}_{44} : \boldsymbol{x}_4{}^{-1} = \boldsymbol{M}_{44} : (\boldsymbol{x}_4{}^{-1} + p_0 \mathbf{1})$ for arbitrary p_0, and p_0 is uniquely determined from the analytically solvable nonlinear equation $\mathrm{Tr}(\boldsymbol{x}_4) = 3$. The described iterative procedure is an integral part of the BEMD method. A fix point and stationary value for $\boldsymbol{x}_4$ is only reached if $\boldsymbol{x}_4$ and $\boldsymbol{M}_{44}$ correspond to each other. During the course of this process the Lagrange parameter is updated in time intervals larger than τ_s needed to determine the updated $\boldsymbol{M}_{44}$. As soon as the physical regime is reached, where the Lagrange parameter remains unaltered, also the mean $\boldsymbol{\Pi}_4$ agrees with $\boldsymbol{x}_4$.

BEMD – NEMD Switch

We presented the BEMD simulation strategy, which demonstrates the application of the GENERIC formalism and the calculation of the friction matrix, and applied it to a rarefied Lennard–Jones gas subjected to shear flow. Our choice of nonequilibrium variables, (8.50), is inspired by Grad's moment method for gases. Multiplostatted equations of motion (8.58)–(8.60), (8.62)–(8.64) – the three former equations simplify to the latter ones for the rarefied gas – were employed to maintain the generalized canonical ensemble, (8.45), (8.56). The friction matrix has been iteratively obtained through (8.48), (8.54) employing the Green–Kubo type expression (8.49), (8.55). BEMD supersedes the conventional ('reference') NEMD simulations concerning efficiency in the weak flow regime, i.e., whenever they are applicable. The range of applicability was discussed in Sect. 8.9.2. We made sure that same conclusions hold for slight variations of the reference algorithm, cf. [383]. For larger shear rates, essentially in the non-Newtonian regime of the rarefied gas, conventional NEMD seems to be a more suitable approach. For our current purposes, the rescaling procedure suffices, particularly for the moderate flow rates under study, where most thermostats (including Nosé-Hoover) give 'identical' results [395]. An upper limit for the NEMD method using the 'Gaussian' thermostat [389] has been discussed in [383]. Profile unbiased thermostats are commonly used at dimensionless shear rates $\Gamma > 5$. From the present investigation we conclude that an efficient way to obtain the zero shear-rate viscosity may be to run a simulation at a very low rate in order to get an estimate for the relaxation time τ to be chosen as the initial value for τ_{init}. Then, simulations can be carried out in the Newtonian regime at dimensionless shear rate $\Gamma = \dot{\gamma}\tau_{\mathrm{init}}$ in the range $0.01 < \Gamma < 0.1$, which constitutes a regime where the presented multiplostatted equations are most efficient. Subsequently, NEMD can be used to calculate material functions at shear rates $\Gamma > 0.1$. This procedure requires an amount of computing time that is at least an order of magnitude smaller than the one required for a complete NEMD run.

Alternatively, if an estimate for the relaxation time corresponding to the nonequilibrium variable is not available, one can proceed from large to smaller field strengths (rates) using NEMD and extracting the microscopic material quantity (pressure tensor $\boldsymbol{\pi}$) as a time average until error bars become significantly large but smaller than, say, 5% at a certain flow (external field) strength, cf. Fig. 8.1 for a schematic drawing. Next, one has to run NEMD and BEMD in parallel and switch to BEMD as soon as both results for $\boldsymbol{\pi}$ coincide at a certain 'critical' rate. Below this rate, BEMD can be safely used to complete the flow curve. Here, we do not need an estimate for the switch rate or critical rate using arguments as those presented in Sect. 8.9.2.

The closer the generalized canonical distribution function represents the correct nonequilibrium distribution function, the more efficient BEMD becomes. The new method can be more usefully applied to more complex, and dense fluids. The generalized canonical distribution used here has been shown in [396] to provide the most accurate closure approximation for the theory of rodlike liquid crystalline polymers (where this distribution is termed 'Bingham distribution', and where the nonequilibrium variable is chosen as $\boldsymbol{\Pi}_4 = \boldsymbol{uu}$ with rod axis $\boldsymbol{u}$). The success of BEMD is intimately related to the suitable choice of relevant variables. For bulk polymers in the molten state such a relevant variable may be the end-to-end distance, the first normal mode coordinate or the tensor of gyration [390], as suggested by experience with theoretical approaches.

8.9.3 DOLLS/SLLOD Analogy with Multiplostatted Equations

Due to the velocity gradient $\boldsymbol{\kappa}^T$ appearing in (8.63) the time evolution equations (8.62)–(8.64) offer similarities with the DOLLS [360] equations, whereas it is known that the SLLOD equations – with $\boldsymbol{\kappa}$ instead of $\boldsymbol{\kappa}^T$ in (8.63) – give an exact description of shear flow arbitrarily far from equilibrium. SLLOD can be also seen as rheological theory [48]. Following the work of Edwards and Dressler [397] the DOLLS algorithm is not 'completely Hamiltonian' because the connection with the underlying Lagrangian problem, via a Legendre transformation, has been severed. Accordingly, the Hamiltonian equations should appropriately be expressed through a non-canonical Poisson bracket. In doing so, Edwards and Dressler and earlier Tuckerman et al. [398] obtain GSLLOD equations (generalized SLLOD, modified by a term which appears for flows for which $\boldsymbol{\kappa}^2$ does not vanish). Following their concept one would supplement (8.63) by a term $-m\,\boldsymbol{\kappa}^2\cdot\boldsymbol{r}_j$, and replace the existing $\boldsymbol{\kappa}^T$ by $\boldsymbol{\kappa}$. Concerning the acceleration $\ddot{\boldsymbol{r}}$ for all the mentioned models, we obtain ($m=1$, no interactions, $\boldsymbol{F}=\boldsymbol{0}$)

$$\ddot{\boldsymbol{r}}_j = \alpha_1(\boldsymbol{\kappa}^T-\boldsymbol{\kappa})\cdot\dot{\boldsymbol{r}}_j + \alpha_2(\boldsymbol{\kappa}^T-\boldsymbol{\kappa})\cdot\boldsymbol{\kappa}\cdot\boldsymbol{r}_j + \alpha_3\,\boldsymbol{\kappa}^2\cdot\boldsymbol{r}_j - \dots\,, \tag{8.65}$$

where the coefficients are listed in Table 8.1. If the macroscopic flow profile $\boldsymbol{\kappa}$ should not influence the inertia of particles through these algorithms, GSLLOD (with $\forall_i\alpha_i = 0$) offers the appropriate structure. Independent of the type of algorithm, flow is driven by additional Lees-Edwards boundary [393] conditions. From equation (8.65) it is obvious that for the case of shear flows ($\boldsymbol{\kappa}^2=\boldsymbol{0}$), SLLOD equals GSLLOD while for elongational flows ($\boldsymbol{\kappa}^T=\boldsymbol{\kappa}$) SLLOD equals DOLLS. Actually, by solving (8.65) for initial conditions $\boldsymbol{r}(0) = (0, y_0\neq 0, 0)$ and $\dot{\boldsymbol{r}}(0) = \alpha\boldsymbol{\kappa}\cdot\boldsymbol{r}(0)$

Table 8.1. Coefficients appearing in the acceleration equation (8.65) for the coordinates for diverse models. Here BEMD stands for the isotropic ($x_4 = 1$) version of the (8.62)–(8.63)

Coefficient	Relevance	BEMD	DOLLS	SLLOD	GSLLOD
α_1	shear	1	1	0	0
α_2	shear	0	1	0	0
α_3	elongation	0	1	1	0

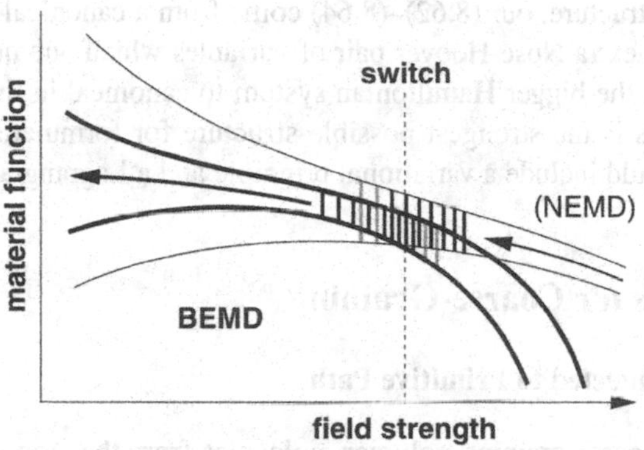

Fig. 8.1. Suggested switch at a critical field strength (flow rate) from NEMD to BEMD, or vice versa, as described in Sect. 8.9.2. For the simple model system considered here, NEMD can be directly employed to obtain material functions from atomistic configurations. More generally, simulations must be performed on several levels of description invoking the choice of relevant variables on each separate level. The error for the material function grows with decreasing field strength because it actually represents a coefficient of proportionality for an inverse deviation from equilibrium (viscosity rather than a component of the pressure tensor)

($\alpha = 0, 1$), we see immediately that SLLOD without appropriate boundary conditions is the only version which does not produce 'aphysical' trajectories for both types of flow. For this simple initial condition we consider a result as aphysical if the y component changes under shear flow, or if it changes sign or keeps it initial value under elongational flow in the absence of forces. The method of rescaling peculiar velocities used here for performing NEMD is aphysical in the above sense, since a one particle analysis is impossible. Considering boundary conditions [393,395,399], the picture and conclusions are slightly different. Boundary conditions (such as Lees-Edwards) transmit forces which, if they are stabilizing a (homogeneous) flow situation ($\dot{r} = \boldsymbol{\kappa} \cdot \boldsymbol{r}$) compatible with these boundary conditions, effectively increase α_3 in (8.65) by one. That is why, in our opinion, the GSLLOD algorithm rather than SLLOD should be appropriate for elongational flows if such boundary conditions are used.

Most importantly, we remind the reader that for the method outlined in this article, we do not need to produce physical trajectories at small and large (shear) rates. Instead, we need to preserve the generalized canonical distribution (8.45) (physical or aphysical), which is the case for our equations (8.62)–(8.63). Our goals are absolutely different from those of DOLLS/SLLOD approaches. Their ambition is to impose significant flow, and this comes with a significant modification of Hamilton's equations of motion. We want to achieve generalized canonical distributions with a minor modification of Hamilton's equations of motion. (Minor means that the modification should introduce errors of the same order of magnitude as those coming with the assumption of a time-scale separation.) For a rarefied gas faced with the Grad level, this limits us to small flow rates, cf. Sect. 8.9.2. Concerning the Hamiltonian structure, our (8.62)–(8.64) come from a canonical symplectic formulation (with an extra Nosé-Hoover pair of variables which one needs to go from microcanonical in the bigger Hamiltonian system to canonical in the system of actual interest). This is the strongest possible structure for formulating Hamiltonian dynamics and should include a variational principle and a Lagrangian formulation.

8.10 Examples for Coarse-Graining

8.10.1 From Connected to Primitive Path

A procedure for coarse-graining polymer molecules from the atomistic level of description (and also FENE chain level) to the reptation level for entangled polymers had been presented in [140]. While this method is based on collapsing a certain number of atoms or monomers into a large unit at their center of mass, the smooth and uniform dependence of the coarse-grained chain on positions of all atoms proposed in [400] is useful if one is interested in a two-way coupling of two levels of description as pointed out in [401]. We just summarize how to explicitely apply coarse-graining from the latter procedure, which is illustrated in Fig. 8.2. The transformation, parametrized by a single parameter, $\mathcal{P}_\xi : \{\boldsymbol{x}^0{}_i\} \to \{\boldsymbol{x}_i\}$ maps a set of $i = 1, .., N$ atomistic (or FENE chain) coordinates of a linear chain to a new set with an equal number of coordinates, called coarse-grained coordinates $\boldsymbol{x}_i$, which define the coarse-grained chain or 'primitive path' $\{\boldsymbol{x}_i\}$ of the atomistic chain. In order to motivate the mapping, we require, that $\mathcal{P}_0$ =Id, i.e., for $\xi = 0$ all information of the atomistic chains is conserved for the coarse-grained chain. The opposite limit reflects a complete loss of information about the atomistic structure, i.e, the projection in the limit $\xi \to \infty$ gives give a straight line (or dot) for arbitrary atomistic configurations. The recommended mapping results from minmization of the energy

$$E \propto \frac{1}{2}\sum_{i=1}^{N}(\boldsymbol{x}_i - \boldsymbol{x}^0{}_i)^2 + \frac{\xi}{2}\sum_{i=1}^{N-1}(\boldsymbol{x}_{i+1} - \boldsymbol{x}_i)^2 , \tag{8.66}$$

for a system of two types of Hookean springs. The first type connects adjacent beads within the primitive chain, the second type connects the beads of the primitive chain with the atomistic beads, and ξ is the ratio between spring coefficients. The mapping from atomistic $\boldsymbol{x}_o$ to coarse-grained coordinates $\boldsymbol{x}$ reads, with the $N \times N$ tri-diagonal matrix $\mathcal{P}^{-1}$ which can be inverted with order N effort (see Sect. 12.7.1):

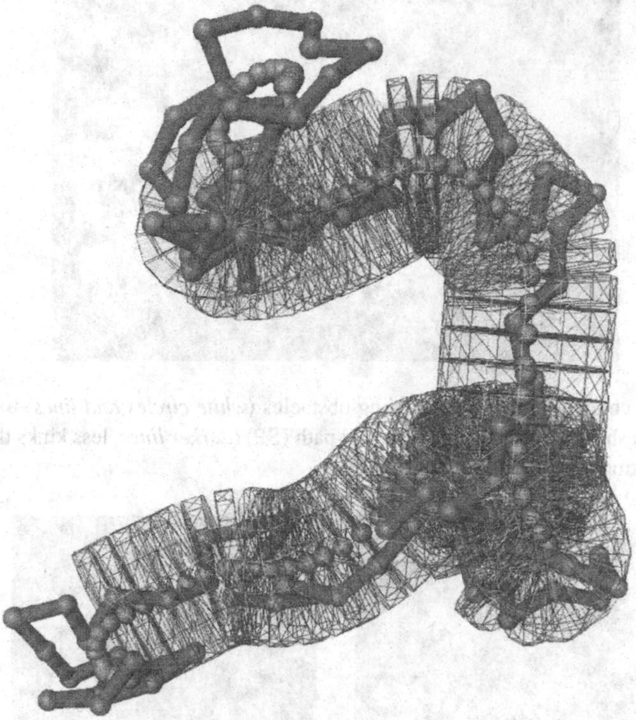

Fig. 8.2. Miscroscopic chain (*dark beads*) and its primitive path (*light beads*, tangential cylinder indicated). The latter is obtained by the mapping $\mathcal{P}_\xi$ via (8.67) for a certain ratio of spring coefficients ξ

$$
\boldsymbol{x}_i = \sum_{j=1}^{N} \mathcal{P}_{ij} \cdot \boldsymbol{x}^0{}_j, \quad \mathcal{P}^{-1} = \begin{pmatrix} 1+\xi & -\xi & 0 & \cdots & \cdots & 0 \\ -\xi & 1+2\xi & -\xi & 0 & \ddots & \vdots \\ 0 & -\xi & 1+2\xi & \ddots & \ddots & \vdots \\ \vdots & \ddots & \ddots & \ddots & -\xi & 0 \\ \vdots & \ddots & 0 & -\xi & 1+2\xi & -\xi \\ 0 & \cdots & \cdots & 0 & -\xi & 1+\xi \end{pmatrix}, \quad (8.67)
$$

for all $i = 1 \ldots N$. iiCoarse-graining!linear polymer

The discrete coarse-graining had been recently analyzed in [401] for wormlike 'atomistic' chains characterized by their squared end-to-end vector $\langle \boldsymbol{R}_{(0)}{}^2 \rangle$ and their tube diameter d_T (i.e., quantities usually tabulated, cf. Sect. 4.5 and Table 4.5). One of the important result of [401] states, that the correct parameter ξ is determined by these two characteristics via

$$
\frac{1}{\xi^{1/2}} \propto \frac{\langle \boldsymbol{R}_{(0)}{}^2 \rangle}{N-1} \frac{1}{d_T^2}, \quad \text{for } \xi^{1/2} \ll N, \quad (8.68)
$$

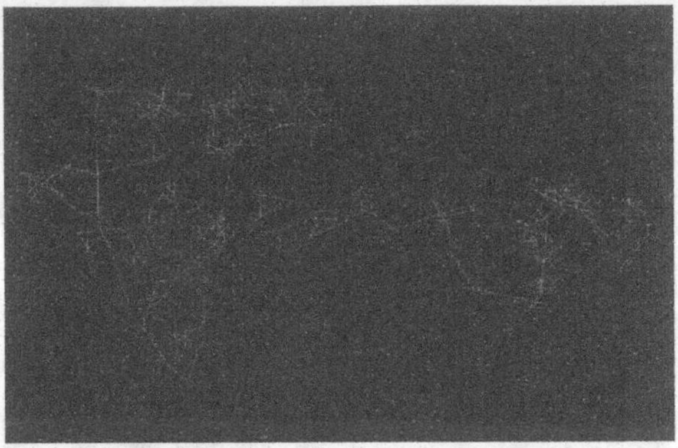

Fig. 8.3. Input configuration (2D) including obstacles (*white circles and lines*) together with the constructed shortest multiple disconnected path (SP) (*darker lines*, less kinks than original chain). Simulation code available in [402]

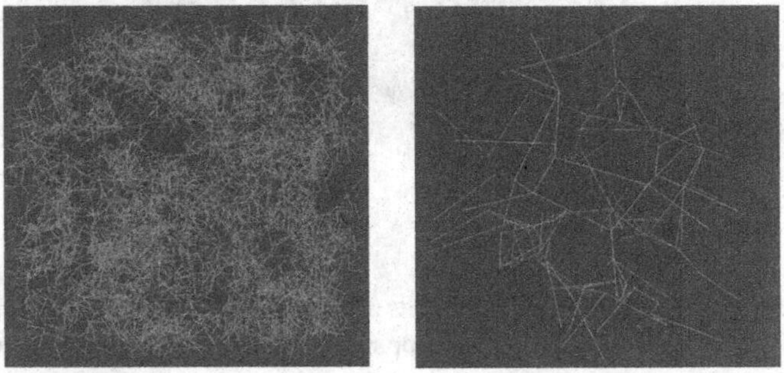

Fig. 8.4. Original chains (*left*) and SP (*right*). Computation done within roughly 1 second for such a system (polymer melt, 10 chains à 500 beads). Simulation code available in [402]

with a prefactor of order unity. In terms of the quantities introduced in Sect. 4.5 this relationship is rewritten as $\xi \propto N_e^2$, for $N \gg N_e$ with the characteristic entanglement length N_e.

8.10.2 From Disconnected to Primitive Path

Rubinstein and Helfand, and later Everaers et al. introduced a concept to extract primitive paths for dense polymeric melts made of linear chains (a multiple disconnected multibead 'path'), where each primitive path is defined as a path connecting the (space-fixed) ends of a polymer under the constraint of non-interpenetration (excluded volume) between primitive paths of different chains, such that the multiple disconnected path fulfills a minimization criterion. In [402] we presented an algorithm which returns a shortest path (SP) and related number of entanglements for a given configuration of a polymeric system in 2 or 3 dimensions, cf. Figs. 8.3, 8.4. Our algorithm uses geometrical operations and provides a – model independent – efficient approximate solution to this challenging problem. Primitive paths are treated as 'infinitely' thin (we further allow for finite thickness to model excluded volume), and tensionless lines rather than multibead chains, excluded volume is taken into account without a force law. This implementation allows to construct a shortest multiple disconnected path (SP) for 2D systems (polymeric chain within spherical obstacles) and an optimal SP for 3D systems (collection of polymeric chains). The number of entanglements is then simply obtained from the SP as either the number of interior kinks (Z_{kinks}), or from the average length of a line segment (Z_{coil}). Further, information about structure and potentially also the dynamics of entanglements is immediately available from the SP. While our algorithm runs to minimize the Euclidean path length, previous implementations carefully [403,404] minimized bond energies, which of course in some cases, must lead to quantitative different results. But still, both approaches tend to produce very comparable path mesh characteristics.

With linear, unanchored, polymers a concept of 'topological equivalence' adapted from knot theory is useless because all paths are topologically equivalent. One can always disentangle paths by pulling one path around the end of the other (or itself). Fortunately, in the limit of large chain lengths N there are several schemes which will eliminate these unlikely distortion processes and create a meaningful definition of topological equivalence. Perhaps the easiest expedience is to artificially make the molecule cyclic by drawing a straight line between the two ends of each polymer. It is evident that the total length of these lines is less than the length of the set of primitive paths, so as $N \to \infty$ the extra lines contribute negligibly to the entanglement net. In our approach, we obtain a shortest path which is independent of the sequence chain of displacements in the limit of infinitely small, impractical, displacements (requiring infinite time) during the shortening process. The solution must be regarded as approximate. The shortening process, where the SP length is strictly decreasing, prevents disentangling of chains by the above-mentioned pull-around.

Next, we shortly describe the algorithm and apply the method to study the 'concentration' dependence of the degree of entanglement in phantom chain systems. As for binary interaction particle dynamics methods, where the driving forces on all beads can be calculated by a simple double loop over particles ('N^2' type implementation, see Sect. 12.4.1 for a simple example), and research focuses on the efficient calculation of forces, we describe here, how to build an 'N^2' version of our algorithm, which is easy to implement. All details about the efficient code are given in the source code attached to [402]. The main procedure acts on a pair of arbitrarily selected, but adjacent, segments (the 'in-subpath') of the current multiple disconnected path, and returns one or more new (connected) segments replacing the two original ones (the 'out-subpath'), cf. Fig. 8.5. Selected pairs of segments are always assigned to a certain original chain. The out-subpath must have the following properties: (i) The coordinates of the limiting nodes of in- and out-subpaths coincide, (ii) its contour length is less than or equal compared with the one of the in-subpath, (ii) The in-subpath can be continuously transformed into the out-subpath without touching any existing object (points in 2D, paths of all chains in 3D – where all means all except the subpath to which the in-path belongs as long as we do not wish to count self-entanglements), (iv) the length of each individual segment of the out-path is less than one quarter of the minimum box length (a method to avoid any complications with periodic boundary conditions).

This procedure alters the shape and number of segments of the current path and returns an updated path which re-enters the procedure. All operations have to take care about the periodicity of the simulation cell. Our implementation of this procedure uses a small positive number (a parameter) as the minimum allowed distance between any two points on segments, rather than recording orientations and connectivity when two segments almost touch each other. Further, single segment vectors are split into two identical segment vectors in order to fulfill (iv). The above procedure is called iteratively until the overall contour length of the multiple connected path does not decrease anymore for any choice of entering pair, and does not decrease upon choosing a different ordering of entering pairs. The final path is the SP. Depending on the actual implementation of this procedure, the conformation of the SP is not completely insensitive to the order we select pairs. The number of entanglements and mean mesh size of the entangled network, however, is quite insensitive to the ordering. The procedure can be used in a Monte-Carlo type fashion, and moves can be rejected according to user-defined criteria, as long as they do not prevent selected pairs from further 'shrinking'.

We ensure property ii) by calculating, for the given pair of segments, all obstacles located in, and all points on lines crossing, the secant area of the pair. To this end we make use of intercept point formulas (point+area and line-area). Once we have a set (say S) of intersecting coordinates, we need to decide which single member of S is relevant for the construction of new segments. If the size of S is zero, the out-path will be a straight line between the fixed ends of the pair. If the size of S is unity, the node between the two segments is moved close to the intercept point, along the straight line connecting intercept point and node. If the size of S is larger than unity, we choose from S the point (with coordinates C) which offers the smallest angle between lines A-B and A-C, where A-B denotes the 'first' segment (of the given pair), B the centered node, and A a fixed node at one of the ends of the pair. This point is then the only remaining member of S, and we proceed as if the size of S was unity. Any of the operations reduces the overall contour length of the subpath and keeps it possibly unchanged only if the node is located close to one or more intercept points. This does not imply, that there is no overall movement anymore. The intersect points themselves move according to the above iterative procedure.

In order to count the number of kinks, artificial segments, just introduced to the data structure to ensure (iv) (relevant for 'small' systems) are removed at the final stage.

While in Sect. 4.4 the critical molecular weight had been measured, here we thus construct a SP and extract the entanglement weight N_e (number of monomers between entanglement points, obtained from the segment lengths of the SP assuming equidistant separations of monomers) and the number of entanglements Z for a chain,

$$Z \equiv \frac{N}{N_e} - \frac{N}{N-1} \,. \tag{8.69}$$

The correction term $N/(N-1)$ is needed to correctly cover the limit of rods, for which we wish to have $Z = 0$ and $N_e = N - 1$. Equation (8.69) usually does not hold if quantities N_e, Z are replaced by their averages over chains. Following [403], N_e is defined from the primitive paths (SP) as the ratio $N_{e,coil} \equiv d/b_{\rm pp}$, with $d \equiv R^2/L$, and $b_{\rm pp} \equiv L/(N-1)$, hence,

$$Z_{coil} \equiv \left(\frac{N}{N-1}\right)\left(\frac{L^2}{R^2} - 1\right), \tag{8.70}$$

where L denotes the mean contour length, and R^2 the mean squared end-to-end distance of a primitive path with N beads, tube diameter d, further $b^2 = db_{\rm pp}$, and $R_{\rm ee}^2 = d^2 Z_{coil}$, or equivalently $N_e = d^2(N-1)/R_{\rm ee}^2$. For rods, $Z_{coil} = 0$.

Since R and the number of beads N are fixed during the construction of primitive paths, the number of entanglements Z per chain and the entanglement molecular weight is determined by the mean contour length L of the primitive path, when (8.70) is used as the definition for Z. This definition prevents defining and extracting kinks or step lengths of primitive paths and assumes that the primitive path is a Gaussian coil with step length $b_{\rm pp}$. Alternatively, we extract Z from the primitive path as the average number of interior kinks,

$$Z_{\rm kinks} \equiv \frac{\#\text{kinks}}{\#\text{chains}}. \tag{8.71}$$

For rods, $Z_{\rm kinks} = 0$. As we will see, both definitions for Z yield very similar values (in an unstrained, equilibrium state) upon defining a kink as a node representing the primitive path (as the present code constructs it), where we do not count nodes which are closer together (in terms of contour length) than twice the line thickness. These points have to be discarded since a 'sharp corner' of the line enclosing a small angle involves more than a single point on the primitive path.

In order to test the output and scaling behaviors, we present in Fig. 8.6 results obtained for (in total roughly 1500) monodisperse systems made of 100 linear, random phantom chains with $N = 20, 40, 60, 80, 100, 150, 200, 200, 250, 300, 500, 1000$ beads, bond lengths $b_0 = 0.5, 0.6, .., 1.3$, linethicknesses $0.05, 0.06, .., 0.13 \times 10^{-5}$, contained in a cubic simulation cell at bead number densities $n = 0.050, 0.075, 0.10, 0.15, 0.20, 0.30, .., 1.30$. For these flexible chains, $b = b_0$. Here, the input configurations are uncorrelated, and linethickness is chosen very small (primitive paths are, of course, still uncrossable), excluded volume is therefore absent in the generation process, and very small (in fact, it is squared linethickness times length of the SP times π) during the analysis. A master curve for all data (Fig. 8.6) for the dimensionless tube diameter $d/(b\sqrt{N}) = \sqrt{N_e/N}$. We have

$$d = (b\sqrt{N})\, f(x), \quad \text{with } x \equiv n_p (b\sqrt{N})^3 N^{(\gamma-1/2)} = (nb^3) N^{\gamma}. \tag{8.72}$$

As outlined above, we have two methods for extracting the tube diameter, or N_e, denoted as $N_{\rm e,coil}$ and $N_{\rm e,kinks}$. Analyzing the data, cf. Fig. 8.7, we find the same, nonzero, exponent $\gamma = 0.82 \pm 0.04$ for both methods, i.e., $x = (nb^3)N^{0.82}$. If we

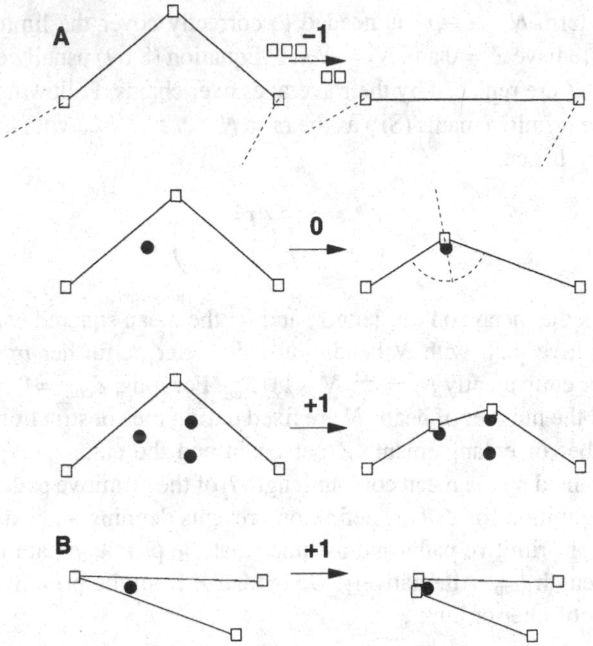

Fig. 8.5. Basic moves during generation of the SP changing the number and contour length of edges. Example A shows two neighboring line segments in a plane with three different numbers of obstacles (compare rows 1-3). These obstacles are 3D line obstacles crossing the area, or 2D point obstacles). Configurations are shown before (*left*) and after (*right*) the basic operation which eventually alters the number of line segments, while reducing the overall contour length of the line. Example B shows the case where, due to numerical precision, artificial, short line segments are introduced to properly take into account uncrossability of chains. These artifical segments do not influence the contour length, and thus leave $Z_{\rm coil}$, (8.70), unaltered, and are ignored when counting Z via the number of segments, cf. (8.71)

regard x as the relevant scaling variable, we can identify a scaling exponent ν' in $x \propto (nb^3)N^{3\nu'-1}$ [405] as $\nu' = 0.610 \pm 0.015$. The classical scaling exponent appearing in the radius of gyration $R \propto N^\nu$ is $\nu = 1/2$ for our artificial phantom chain system. In this case both exponents differ since concentration effects solely enter the tube diameter, mesh size, or variable x, but not the statistics of random paths. We therefore cannot adapt scaling theory, cf. [405], for physical, semidilute polymer solutions and melts under athermal or theta solvent conditions. While $\nu = 1/2$ corresponds to theta solvents, $\nu = 0.588$ is expected for athermal solvents obeying self-avoiding walk statistics.

Specifically, for $x \geq 17$ (kinks) and $x \geq 30$ (coils), we have $f(x) \propto x^{-0.60\pm0.02} \approx x^{-1/(2\gamma)}$ for both quantities, which reflects the fact that d becomes independent of chain length. For x smaller than these thresholds, $f = 1$, such that $N_e = N - 1$ and $Z = 0$ below the crossover, in agreement with our definitions. From the master plot, we obtain, for infinitely long chains, the prefactors: $N_e = d^2/b^2 \propto (nb^3)^{-1/\gamma}$, more precisely, $N_e = 14\,(nb^3)^{-1.22}$ (kinks) and $N_e = 50\,(nb^3)^{-1.22}$ (coils). Different

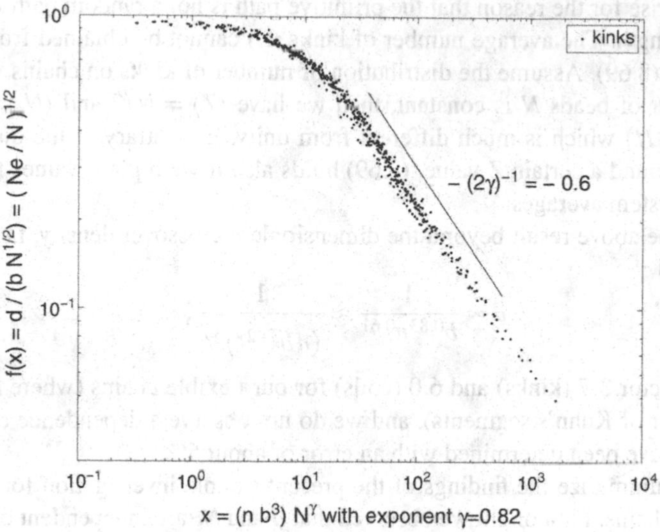

Fig. 8.6. Master plot for roughly 1500 configurations for different random path systems characterized by number of steps $N-1$, step lengths $b_0 = b$, and step number densities n. Shown is the dimensionless tube diameter $d/(b\sqrt{N})$ vs. the dimensionless density $n_p(b\sqrt{N})^3 N^{\gamma-1/2}$, with $n_p = n/N$. A value $\gamma = 0.82 \pm 0.04$, determined via Fig. 8.7, is needed to shift all data onto a single 'line'. The SP is completely unentangled, if all primitive paths are straight lines

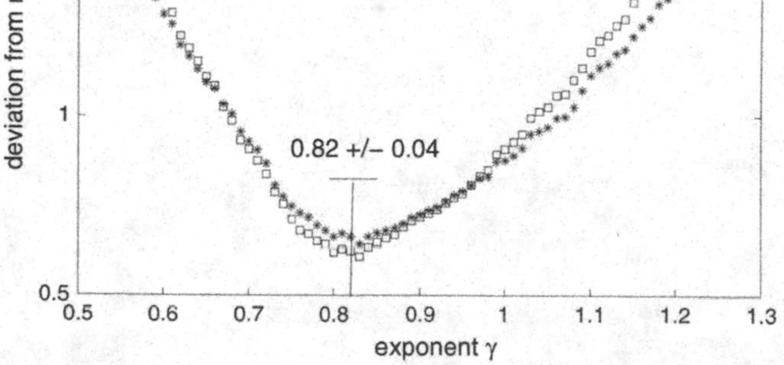

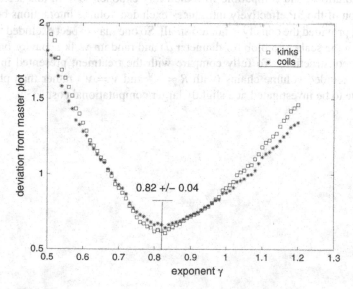

Fig. 8.7. Standard deviation vs. unknown exponent γ for phantom chain N_e data (collected in bins vs. dimensionless length scale $x = (nb^3)N^\gamma$). The minimum standard deviation corresponds to an exponent γ for the best representation in a master plot, cf. Fig. 8.6. Adapted from [402]

prefactors arise for the reason that the primitive path is not a random path with constant step length. The average number of kinks (Z) cannot be obtained from N_e via the formula (8.69). Assume the distribution of number of kinks on chains with constant number of beads N is constant, then we have $\langle Z\rangle = N/2$ and $\langle N_e\rangle \langle Z\rangle /N \approx (1/2)\ln(1+N)$ which is much different from unity. In contrary, if the distribution is peaked around a certain Z value, (8.69) holds also if we replace values for single chains by system averages.

Using the above result beyond the dimensionless crossover density, i.e., for $x > 100$, we find

$$d \propto \frac{1}{b^{0.83}n^{0.61}} \approx \frac{1}{(nb^{3-2\gamma})^{\frac{1}{2\gamma}}}, \tag{8.73}$$

with a prefactor 3.7 (kinks) and 6.0 (coils) for our flexible chains (where N_e stands for a number of Kuhn's segments), and we do not observe a dependence on N. The exponents have been determined with an error of about 5%.

Let us summarize the findings of the present sample investigation for a system of random chains. First of all, we observed that d and N_e are independent of N. With the general representation $d = (b\sqrt{N})f(x)$, the dimensionless quantity $x = (nb^3)N^\gamma$ involving an exponent γ, and the observed power-law behavior for $f(x)$ at large x, we immediately arrive at $f(x) \propto x^{-1/(2\gamma)}$, $N_e \propto (nb^3)^{-1/\gamma}$, and $d \propto n^{-1/(2\gamma)}b^{1-3/(2\gamma)}$. We found $\gamma = 0.82 \pm 0.04$. The observed scaling implies that only the exponent $\nu' = (1+\gamma)/3$ matters. Its numerical value is not too far from 0.588, the value expected for athermal solutions, and compatible with the Flory exponent 0.6. In this section, the construction of the SP effectively introduces excluded volume interactions between the chains, provided the density is not too small. So one may expect excluded volume behavior on the scale of a blob (of diameter d) and random-walk behavior on larger scales (by construction). To fully compare with the treatment presented in [405], however, excluded volume chains (with $R \propto N^\nu$ and $\nu = \nu'$) rather than phantom chains have to be investigated at a slightly larger computational cost.

Part II

Theory & Computational Recipes

9

Equilibrium Statistics: Monte Carlo Methods

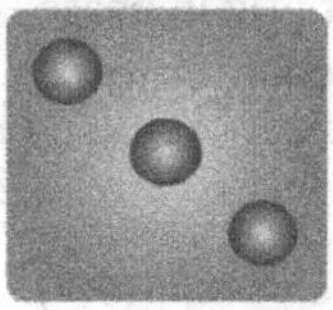

Monte Carlo methods use random numbers, or 'random' sequences, to sample from a known shape of a distribution, or to extract distribution by other means. and, in the context of this book, to (i) generate representative equilibrated samples prior being subjected to external fields, or (ii) evaluate high-dimensional integrals. Recipes for both topics, and some more general methods, are summarized in this chapter. It is important to realize, that Monte Carlo should be as artificial as possible to be efficient and elegant. Advanced Monte Carlo 'moves', required to optimize the speed of algorithms for a particular problem at hand, are outside the scope of this brief introduction. One particular modern example is the wavelet-accelerated MC sampling of polymer chains [406].

In physics, Monte Carlo methods are commonly used to treat the equilibrium statistics of simple and complex systems. Since the study of systems far from equilibrium requires full knowledge about equilibrium properties, and since we did not spent much space to describe equilibrium properties of the nonequilibrium systems discussed so far, this section offers a background on how to calculate equilibrium statistics, phase behavior of Hamiltonian systems. The treatment is kept sufficiently general in order to cover Monte Carlo methods as solvers for high-dimensional integration. High-dimensional integrals occur frequently in a huge number of diverse applications and areas including financial mathematics and data recognition.

According to classical statistical mechanics expectation values of observables $A(\boldsymbol{x})$, where $\boldsymbol{x}$ is a phase space coordinate of the system, can be computed in the canonical ensemble

$$\langle A \rangle = \int A(\boldsymbol{x})\, p(\boldsymbol{x})\, \mathrm{d}\boldsymbol{x}$$

$$p(\boldsymbol{x}) = \frac{1}{Z_\beta} e^{-\beta H(\boldsymbol{x})} , \tag{9.1}$$

where $H(\boldsymbol{x})$ denotes the Hamilton function of the system, Z_β the partition function, $\beta = (k_B T)^{-1}$ the inverse absolute temperature and k_B Boltzmann's constant. The partition function

$$Z_\beta = \int e^{-\beta H(\boldsymbol{x})} d\boldsymbol{x} \tag{9.2}$$

ensures proper normalization $\langle 1 \rangle = 1$ and yields, e.g., the free energy $F(\beta) = -k_B T \ln Z_\beta$. The main idea of the Monte Carlo method is to treat this problem as a problem for numerical integration, where, in view of the usually high dimension of the integral (dimension is proportional to the degrees of freedom $\propto N$ with the number of particles N for a many-particle system), an equidistant or regular grid in phase space cannot be used to compute the integral (cf. Table 9.1), but grid coordinates $\boldsymbol{x}_i$ have to be statistically selected, to obtain

$$\langle A \rangle = \frac{1}{M} \sum_{i=1}^{M} A(\boldsymbol{x}_i)\, p(\boldsymbol{x}_i) \tag{9.3}$$

in the limit $M \to \infty$. Since the distribution function usually varies over several orders of magnitude – $H(\boldsymbol{x})$ is an extensive variable, $H(\boldsymbol{x}) \propto N$ –, a regular phase space grid is often inefficient.

Table 9.1. Convergence behaviors for N function evaluations in d dimensions

Method	Scaling Behavior	
Trapezoidal	$N^{-2/d}$	regular grid in coordinate space
Simpson	$N^{-4/d}$	regular grid in coordinate space
Standard pseudorandom Monte Carlo	$N^{-1/2}$	(independent of d)
Quasi Monte Carlo	$N^{-1} \log^D N$	(for some D, realized for $N > e^d$)

9.1 Expectation Values, Metropolis Monte Carlo

The Metropolis algorithm aims to calculate $\langle A \rangle$ without the need to calculate Z_β. It is based on the idea to prefer phase space coordinates located in the relevant part of the phase space, which depends on temperature and further independent thermodynamic variables: 'importance sampling'. This is achieved by recursively creating a Markov chain ('one-step memory') of states $\boldsymbol{x}_i$: $\boldsymbol{x}_i \to \boldsymbol{x}_{i+1} \to \boldsymbol{x}_{i+2} \to \ldots$ using appropriate transition probabilities $w(\boldsymbol{x} \to \boldsymbol{x}')$ between states $\boldsymbol{x}$ and $\boldsymbol{x}'$. These probabilities must be chosen such that after a 'large' number of steps the states $\boldsymbol{x}$ are distributed according to the given probability distribution $p(\boldsymbol{x})$. It is achieved, for example, by fulfilling the so called condition of detailed balance

$$p(\boldsymbol{x})w(\boldsymbol{x} \to \boldsymbol{x}') = p(\boldsymbol{x}')w(\boldsymbol{x}' \to \boldsymbol{x}), \tag{9.4}$$

where it is important to realize that Z_β effectively drops out. A particular solution to (9.4) is the Metropolis scheme (see Sect. 12.3.2 for the implementation)

$$w(\boldsymbol{x} \to \boldsymbol{x}') = \min(1, \mathrm{e}^{-\beta \Delta H}), \quad \Delta H \equiv H(\boldsymbol{x}') - H(\boldsymbol{x}). \tag{9.5}$$

For $M \to \infty$ the phase space coordinates $\boldsymbol{x}_i$ obtained using a jump probability fulfilling (9.4) are distributed according to $p(\boldsymbol{x})$, and the expectation value $\langle A \rangle$ is evaluated as arithmetic average $\langle A \rangle = \frac{1}{M} \sum A(\boldsymbol{x}_i)$. This approach usually does not yield information about the partition function Z or free energy F, even though the partition function can be expressed itself as an expectation value $Z_\beta = 1/\langle \mathrm{e}^{\beta H} \rangle$, because, for the evaluation of $\langle \mathrm{e}^{\beta H} \rangle$ states with high energy H are relevant, but rarely visited. On the other hand, rare visits of high energy states are the key for understanding the efficiency of the Metropolis scheme for calculating 'conventional' expectation values such as the mean energy, pressure, or the end-to-end distance of polymers. A positive aspect of the Monte Carlo method is the flexibility in choosing the step size for a move $\boldsymbol{x} \to \boldsymbol{x}'$. The optimum step, and step size depends on the system under study. For dense liquids, due to excluded volume, random configurations are usually very unlikely to be statistically relevant, while small displacements of a statistically relevant configuration will be often relevant, too. For the Ising model, a single step may be the reorientation of a single spin, or the reorientation of a cluster of spins, and the conventional game is to search for most efficient implementations. It is important to realize that the Monte Carlo trajectory through phase space can be interpreted as resulting from an underlying dynamics obeying a master equation (loss- and gain-equation) for the (time dependent) probability distribution. The equation of change for $p(\boldsymbol{x},t)$ reads

$$\frac{d}{dt} p(\boldsymbol{x},t) = -\sum_{\boldsymbol{x}'} w(\boldsymbol{x} \to \boldsymbol{x}') p(\boldsymbol{x},t) + \sum_{\boldsymbol{x}'} w(\boldsymbol{x}' \to \boldsymbol{x}) p(\boldsymbol{x}',t). \tag{9.6}$$

It is obvious that $p(\boldsymbol{x})$, which fulfills the condition of detailed balance (9.4) is the stationary solution of the master equation (9.6). Of course, the meaning of time is not a physical one, a time unit corresponds to a single Monte Carlo step, and we have freedom to choose the Monte Carlo step. Dynamic correlations, however, are important to quantify the precision of the obtained results, and eventually have a physical meaning, if the configuration space $\boldsymbol{x}$ is a subspace of the space of slow variables, which are coupled to fast degrees of freedom (acting as a 'heat bath').

We have seen that the Metropolis scheme is not useful for calculating Z_β. While expectation values often contain sufficient information to validate models, the above-mentioned approach still has strong deficiencies. First of all, a new simulation has to be performed for each temperature. Second, low energy states are visited more often than potentially needed to extract the relevant information about their effective contribution to expectation values. Third, close to phase transitions (close to critical temperatures), the Metropolis scheme seriously slows down, if not sophisticated moves are introduced which overcome the problem of long-range correlations.

9.2 Normalization Constants, Partition Function

In this section we shortly summarize methods to directly compute the partition function Z_β, or equally, a class of high dimensional integrals parameterized by temperature. In particular, the density-of-states Monte Carlo method overcomes the problem with a slow-down close to critical temperatures since Z_β is obtained from the density of states which is independent of temperature.

To be more general, and to also match with the nomenclature used by non-physicists, let us consider a d-dimensional integral over 'arbitrary' function $q(\boldsymbol{x},\beta)$, $\boldsymbol{x} \in \Re^d$ parameterized by $0 \le \beta \le 1$

$$Z_\beta \equiv \int q(\boldsymbol{x},\beta)\, d\boldsymbol{x}\,. \tag{9.7}$$

The normalized density, fulfilling $\int p\, d\boldsymbol{x} = 1$, explicitly reads

$$p(\boldsymbol{x},\beta) = \frac{q(\boldsymbol{x},\beta)}{Z_\beta}\,. \tag{9.8}$$

A particular q is the so-called 'geometric path' between functions q_0 and q_1:

$$q(\boldsymbol{x},\beta) = q_0(\boldsymbol{x})^{1-\beta}\, q_1(\boldsymbol{x})^{\beta}\,. \tag{9.9}$$

We see, that the canonical distribution can be considered as a special geometric (temperature) path if we choose

$$\begin{aligned} q_0(\boldsymbol{x}) &\overset{\text{can}}{=} 1, \quad q_1(\boldsymbol{x}) \overset{\text{can}}{=} \mathrm{e}^{-H(\boldsymbol{x})}\,, \\ Z_\beta &\overset{\text{can}}{=} \int \mathrm{e}^{-\beta H(\boldsymbol{x})}\, d\boldsymbol{x}, \quad Z_0 \overset{\text{can}}{=} \int 1\, d\boldsymbol{x} = V\,. \end{aligned} \tag{9.10}$$

Here, we introduce the notation $\overset{\text{can}}{=}$ whenever the particular choice of a canonical (can) distribution for q_1, q_0 of (9.10) is made to simplify a more general expression, and V denotes 'volume'. Of course it is then identical calculating Z_1 for all β, or, for all β, Z_β at fixed $\beta = 1$. From Z_β one obtains free energy F, mean energy $\langle H\rangle$,

$$F \equiv -\frac{1}{\beta}\ln Z_\beta, \quad \langle H\rangle = -\frac{d\ln Z_\beta}{d\beta}\,, \tag{9.11}$$

as well as entropy $S = \beta(\langle H\rangle - F)$, heat capacity $C_V = \beta^2(\langle H^2\rangle - \langle H\rangle^2)$ etc.. The canonical distribution maximizes $S = -\int p \ln p\, d\mathbf{x} = -\langle \ln p\rangle$ under the constraint $\langle H\rangle$ =constant and $\int p\, d\mathbf{x} = 1$. Notice, that β appears to be inverse temperature, but will be eventually used below as a pure, 'unphysical', interpolating parameter where $0 \le \beta \le 1$ holds. For the same reason, we will be only interested in calculating Z_1. Once we have a method to compute Z_1 for given hamiltonian $H(\boldsymbol{x})$, we also know how to compute Z_β by just multiplying this hamiltonian by β. Except for the case of density of states Monte Carlo in Sect. 9.3, each new β requires a new, independent, calculation.

9.2.1 Standard Monte Carlo

Standard Monte Carlo consists of choosing an arbitrary, eventually optimized, and normalized $P(\boldsymbol{x})$ to obtain Z_β from M realizations I compatible with P:

$$Z_\beta = \int \frac{q(\boldsymbol{x},\beta)}{P(\boldsymbol{x})} P(\boldsymbol{x})\, d\boldsymbol{x} = \langle I \rangle_P \text{ with } I = \frac{q(\boldsymbol{x},\beta)}{P(\boldsymbol{x})} \,. \tag{9.12}$$

The error for the calculated Z_β is estimated 'on the fly' as

$$\sigma^2 = \frac{1}{M}\left(\langle I^2 \rangle_P - \langle I \rangle_P^2\right) . \tag{9.13}$$

This formula is derived making use of the law of large numbers and offers the reason why the standard deviation σ decreases with $M^{-1/2}$ independent of dimension d. Two special choices for P, denoted as 'uniform' and 'canonical', are

A) P uniform (constant)
(use uniform pseudo or quasi random generator to approximately realize P)

$$P = \frac{1}{\int 1\, d\boldsymbol{x}} = V^{-1} \rightarrow Z_\beta = V \langle q \rangle_P \overset{\text{can}}{=} V \left\langle e^{-\beta H} \right\rangle_{\text{uniform}} ; \tag{9.14}$$

B) $P \propto q$ (canonical)
(use Metropolis, rejection method, inversion etc. to approximately realize P)

$$P = \frac{q}{Z_\beta} \rightarrow Z_\beta = \frac{V}{\langle q^{-1} \rangle_P} \overset{\text{can}}{=} \frac{V}{\left\langle e^{\beta H} \right\rangle_{\text{canonical}}} . \tag{9.15}$$

The standard Monte Carlo method still requires one run for each β. For the uniform case (A) it is straightforward to obtain realizations. Using a build-in pseudo random generator is upon the approximate solutions. For case (A), we miss the 'relevant' low energy regions, which is bad for above-mentioned expectation values, but good for calculating Z_β if uniform sampling in $\boldsymbol{x}$ is cheap, and if all energy levels are reached in a comparable fashion (not so for dense fluids). Concerning case (C) and more general cases, it is often difficult to realize P, which prevents calculating Z_β to desired accuracy. Just for the one-dimensional case ($d = 1$) we would like to mention the variable transformation method to obtain realizations y distributed according to $P(y)$ using uniformly distributed random numbers with $p(x) = V^{-1}$, $x \in [0,V]$. By noticing the identity

$$\int p(x)\, dx = 1 = \int P(y)\, dy = \int p(y(x)) x'(y)\, dy \tag{9.16}$$

we need to solve the differential equation $x'(y) = V P(y)$to obtain $x(y)$, and if we further invert x analytcally or numerically to obtain $y(x)$, this constitutes the rule to obtain P-distributed realizations $y(\boldsymbol{x})$ from random numbers $\boldsymbol{x}$.

9.2.2 Direct Importance Sampling

Direct importance sampling is based on the following basic identity, valid for arbitrary $\alpha(\boldsymbol{x})$

$$\frac{Z_1}{Z_0} = \frac{\int q_1\,d\boldsymbol{x}}{\int q_0\,d\boldsymbol{x}} = \frac{\int \alpha q_1 q_0\,d\boldsymbol{x} \int q_1\,d\boldsymbol{x}}{\int q_0\,d\boldsymbol{x} \int \alpha q_0 q_1\,d\boldsymbol{x}} = \frac{\langle \alpha q_1 \rangle_{p_0}}{\langle \alpha q_0 \rangle_{p_1}} \overset{\text{can}}{=} \frac{\langle \alpha e^{-H} \rangle_{\text{uniform}}}{\langle \alpha \rangle_{\text{canonical}}} . \tag{9.17}$$

Special cases:

A) $\alpha = 1/q_0 \overset{\text{can}}{=} 1$ (uniform)
B) $\alpha = 1/(q_0 q_1) \overset{\text{can}}{=} e^H$ (canonical)
C) $\alpha = \min(1/q_0, 1/q_1)$ (acceptance ratio method)
D) $\alpha = q_{\frac{1}{2}}/(q_0 q_1)$ with $q_{\frac{1}{2}}$ in between q_0 and q_1
(bridge sampling, umbrella sampling, thermodynamic integration)

$$\frac{Z_1}{Z_0} = \frac{\left\langle q_{\frac{1}{2}}/q_0 \right\rangle_{p_0}}{\left\langle q_{\frac{1}{2}}/q_1 \right\rangle_{p_1}} \overset{\text{can}}{=} \frac{\left\langle q_{\frac{1}{2}} \right\rangle_{\text{uniform}}}{\left\langle q_{\frac{1}{2}} e^H \right\rangle_{\text{canonical}}} \tag{9.18}$$

Functions $q_{\frac{1}{2}}$ and q_0, q_1 should possess overlap. Otherwise refine by using several spans, or perform iterations with several '$q_{\frac{1}{2}}$'. Optimal path (weighted harmonic mean):

$$q_{\frac{1}{2}}(\boldsymbol{x}) = \frac{p_0 p_1}{s_0 p_0(\boldsymbol{x}) + s_1 p_1(\boldsymbol{x})}, \quad s_t = n_t/(n_0 + n_1) . \tag{9.19}$$

This α is not directly usable since it depends on Z_1/Z_0. Iterative schemes have been proposed, e.g., in [407], which can reduce the simulation error by orders of magnitude when compared to the conventional importance sampling method. Considering infinitely many spans $q_\varepsilon, q_{2\varepsilon}, \ldots q_{1-\varepsilon}$, bridge sampling equals path sampling.

9.2.3 Path Sampling

Path sampling is based on basic identity

$$\frac{d}{d\beta} \ln Z_\beta = \int \frac{1}{Z_\beta} \frac{d}{d\beta} q(\boldsymbol{x}, \beta)\,d\boldsymbol{x} = \left\langle \frac{d}{d\beta} \ln q(\boldsymbol{x}, \beta) \right\rangle_p \overset{\text{can}}{=} -\langle H(\boldsymbol{x}) \rangle_p , \tag{9.20}$$

where definition (9.8) was used. Integrating (9.20) over β from 0 to 1 yields, for the 'physical' case,

$$\ln \frac{Z_1}{Z_0} \overset{\text{can}}{=} -\int H(\boldsymbol{x}) p(\boldsymbol{x}, \beta)\,d\boldsymbol{x}\,d\beta = -\int \frac{H(\boldsymbol{x})}{P(\beta)} p(\boldsymbol{x}; \beta)\,d\boldsymbol{x}\,d\beta , \tag{9.21}$$

This expression involves a joint distribution $p(\boldsymbol{x}; \beta) = p(\boldsymbol{x}, \beta) P(\beta)$, and $P(\beta)$ which may be, for example, uniform. In practise, the goal is to find the optimum path $P(\beta)$.

9.3 Density of States Monte Carlo (DSMC)

Density of states Monte Carlo takes an orthogonal 'view' to the problem of high-dimensional integration. It is based on the identity, with $Q(E) = \mathrm{e}^{-E}$

$$\frac{Z_\beta}{Z_0} = \frac{\int q(x,\beta)\,dx}{Z_0} \overset{\text{can}}{=} \frac{\int Q(H(x))^\beta\,dx}{Z_0}$$

$$= \int n(E)\mathrm{e}^{-\beta E}\,dE = \left\langle Q^\beta(E)\right\rangle_{n(E)}, \tag{9.22}$$

where $n(E)$, due to (9.22), is defined as a normalized density of states, $\int n(E)\,dE = 1$. The appeal of this method, introduced in [408], lies in the fact, that $n(E)$ is independent of β. Thus Z_β for all β is obtained from a single 'athermal' simulation. For this reason the method cannot suffer from a slow down close to critical temperature, for example. The method converts the high-dimensional integration to a one-dimensional one, in its simplest form, where we have a single Lagrange multiplier (β). The method is much different in spirit compared with the previous ones mentioned in this section. Here, the distribution $n(E)$ is not known at all a priori! For all above examples, at least we knew the shape of the distribution function in advance. In order to iteratively converge to the correct density of states, a procedure based on the following consideration had been proposed [408, 409]: Assume, the histogram $h(E)$ of visited states is 'flat', while performing a Monte Carlo simulation which realizes a certain, given, $p(E)$, then $h(E) \propto p(E)n(E)$ with probability p visiting energy level E, and $n(E) \propto p(E)^{-1}$. The trick is now to obtain a flat histogram, where 'flat' means 'mostly constant', or 'not varying much within the possible energy range', and has to be quantified. The following explicit simulation scheme had been proposed, which converges, and also fulfills a detailed balance criterion for the density of states in the limit of small 'loop parameter' f. A large loop parameter enforces the phase space trajectory to reach all energy levels quickly, which is just the opposite of what the conventional Metropolis scheme of Sect. 9.1 achieves.

DSMC Algorithm

Perform Metropolis Monte Carlo with time-dependent transition probability $w(\boldsymbol{x} \to \boldsymbol{x}')(t) = \min(1, p(H(\boldsymbol{x}'),t)/p(H(\boldsymbol{x}),t))$.

- Start ($i = 1$):
  $p(E,t) = f^{-h_i(E,t)}$ (h_i is the histogram recorded in round i up to time t) with 'large' $f = e$ to rapidly explore the whole E range, and converge towards a 'flat' $h_i(E,t_{\max})$ (achieved at time $t_{\max}$). Set $p_i(E) \equiv p(E,t_{\max})$.
- Iteration step ($i+1$):
  $p(E,t) = p_i(E) * f^{-h_{i+1}(E)/(i-1)}$ to obtain for large i a steady, polished, $p(E)$ and to approximately fulfill detailed balance in the corresponding Metropolis scheme.

The method is easy to implement, and it is possible to extend the method by using parallel runs with different, overlapping E windows. A 'drawback' of density of states Monte Carlo seems to lie in the limited range of accessible expectation values $\langle A \rangle$ with $A = A(E)$. Histogram recording becomes memory consuming if generalized canonical distributions are considered (involving more than a single Lagrange parameter). However, one can record realizations A and obtain $\langle A \rangle$ in a postprocessing step, or further explore the numerical precision of the following, trivial, identity

$$\left\langle A^{n\beta} \right\rangle = \frac{\tilde{Z}_\beta}{Z_\beta}, \quad \tilde{Z}_\beta \equiv \int e^{-\beta \tilde{H}(\boldsymbol{x})} \, d\boldsymbol{x}, \quad \tilde{H}(\boldsymbol{x}) \equiv H(\boldsymbol{x}) - n \ln A(\boldsymbol{x}) \,, \tag{9.23}$$

which requires performing two independent simulations, with two different hamiltonians H and $\tilde{H}$ (but otherwise identical) to obtain $\left\langle A^{n\beta} \right\rangle$ for all β and a single n. To obtain $\langle A \rangle$ for several β, which may the most classical task, we would need to run several simulations with fixed $n = 1/\beta$, which must be compared with the conventional Metropolis scheme of Sect. 9.1.

Further Extensions

The density of states Monte Carlo method had been recently extended to compute phase diagrams of dense Lennard–Jones fluids and binary glasses with N particles in a volume V, characterized by a two-body interaction potential $U(\boldsymbol{x})$ in [410, 411]. Here, use is made of the identity (fixed N, V)

$$\beta(E) = k_B \frac{\partial S}{\partial E}|_V = \frac{\partial \ln \Omega(N,V,E)}{\partial E} \,, \tag{9.24}$$

where Ω denotes the microcanonical partition function. Integration over E yields

$$\ln \Omega(E) = \int_{E_0}^{E} \beta(E') \, dE' \,. \tag{9.25}$$

Further, an expression for the configurational temperature [412, 413],

$$\beta(E) = \frac{\langle -\sum_i \nabla_i \boldsymbol{F}^i \rangle}{\langle \sum_i |\boldsymbol{F}^i|^2 \rangle} = \frac{a(E)}{b(E)} \tag{9.26}$$

is employed, which is obtained using a phase point transformation, assuming that at constant E, the phase space is uniformly occupied, while U is smooth. Histograms a, b, h, Ω are evaluated and Ω is used as guide for a walker in energy space. Further, the known probability of observing a configuration $\boldsymbol{x}$ having total potential energy $U(\boldsymbol{x})$ is used as guide for a walker in configuration space. Finally, one obtains a smooth, since integrated, distribution Ω from a and b. In all other respects, the above listed algorithm is adapted for this application.

9.4 Quasi Monte Carlo

The strikingly simple idea of quasi Monte Carlo is to replace pseudo random numbers by a quasi random sequence which is 'known to be uniform' in high dimensional $\boldsymbol{x}$ space (exhibiting low 'discrepancy'). Quasi Monte Carlo methods use deterministic samples at points that belong to low discrepancy sequences and approximate the integrals by the arithmetic average of N function evaluations. According to the KoksmaHlawka inequality their worst case error is of order $(1/N) * \ln^d N$; where d denotes the dimension. Since this term becomes huge when N is fixed and d is large, as sometimes happens in practice, traditionally, there has been a certain degree of concern about Quasi Monte Carlo [414] presents sufficient conditions for fast Quasi Monte Carlo convergence which apply to isotropic and non-isotropic problems. It is shown that the convergence rate of Quasi Monte Carlo is of order $N^{-1+p/\ln N^{1/2}}$ with $p \geq 0$. Compared to the expected rate $N^(-1/2)$ of Monte Carlo it shows the superiority of Quasi Monte Carlo. To understand the success of Quasi Monte Carlo in some applications, also the notion of effective dimension has been introduced, cf. [415].

One example is the 'Richtmeyer sequence' $\boldsymbol{x}^{(n)} \in [0,1]^d$ with $x_\mu^{(n)} = n\sqrt{\mathcal{P}_\mu}$ mod 1 with prime number $\mathcal{P}$. Other often used sequences are the so called Faure and Niedermeyer sequences. There is no exhaustive knowledge about the overall efficiency of quasi Monte Carlo methods for computing high dimensional integrals. Quasi Monte Carlo is certainly good for low-dimensional integration, but supersedes pseudo Monte Carlo only if $N > \mathrm{e}^d$ (huge for large d). It is yet an empirical observation that quasi random numbers should not be used for solving stochastic differential equations (Langevin equations) [416].

10

Irreducible and Isotropic Cartesian Tensors

$$\Delta^{(2)}_{\mu\nu,\lambda\gamma}$$

In this chapter we summarize definitions and properties of cartesian, anisotropic, irreducible and isotropic tensors and related tensor operators. In several aspects more exhaustive, detailed treatments, eventually using a different notation, can be found in [4, 82, 417–420]. We present rules and properties using tensor product notation and tend to avoid component notation (except in footnotes). The formulas presented in this chapter help to evaluate tensor operators (differentiation, integration) without performing a differentiation or an integral (cf. Sect. 10.5 and (10.68)), to rewrite arbitrary tensors of rank l made of unit vectors $\boldsymbol{u}$ in terms of the dyadics $\boldsymbol{u}_{(k)} \equiv \boldsymbol{uu..u}$ of rank $k \leq l$, to rewrite anisotropic tensors $\boldsymbol{u}_{[l]} \equiv \overline{\boldsymbol{uu..u}}$ of rank l in terms of $\boldsymbol{u}_{(k)}$ of rank $k \leq l$ (using (10.14)), and vice versa (recursively using (10.67)). This sets us in position to write down (coupled) moment equations starting from a given differential equation for (orientational) distribution functions in Chap. 11, and to write down approximate sets of coupled equations for moments of the distribution function.

10.1 Notation

Let $\boldsymbol{T}^l$ be an arbitrary tensor of rank l. Low rank tensors are scalars (rank 0) and vectors (rank 1). Components of $\boldsymbol{T}^l$ are denoted as $(\boldsymbol{T}^l)_{\mu_1\mu_2..\mu_l} = T^l_{\mu_1\mu_2..\mu_l}$, and a bold face index $\boldsymbol{\mu}$ stands for the ordered set of indices $\mu_1\mu_2..\mu_l$. Examples clarifying notation used in this book:

$$(\boldsymbol{T}^l)_{\boldsymbol{\mu}} = T^l_{\mu_1\mu_2..\mu_l} = T^l_{\boldsymbol{\mu}}, \quad \text{rank l}$$
$$(\boldsymbol{T}^l\boldsymbol{T}^m)_{\boldsymbol{\mu}} = T^l_{\mu_1\mu_2..\mu_l}T^m_{\mu_{l+1}\mu_{l+2}..\mu_{l+m}}, \quad \text{rank l+m}$$

$$
\begin{aligned}
(\boldsymbol{T}^l\cdot\boldsymbol{T}^m)_{\boldsymbol{\mu}} &= T^l_{\mu_1\mu_2..\mu_{l-1}\lambda}T^m_{\lambda\mu_l..\mu_{l+m-2}}, \quad \text{rank l+m−2}\\
(\boldsymbol{T}^l:\boldsymbol{T}^m)_{\boldsymbol{\mu}} &= T^l_{\mu_1\mu_2..\mu_{l-2}\lambda\gamma}T^m_{\gamma\lambda\mu_{l-1}..\mu_{l+m-4}}, \quad \text{rank l+m−4}\\
(\boldsymbol{T}^l\odot^k\boldsymbol{T}^m)_{\boldsymbol{\mu}} &= T^l_{\mu_1\mu_2..\mu_{l-k}\lambda_1..\lambda_k}T^m_{\lambda_k..\lambda_1\mu_{l-k+1}..\mu_{l+m-2k}}, \quad \text{rank l+m−2k}\\
(\boldsymbol{u}\times\boldsymbol{T}^l)_{\boldsymbol{\mu}} &= \varepsilon_{\mu_1\lambda\kappa}u_\lambda T^l_{\kappa\mu_2..\mu_l}\\
(\boldsymbol{1}^{(l)})_{\boldsymbol{\mu\nu}} &\equiv \prod_{i=1}^{l}\delta_{\mu_i\nu_i},\quad \boldsymbol{1}^{(1)}=\boldsymbol{1},\quad 1^{(1)}_{\mu\nu}=\delta_{\mu\nu}
\end{aligned}
\tag{10.1}
$$

where summation ($\Sigma^3_{\mu_l=1}$) over repeating indices is always understood (we use the Einstein summation convention). Contraction over indices has to be always performed in a strict, ordered, way. In the above, $\delta_{\mu\nu}$ is the Kronecker symbol, the components of the 3×3 unity matrix $\boldsymbol{1}$, i.e., $\delta_{\mu\nu}=1$ if $\mu=\nu$, and $\delta_{\mu\nu}=0$ if $\mu\neq\nu$. Notice, $\delta_{\mu\mu}=3$ due to the summation convention. The symbols $\varepsilon_{\mu\nu\lambda}$ are the components of the total antisymmetric tensor $\boldsymbol{\varepsilon}$ of rank 3, with $\varepsilon_{\mu\nu\lambda}=1\ (-1)$ if (μ,ν,λ) is an even (odd) permutation of $(1,2,3)$, respectively, and $\varepsilon_{\mu\nu\lambda}=0$ if two or more of its three indices are identical. There are some very basic 'rules' for evaluating expressions containing $\delta_{\mu\nu}$ and $\boldsymbol{\varepsilon}$ such as $\delta_{\mu\nu}T_\nu=T_\mu$, $\delta_{\mu\nu}\delta_{\nu\lambda}=\delta_{\mu\lambda}$, $\varepsilon_{\mu\nu\lambda}=-\varepsilon_{\nu\mu\lambda}$, $\varepsilon_{\mu\nu\lambda}T^{\text{sym}}_{\nu\lambda}=0$, $\varepsilon_{\mu\nu\kappa}\varepsilon_{\kappa\lambda\gamma}=\delta_{\mu\lambda}\delta_{\nu\gamma}-\delta_{\mu\gamma}\delta_{\nu\lambda}$, which implies $\varepsilon_{\mu\nu\kappa}\varepsilon_{\kappa\nu\gamma}=-2\delta_{\mu\gamma}$. Further, $\varepsilon_{\mu\nu\lambda}u_\nu v_\lambda=(\boldsymbol{u}\times\boldsymbol{v})_\mu$. The superscript 'symm' denotes the symmetrized, and normalized tensor, $T^{\text{symm}}_{\mu\nu}=(T_{\mu\nu}+T_{\nu\mu})/2$ and accordingly for higher order tensors. A k-fold contraction is denoted by the symbol $\odot^k$; for $k=1$ and $k=2$ we still prefer to use the classical notation '$\cdot$' and '$:$', rather than $\odot^1$, and $\odot^2$, respectively.

We introduce the rank l tensors $\boldsymbol{u}_{(l)}$, the 'anisotropic' tensors $\boldsymbol{u}_{[l]}$ (l-fold symmetric traceless dyadic product of unit vectors $\boldsymbol{u}$; the $\boldsymbol{u}_{[l]}$ are then called 'irreducible' or 'anisotropic') and the corresponding irreducible tensor $\boldsymbol{Q}_{[l]}$ via

$$
\begin{aligned}
\boldsymbol{u}_{(l)} &\equiv \boldsymbol{uu}..\boldsymbol{u}\,,\\
\boldsymbol{u}_{[l]} &\equiv \overline{\boldsymbol{u}_{(l)}} = \overline{\boldsymbol{uu}\ldots\boldsymbol{u}}\,,\\
\boldsymbol{a}_{(l)} &\equiv \langle\boldsymbol{u}_{(l)}\rangle = \langle\boldsymbol{uu}\ldots\boldsymbol{u}\rangle\,,\\
\boldsymbol{a}_{[l]} &\equiv \langle\boldsymbol{u}_{[l]}\rangle = \left\langle\overline{\boldsymbol{uu}\ldots\boldsymbol{u}}\right\rangle, \quad \text{alignment tensor of rank } l\,,\\
\boldsymbol{Q}_{[l]} &\equiv Q^l\,\boldsymbol{u}_{[l]} = \overline{\boldsymbol{QQ}\ldots\boldsymbol{Q}}\,,
\end{aligned}
\tag{10.2}
$$

The hard brackets '[]' symbolize an irreducible quantity, the soft brackets a simple dyadic product, 'a' stands for 'averaged' quantity.

Definitions of the operators $\nabla,\mathcal{L},\frac{\partial}{\partial\boldsymbol{u}},\Delta$ are given in Sect. 10.3, and repeated in the Sect. 10.5.

10.2 Anisotropic (Irreducible) Tensors

How to construct, for given $\boldsymbol{u}_{(l)}$ of rank l, the anisotropic, irreducible tensors $\boldsymbol{u}_{[l]}$: A naive, illustrative, solution is to make an ansatz (below with int($l/2$) parameters $\alpha_{..}$)

using solely (!) symmetric tensors made of $\boldsymbol{u}$ and the unity matrix $\mathbf{1}$, and to require a single, arbitrarily chosen, trace (repeating two indices) to vanish. This works as follows (apart from the trivial relationship for $l = 1$: $\boldsymbol{u}_{[1]} = \boldsymbol{u}$), where $\boldsymbol{i}$ symbolizes a 'half unity matrix' with the property $\mathbf{1} = \boldsymbol{ii}$, or $\delta_{\mu\nu} = i_\mu i_\nu$:

- $l = 2$

$$\begin{aligned}
\boldsymbol{u}_{[2]} &= \boldsymbol{u}_{(2)}^{\text{sym}} - \alpha \mathbf{1}^{\text{sym}} = \boldsymbol{u}_{(2)} - \alpha\, \boldsymbol{ii}, \\
0 = \text{Tr}(\boldsymbol{u}_{[2]}) &= u_\lambda u_\lambda - \alpha\, \delta_{\lambda\lambda} = 1 - 3\alpha \rightarrow \alpha = \frac{1}{3}, \\
\rightarrow \boldsymbol{u}_{[2]} &= \boldsymbol{u}_{(2)} - \frac{1}{3}\mathbf{1} \\
\Leftrightarrow \overline{\boldsymbol{uu}} &= \boldsymbol{uu} - \frac{1}{3}\mathbf{1}
\end{aligned} \tag{10.3}$$

- $l = 3$

$$\begin{aligned}
\boldsymbol{u}_{[3]} &= \boldsymbol{u}_{(3)}^{\text{sym}} - \alpha \{\boldsymbol{u}\mathbf{1}\}^{\text{sym}}, \\
&= \boldsymbol{u}_{(3)} - \alpha \frac{1}{3}(\boldsymbol{iiu} + \boldsymbol{iui} + \boldsymbol{uii}), \\
0 = \text{Tr}(\boldsymbol{u}_{[3]}) &= \boldsymbol{u} - \alpha \frac{1}{3}(3\boldsymbol{u} + \boldsymbol{u} + \boldsymbol{u})) = \left(1 - \frac{5}{3}\alpha\right)\boldsymbol{u} \rightarrow \alpha = \frac{3}{5}, \\
\rightarrow \boldsymbol{u}_{[3]} &= \boldsymbol{u}_{(3)} - \frac{3}{5}\{\boldsymbol{u}\mathbf{1}\}^{\text{sym}}.
\end{aligned} \tag{10.4}$$

- $l = 4$

$$\begin{aligned}
\boldsymbol{u}_{[4]} &= \boldsymbol{u}_{(4)}^{\text{sym}} - \alpha_1 (\boldsymbol{u}_{(2)}\mathbf{1})^{\text{sym}} - \alpha_2 \{\mathbf{11}\}^{\text{sym}}, \\
&= \boldsymbol{u}_{(4)} - \alpha_1 \frac{1}{6}(\boldsymbol{iiu}_{(2)} + \boldsymbol{iuiu} + \ldots) - \alpha_2 \frac{1}{3}(\mathbf{11} + \boldsymbol{i}\mathbf{1}\boldsymbol{i} + \boldsymbol{iiii}), \\
0 = \text{Tr}(\boldsymbol{u}_{[4]}) &= \boldsymbol{u}_{(2)} - \alpha_1 \frac{1}{6}(3\boldsymbol{u}_{(2)} + 4\boldsymbol{u}_{(2)} + \mathbf{1}) - \alpha_2 \frac{1}{3}(3\mathbf{1} + 2\mathbf{1}), \\
&= \left(1 - \frac{7}{6}\alpha_1\right)\boldsymbol{u}_{(2)} - \left(\frac{1}{6}\alpha_1 + \frac{5}{3}\alpha_2\right)\mathbf{1} \rightarrow \alpha_1 = \frac{6}{7},\ \alpha_2 = -\frac{3}{35}, \\
\rightarrow \boldsymbol{u}_{[4]} &= \boldsymbol{u}_{(4)} - \frac{6}{7}\{\boldsymbol{u}_{(2)}\mathbf{1}\}^{\text{sym}} + \frac{3}{35}\{\mathbf{11}\}^{\text{sym}}.
\end{aligned} \tag{10.5}$$

Notice, that the same strategy can be used (and easily implemented on a computer) for any tensor made of $\boldsymbol{u}$'s and constants such as $\boldsymbol{i}$'s and $\boldsymbol{\varepsilon}$, where the ansatz just has to have the correct rank. Examples are $\boldsymbol{u} \times \boldsymbol{u}_{[l]}$ (rank l) or $\nabla \boldsymbol{u}_{[l]}$ (rank $l+1$), $\boldsymbol{u}_{[l]}\boldsymbol{u}_{[m]}$ (rank $l+m$), $\boldsymbol{u}_{[l]} \cdot \boldsymbol{u}_{[m]}$ (rank $l+m-2$) etc.[1] In particular, all terms on the right

[1] Another explicit example: $\overline{\boldsymbol{huu}} = (\boldsymbol{huu} + \boldsymbol{uhu} + \boldsymbol{uuh})/3 - (\mathbf{1}\boldsymbol{h} + \boldsymbol{ihi} + \boldsymbol{h}\mathbf{1})/15 - 2\boldsymbol{u} \cdot \boldsymbol{h}(\mathbf{1}\boldsymbol{u} + \boldsymbol{iui} + \boldsymbol{u}\mathbf{1})/15$ is symmetric and traceless, thus anisotropic. See Chapt. 10 for the $\boldsymbol{\Delta}^{(3)}$-operator, which has the property $\overline{\boldsymbol{huu}} = \boldsymbol{\Delta}^{(3)} \odot^3 \boldsymbol{huu}$.

hand sides of the above equations can be made anisotropic, and finally, we can also invert the equations to express any tensor in terms of anisotropic tensors $\boldsymbol{u}_{[l]}$. As soon as we would have reached this state, we could use the rules of the tables of Sect. 10.5 to perform any derivations. However, there is a direct route which prevents solving a system of equations to obtain the irreducible part of any given tensor (in terms of $\boldsymbol{u}$'s).

10.3 Differential Operators (∇, $\mathcal{L}$ etc.)

Consider a vector $\boldsymbol{Q} = Q\boldsymbol{u}$, unit vector $\boldsymbol{u}$, and the norm Q (length) of $\boldsymbol{Q}$, i.e., $Q \equiv |\boldsymbol{Q}|$ and $|\boldsymbol{u}| = 1$. The differential operators ∇ (nabla operator), Δ (Laplace operator), $\frac{\partial}{\partial \boldsymbol{u}}$ (gradient on unit sphere), and $\mathcal{L}$ (angular operator) are defined as follows

$$
\begin{aligned}
\nabla &\equiv \frac{\partial}{\partial \boldsymbol{Q}}, \quad \Delta \equiv \nabla \cdot \nabla , \\
\frac{\partial}{\partial \boldsymbol{u}} &\equiv Q(\boldsymbol{1} - \boldsymbol{u}_{(2)}) \cdot \nabla , \\
\mathcal{L} &\equiv \boldsymbol{Q} \times \nabla = \boldsymbol{u} \times \frac{\partial}{\partial \boldsymbol{u}} .
\end{aligned} \tag{10.6}
$$

The latter identity is proven in the footnote.[2] Using basic identities such as $\nabla Q = \boldsymbol{u}$ and $\boldsymbol{u} \cdot \nabla = (\partial \boldsymbol{Q}/\partial Q) \cdot \nabla = \partial/\partial Q$ allows to obtain a number of identical representations (splitting into radial and orientational part) for the nabla operator,

$$
\begin{aligned}
\nabla &= \boldsymbol{u}_{(2)} \cdot \nabla + (\boldsymbol{1} - \boldsymbol{u}_{(2)}) \cdot \nabla \\
&= \boldsymbol{u} \frac{\partial}{\partial Q} + \frac{1}{Q} (\boldsymbol{1} - \boldsymbol{u}_{(2)}) \cdot \frac{\partial}{\partial \boldsymbol{u}} \\
&= (\nabla Q) \frac{\partial}{\partial Q} + (\nabla \boldsymbol{u}) \cdot \frac{\partial}{\partial \boldsymbol{u}} \\
&= \boldsymbol{u} \frac{\partial}{\partial Q} - \frac{1}{Q} \boldsymbol{u} \times \mathcal{L} .
\end{aligned} \tag{10.7}
$$

The angular operator $\mathcal{L}$ solely acts on the orientational part, while the term $\propto \partial/\partial Q$ on the rhs of (10.7) acts on the radial part of the operator argument. Further, one has [3]

$$
\boldsymbol{u} \times \mathcal{L} = -(\boldsymbol{1} - \boldsymbol{u}_{(2)}) \frac{\partial}{\partial \boldsymbol{u}}, \quad \boldsymbol{u} \cdot \mathcal{L} = \boldsymbol{u} \cdot \frac{\partial}{\partial \boldsymbol{u}} = 0 . \tag{10.8}
$$

---

[2] Proof of (10.7): $(\boldsymbol{u} \times \partial/\partial \boldsymbol{u})_\mu = Q\varepsilon_{\mu\nu\lambda} u_\nu (\delta_{\lambda\gamma} - u_\lambda u_\gamma) \nabla_\gamma = Q\varepsilon_{\mu\nu\lambda} u_\nu \delta_{\lambda\gamma} \nabla_\gamma = Q(\boldsymbol{u} \times \nabla)_\mu = (\boldsymbol{Q} \times \nabla)_\mu$ where we used the antisymmetry of $\boldsymbol{\varepsilon}$, cf. previous footnote.

[3] To let the reader feel comfortable with the short notation, here is the proof of (10.8) – all tiny steps – in component notation: $(\boldsymbol{u} \times \mathcal{L})_\mu = \varepsilon_{\mu\nu\lambda} u_\nu \mathcal{L}_\lambda = u_\nu \varepsilon_{\mu\nu\lambda} \varepsilon_{\lambda\kappa\gamma} u_\kappa \partial/\partial u_\gamma = u_\nu(\delta_{\mu\kappa}\delta_{\nu\gamma} - \delta_{\mu\gamma}\delta_{\nu\kappa}) u_\kappa \partial/\partial u_\gamma = u_\nu u_\mu \partial/\partial u_\nu - u_\nu u_\nu \partial/\partial u_\mu = u_\nu u_\mu \partial/\partial u_\nu - \partial/\partial u_\mu = u_\nu u_\mu \partial/\partial u_\nu - \delta_{\mu\nu} \partial/\partial u_\nu = -(\delta_{\mu\nu} - u_\mu u_\nu) \partial/\partial u_\nu$, where $u^2 = u_\lambda u_\lambda = 1$ and a property of $\boldsymbol{\varepsilon}$ from the foregoing footnote has been used.

The corresponding radial–orientation splitting for the Laplace operator reads

$$\Delta = \Delta_Q + \frac{1}{Q^2}\mathcal{L}^2, \quad \Delta_Q = \frac{\partial^2}{\partial Q^2} + \frac{2}{Q}\frac{\partial}{\partial Q}. \tag{10.9}$$

In order to a perform an exercise which just applies results summarized in this section, let us explicitly make sure that $\mathcal{L}^2 = \frac{\partial}{\partial \boldsymbol{u}} \cdot \frac{\partial}{\partial \boldsymbol{u}}$. We will use (10.24), some expressions from the tables in Sect. 10.5, and the fact that any trace of $\boldsymbol{Q}_{[l]}$ vanishes,

$$\begin{aligned}
\frac{\partial}{\partial \boldsymbol{u}} \cdot \frac{\partial}{\partial \boldsymbol{u}} \boldsymbol{Q}_{[l]} &= \frac{\partial}{\partial \boldsymbol{u}} \cdot \left((l+1)\boldsymbol{u}\boldsymbol{Q}_{[l]} - (2l+1)\frac{1}{Q}\boldsymbol{Q}_{[l+1]} \right) \\
&= (l+1)(\frac{\partial}{\partial \boldsymbol{u}} \cdot \boldsymbol{u})\boldsymbol{Q}_{[l]} - \frac{2l+1}{Q}\frac{\partial}{\partial \boldsymbol{u}} \cdot \boldsymbol{Q}_{[l]} \\
&= 2(l+1)\boldsymbol{Q}_{[l]} - \frac{2l+1}{Q^2}\left((l+2)\boldsymbol{Q} \cdot \boldsymbol{Q}_{[l+1]} - (2l+3)\mathrm{Tr}(\boldsymbol{Q}_{[l+2]}) \right) \\
&= \left(2(l+1) - \frac{(2l+1)(l+2)(l+1)}{2l+1} \right) \boldsymbol{Q}_{[l]} \\
&= -l(l+1)\boldsymbol{Q}_{[l]} = \mathcal{L}^2 \boldsymbol{Q}_{[l]} \,.
\end{aligned} \tag{10.10}$$

The same result can be obtained using properties of the isotropic tensor $\boldsymbol{\Delta}^{(l,1,l)}$ (10.18), cf. Sect. 10.4.1, (10.14).

10.4 Isotropic Tensors

The isotropic tensor $\boldsymbol{\Delta}^{(l)}$ of rank $2l$ is defined by the property, that it projects an arbitrary tensor $\boldsymbol{T}^l$ of rank l to its irreducible, symmetric traceless part,

$$\boldsymbol{\Delta}^{(l)} \odot^l \boldsymbol{T}^l = \overline{\boldsymbol{T}^l} \,. \tag{10.11}$$

Therefore, $\boldsymbol{\Delta}^{(l)}$ is a projector, obeying

$$\boldsymbol{\Delta}^{(l)} \odot \boldsymbol{\Delta}^{(l)} = \boldsymbol{\Delta}^{(l)}. \tag{10.12}$$

A special case is – notice the different subscripts (l) and $([l])$, cf. definitions (10.2) –

$$\boldsymbol{\Delta}^{(l)} \odot^l \boldsymbol{u}_{(l)} = \boldsymbol{u}_{[l]} \,. \tag{10.13}$$

Obviously, the equations (10.3-10.4) provide first implicit representations for $\boldsymbol{\Delta}^{(2)}$, and $\boldsymbol{\Delta}^{(3)}$, while $\boldsymbol{\Delta}^{(1)} = \mathbf{1}$, but as we will see below, there is also a strategy for iteratively constructing $\boldsymbol{\Delta}^{(l)}$, and thus $\boldsymbol{u}_{[l]}$. The $\boldsymbol{\Delta}^{()}$'s will be constructed using l Kronecker symbols for each term, requiring symmetry with respect to the first and last l components independently, and vanishing traces. Useful properties for $\boldsymbol{\Delta}^{(l)}$ can be found in [82].

10.4.1 Construction of the Isotropic Tensors $\boldsymbol{\Delta}^{(l)}$

The explicit recursion formula for all $l \geq 1$ reads, for the components ($\boldsymbol{\mu}$ abbreviates the indices $\mu_1\mu_2..\mu_l$ and $\boldsymbol{\nu}$ the indices $\nu_1\nu_2..\nu_l$)

$$\Delta^{(l)}_{\boldsymbol{\mu},\boldsymbol{\nu}} = \frac{1}{l}\left(\sum_{i=1}^{l}\delta_{\mu_i\nu_1}\Delta^{(l-1)}_{\times^i(\boldsymbol{\mu}),\times^1(\boldsymbol{\nu})}\right) - \frac{2}{l(2l-1)}\left(\sum_{i=1}^{l-1}\sum_{j=i+1}^{l}\delta_{\mu_i\mu_j}\Delta^{(l-1)}_{\times^1(\times^i_{\mu_1}(\times^j_{\nu_1}(\boldsymbol{\mu}))),\times^1(\boldsymbol{\nu})}\right), \tag{10.14}$$

with $\boldsymbol{\Delta}^{(0)} = 1$ to start the recursion. We introduced an 'exchange operator' $\times()$ to keep the notation short, and to allow for immediate implementation in a symbolic programming language,

$$\begin{aligned}\boldsymbol{\mu} &= \mu_1\mu_2..\mu_{i-1}\mu_i\mu_{i+1}..\,,\\ \times^i(\boldsymbol{\mu}) &= \mu_1\mu_2..\mu_{i-1}\quad\mu_{i+1}..\,,\\ \times^i_\nu(\boldsymbol{\mu}) &= \mu_1\mu_2..\mu_{i-1}\,\nu\,\mu_{i+1}..\,,\end{aligned} \tag{10.15}$$

With the formula (10.14) at hand – we skip our proof, and have not seen a similar perfectly explicit formula elsewhere in the literature although it should probably exist –, we can let the computer generate anisotropic tensors without solving the system of equations as for our derivation (10.4). For $l = 1$ and $l = 2$, (10.14) evaluates as

$$\Delta^{(1)}_{\mu_1,\nu_1} = \delta_{\mu_1\nu_1}\Delta^{(0)}_{\times^1(\mu_1),\times^1(\nu_1)} - \sum_{i=1}^{0}.. = \delta_{\mu_1\nu_1}\Delta^{(0)} = \delta_{\mu_1\nu_1} \tag{10.16}$$

and, using $\times^1(\boldsymbol{\nu}) = \nu_2$, $\times^2_{\nu_1}(\boldsymbol{\mu}) = \mu_1\nu_1$, etc.,

$$\begin{aligned}\Delta^{(2)}_{\boldsymbol{\mu},\boldsymbol{\nu}} &= \frac{1}{2}\left(\sum_{i=1}^{2}\delta_{\mu_i\nu_1}\Delta^{(1)}_{\times^i(\boldsymbol{\mu}),\nu_2}\right) - \frac{1}{3}\left(\delta_{\mu_1\mu_2}\Delta^{(1)}_{\nu_1,\nu_2}\right),\\ &= \frac{1}{2}\left(\delta_{\mu_1\nu_1}\delta_{\mu_2\nu_2} + \delta_{\mu_2\nu_1}\delta_{\mu_1\nu_2}\right) - \frac{1}{3}\delta_{\mu_1\mu_2}\delta_{\nu_1\nu_2}\,,\end{aligned} \tag{10.17}$$

which equals (2.11) in [82].

10.4.2 Generalized Cross Product $\boldsymbol{\Delta}^{(l,1,l)}$

The following isotropic tensor $\boldsymbol{\Delta}^{(l,1,l)}$ of rank $2l+1$

$$\Delta^{(l,1,l)}_{\boldsymbol{\mu},\lambda,\tilde{\boldsymbol{\mu}}} \equiv \Delta^{(l)}_{\boldsymbol{\mu},\boldsymbol{\nu}}\,\varepsilon_{\nu_l\lambda\gamma}\Delta^{(l)}_{\times^l_\gamma(\boldsymbol{\nu}),\tilde{\boldsymbol{\mu}}}\,, \tag{10.18}$$

explicitly constructed using (10.14), defines a generalized (irreducible) cross product. With $\boldsymbol{T}^l$ being an arbitrary tensor of rank l,

$$(\boldsymbol{b}\times\boldsymbol{T}^l)_{\boldsymbol{\mu}} \equiv \Delta^{(l,1,l)}_{\boldsymbol{\mu},\lambda,\tilde{\boldsymbol{\mu}}} b_\lambda \overline{\boldsymbol{T}^l}_{\tilde{\boldsymbol{\mu}}} = -(\overline{\boldsymbol{b}\cdot\boldsymbol{\varepsilon}\cdot\overline{\boldsymbol{T}^l}})_{\boldsymbol{\mu}}\,, \tag{10.19}$$

which simplies if $\boldsymbol{T}^l$ is anisotropic, in particular, for $\boldsymbol{T}^l = \boldsymbol{u}_{[l]}$, to

$$\boldsymbol{b}\times\boldsymbol{u}_{[l]} \equiv \overline{\boldsymbol{b}\cdot(-\boldsymbol{\varepsilon})\cdot\boldsymbol{u}_{[l]}}\,. \tag{10.20}$$

since $\boldsymbol{u}_{[l]}$ is symmetric and traceless; the rule is then to replace $\times$ by $\cdot(-\boldsymbol{\varepsilon})\cdot$. In lowest order we simply have $\Delta^{(l,1,l)}_{\mu,\lambda,\nu} = \varepsilon_{\mu,\lambda,\nu}$. Notice the property

$$\Delta^{(l,1,l)}_{\boldsymbol{\mu},\lambda,\boldsymbol{\nu}} = -\Delta^{(l,1,l)}_{\boldsymbol{\nu},\lambda,\boldsymbol{\mu}}\,. \tag{10.21}$$

10.4.3 Generalized Tensor $\boldsymbol{\Delta}^{(l,k,l)}$

In order to further generalize, [82] introduced the notation $\boldsymbol{\Delta}^{(l,k,l)}$ with $\boldsymbol{\Delta}^{(l,0,l)} = \boldsymbol{\Delta}^{(l)}$ (defined in (10.14)), and $\boldsymbol{\Delta}^{(l,1,l)}$ (defined in (10.18)).

$$\Delta^{(l,2,l)}_{\boldsymbol{\mu},\lambda\kappa,\tilde{\boldsymbol{\mu}}} = \Delta^{(l)}_{\boldsymbol{\mu},\boldsymbol{\nu}} \Delta^{(2)}_{\nu_l\gamma,\lambda\kappa} \Delta^{(l)}_{\times'_\gamma(\boldsymbol{\nu}),\tilde{\boldsymbol{\mu}}}\,. \tag{10.22}$$

For $l=1$, one has $\Delta^{(1,2,1)}_{\mu,\lambda\kappa,\nu} = \Delta^{(2)}_{\mu\nu,\lambda\kappa}$.

10.4.4 Implications (Summary)

Using the definitions (10.6), (10.2) and the $\boldsymbol{\Delta}^{(l,k,l)}$-operators of the following sections, we can derive all results listed in the tables of the subsequent Sect. 10.5, and some further very useful relationships (more advanced rules can be found in [82, 417, 418]).

$$\mathrm{Tr}(\boldsymbol{Q}_{[l]}) = 0,\quad \boldsymbol{Q}^{\mathrm{sym}}_{[l]} = \boldsymbol{Q}_{[l]}\,, \tag{10.23}$$

$$\boldsymbol{u}_{[l]} = \frac{2l+1}{l+1}\boldsymbol{u}\cdot\boldsymbol{u}_{[l+1]}\,, \tag{10.24}$$

$$\boldsymbol{u}_{[l-1]} = \frac{2l-1}{l}\boldsymbol{u}\cdot\boldsymbol{u}_{[l]}\,, \tag{10.25}$$

$$\boldsymbol{u}_{[l]}\boldsymbol{u} = \boldsymbol{u}_{[l+1]} + \frac{l}{2l+1}\boldsymbol{\Delta}^{(l)}\odot^{l-1}\boldsymbol{u}_{[l-1]}\,, \tag{10.26}$$

$$\begin{aligned}\boldsymbol{u}_{[l]}\boldsymbol{u}_{[2]} &= \boldsymbol{u}_{[l+2]} + \frac{2l}{2l+3}\boldsymbol{\Delta}^{(l,2,l)}\odot^l\boldsymbol{u}_{[l]}\\ &\quad+\frac{l(l-1)}{(2l+1)(2l-1)}\boldsymbol{\Delta}^{(l)}\odot^{l-2}\boldsymbol{u}_{[l-2]}\,,\end{aligned} \tag{10.27}$$

$$\boldsymbol{u}^2_{[l]} \equiv \boldsymbol{u}_{[l]}\odot^l\boldsymbol{u}_{[l]} = \frac{l!}{(2l-1)!!}\,, \tag{10.28}$$

$$
\begin{aligned}
\boldsymbol{u}_{[l]} \odot^l \boldsymbol{u}_{[l+k]} &= \frac{u^2_{[l+k]}}{u^2_{[l]}} \boldsymbol{u}_{[k]} = \frac{(l+k)!(2k-1)!!}{k!(2l+2k-1)!!} \boldsymbol{u}_{[k]} , && (10.29)\\
\nabla \boldsymbol{Q}_{[l]} &= Q^{l-1}(l\boldsymbol{u} - \boldsymbol{u} \times \boldsymbol{\mathcal{L}})\boldsymbol{u}_{[l]} , \quad (*) && (10.30)\\
\nabla \boldsymbol{u}_{[l]} &= -\frac{1}{Q} \boldsymbol{u} \times \boldsymbol{\mathcal{L}} \boldsymbol{u}_{[l]} , && (10.31)\\
\frac{\partial}{\partial \boldsymbol{u}} \cdot \boldsymbol{Q}_{[l]} &= (l+1)\, Q^l \boldsymbol{u} \cdot \boldsymbol{u}_{[l]} = \frac{l(l+1)}{2l-1} Q^l \boldsymbol{u}_{[l-1]} , && (10.32)\\
\frac{\partial}{\partial \boldsymbol{u}} \boldsymbol{u}_{[l]} &= \frac{\partial \boldsymbol{u}_{[l]}}{\partial \boldsymbol{u}} = (l+1)\boldsymbol{u}\boldsymbol{u}_{[l]} - (2l+1)\boldsymbol{u}_{[l+1]} , && (10.33)\\
\boldsymbol{\mathcal{L}} \cdot \boldsymbol{u}_{[l]} &= \boldsymbol{\varepsilon} : \boldsymbol{u}_{[l]} = 0 , && (10.34)\\
\boldsymbol{\mathcal{L}} \boldsymbol{u}_{[l]} &= -(2l+1)\boldsymbol{u} \times \boldsymbol{u}_{[l+1]} = -l\, \boldsymbol{u}_{[l]} \odot^l \boldsymbol{\Delta}^{(l,1,l)} , && (10.35)\\
\boldsymbol{\mathcal{L}}^2 \boldsymbol{u}_{[l]} &= -l(l+1)\boldsymbol{u}_{[l]} . && (10.36)
\end{aligned}
$$

In (10.28), we have defined the scalar $u^2_{[l]} \equiv \boldsymbol{u}_{[l]} \odot^l \boldsymbol{u}_{[l]}$ (l-fold contraction), which must not be confused with the squared tensor $\boldsymbol{u}^2_{[l]} \equiv \boldsymbol{u}_{[l]} \cdot \boldsymbol{u}_{[l]}$ (single contraction, resulting in a tensor of rank $2(l-1)$.

10.5 Differential Operations (Tabular Form)

Operator Symbol	Operator Defintion	Operator Argument Q	$\boldsymbol{Q}$	$\boldsymbol{u}$
∇	$\frac{\partial}{\partial \boldsymbol{Q}}$	$\boldsymbol{u}$	$\boldsymbol{1}$	$\frac{1}{Q}(\boldsymbol{1} - \boldsymbol{u}_{(2)})$
$\frac{\partial}{\partial \boldsymbol{u}}$	$Q(\boldsymbol{1} - \boldsymbol{u}_{(2)}) \cdot \nabla$	0	$Q(\boldsymbol{1} - \boldsymbol{u}_{(2)})$	$\boldsymbol{1} - \boldsymbol{u}_{(2)}$
$\frac{\partial}{\partial Q}$	$\frac{\partial}{\partial Q}$	1	$\boldsymbol{u}$	0
$\boldsymbol{\mathcal{L}}$	$\boldsymbol{u} \times \frac{\partial}{\partial \boldsymbol{u}}$	0	$-\boldsymbol{\varepsilon} \cdot \boldsymbol{Q}$	$-\boldsymbol{\varepsilon} \cdot \boldsymbol{u}$
Δ	$\nabla \cdot \nabla$	$\frac{2}{Q}$	0	$-\frac{2}{Q^2}\boldsymbol{u}$
$\boldsymbol{\mathcal{L}}^2$	$\boldsymbol{\mathcal{L}} \cdot \boldsymbol{\mathcal{L}}$	0	$-2\boldsymbol{Q}$	$-2\boldsymbol{u}$
$\nabla \cdot$		–	3	$\frac{2}{Q}$
$\frac{\partial}{\partial \boldsymbol{u}} \cdot$		–	$2Q$	2
$\boldsymbol{\mathcal{L}} \cdot$		–	$\boldsymbol{0}$	$\boldsymbol{0}$

Operator Symbol	Operator Argument $\boldsymbol{Q}_{[l]}$	$\boldsymbol{Q}_{[2]}$
$\frac{\partial}{\partial Q}$	$lQ^{l-1}\boldsymbol{u}_{[l]}$	$2Q\boldsymbol{u}_{[2]}$
∇	$Q^{l-1}(l\boldsymbol{u}-\boldsymbol{u}\times\mathcal{L})\boldsymbol{u}_{[l]}$	$Q(2\boldsymbol{u}-\boldsymbol{u}\times\mathcal{L})\boldsymbol{u}_{[2]}$
$\frac{\partial}{\partial \boldsymbol{u}}$	$(l+1)\boldsymbol{u}\boldsymbol{Q}_{[l]}-(2l+1)\frac{1}{Q}\boldsymbol{Q}_{[l+1]}$	$3\boldsymbol{u}\boldsymbol{Q}_{[2]}-5\frac{1}{Q}\boldsymbol{Q}_{[3]}$
$\mathcal{L}$	$Q^l\mathcal{L}\boldsymbol{u}_{[l]}$	$Q^2\mathcal{L}\boldsymbol{u}_{[2]}$
$\triangle$	0	0
$\mathcal{L}^2$	$-l(l+1)\boldsymbol{Q}_{[l]}$	$-6\boldsymbol{Q}_{[2]}$
$\nabla\cdot$	$Q^{l-1}(\frac{l^2}{2l-1}\boldsymbol{u}_{[l-1]}+(2l+1)\boldsymbol{u}_{[2]}:\boldsymbol{u}_{[l+1]})$	$\frac{10}{3}\boldsymbol{Q}$
$\frac{\partial}{\partial \boldsymbol{u}}\cdot$	$\frac{l(l+1)}{2l-1}Q^l\boldsymbol{u}_{[l-1]}$	$2Q^2\boldsymbol{u}$
$\mathcal{L}\cdot$	0	0

Operator Symbol	Operator Argument $\boldsymbol{u}_{[l]}$	$\boldsymbol{u}_{[2]}$
∇	$-\frac{1}{Q}\boldsymbol{u}\times\mathcal{L}\boldsymbol{u}_{[l]}$	$-\frac{1}{Q}\boldsymbol{u}\times\mathcal{L}\boldsymbol{u}_{[2]}=\frac{1}{Q}(-2\boldsymbol{u}_{(3)}+\mathbf{1}\boldsymbol{u}+\boldsymbol{iui})$
$\frac{\partial}{\partial \boldsymbol{u}}$	$(l+1)\boldsymbol{u}\boldsymbol{u}_{[l]}-(2l+1)\boldsymbol{u}_{[l+1]}$	$3\boldsymbol{u}\boldsymbol{u}_{[2]}-5\boldsymbol{u}_{[3]}=-2\boldsymbol{u}_{(3)}+\mathbf{1}\boldsymbol{u}+\boldsymbol{iui}$
$\mathcal{L}$	$-(2l+1)\boldsymbol{u}\times\boldsymbol{u}_{[l+1]}$	$-5\boldsymbol{u}\times\boldsymbol{u}_{[3]}$
$\triangle$	$-\frac{1}{Q^2}l(l+1)\boldsymbol{u}_{[l]}$	$-\frac{6}{Q^2}\boldsymbol{u}_{[2]}$
$\mathcal{L}^2$	$-l(l+1)\boldsymbol{u}_{[l]}$	$-6\boldsymbol{u}_{[2]}$
$\nabla\cdot$	$\frac{2l+1}{Q}\boldsymbol{u}_{[2]}:\boldsymbol{u}_{[l+1]}$	$2\boldsymbol{u}$
$\frac{\partial}{\partial \boldsymbol{u}}\cdot$	$\frac{l(l+1)}{2l-1}\boldsymbol{u}_{[l-1]}$	$2\boldsymbol{u}$
$\mathcal{L}\cdot$	0	0

In calculations, component notation should be always used. $(\mathcal{L}\boldsymbol{u}_{[2]})_{\mu\nu\lambda}=\varepsilon_{\mu\nu\kappa}u_\kappa u_\lambda+\varepsilon_{\mu\kappa\lambda}u_\kappa u_\nu$, is one example which is not immediately rewritten in the simple tensor product notation. However, we prefer to have short notatios in this chapter.

10.6 Nematic Order Parameters

The 2nd rank alignment tensor can be characterized by the amount of distinct eigenvalues, as described in Sect. 7.4. For 1, 2, and 3 distinct eigenvectors, it describes a state of isotropic, uniaxial, and biaxial symmetry, respectively.

10.6.1 Uniaxial Phase

With the ordinary nematic order parameters S_l and director $\boldsymbol{n}$ (with $|\boldsymbol{n}|=1$), both, as well as the alignment tensor $\boldsymbol{a}_{[l]}$ of the nematic phase defined through

$$\boldsymbol{a}_{[l]} \equiv \langle \boldsymbol{u}_{[l]} \rangle = S_l \boldsymbol{n}_{[l]}, \quad \text{with } \boldsymbol{n}_{[l]} \equiv \overline{\boldsymbol{nn..n}}\ , \tag{10.37}$$

we can use the above relationships (10.3-10.4) to immediately arrive at the following sample identities for non-anisotropic tensor, still often prefered in the literature,

$$\begin{aligned}
\boldsymbol{a}_{(2)} &= S_2 \boldsymbol{nn} + \frac{1-S_2}{3}\mathbf{1}\,, \\
\boldsymbol{a}_{(3)} &= S_3 \boldsymbol{nnn} + \frac{3}{5}(S_1 - S_3)(\boldsymbol{n}\mathbf{1})^{\text{symm}}\,, \\
\boldsymbol{a}_{(4)} &= S_4 \boldsymbol{nnnn} + \frac{6}{7}(S_2 - S_4)\{\boldsymbol{nn}\mathbf{1}\}^{\text{sym}} \\
&\quad + \frac{7-10S_2+3S_4}{35}\{\mathbf{11}\}^{\text{sym}}\,.
\end{aligned} \tag{10.38}$$

The generalization of (10.28) becomes

$$\boldsymbol{u}_{[l]} \odot^l \boldsymbol{n}_{[l]} = \frac{l!}{(2l-1)!!} P_l(\boldsymbol{u}\cdot\boldsymbol{n})\,, \tag{10.39}$$

with the lth order Legrende polynomial P_l. In particular, we have, by combining (10.28) with (10.39),

$$S_l = \langle P_l(\boldsymbol{u}\cdot\boldsymbol{n})\rangle\,, \tag{10.40}$$

an expression for the order parameter in terms of an expectation value. If the alignment tensors are available, which is the typical case when performing simulations where the alignment tensors are obtained as time averages, and the director is not known a priori, we can calculate the squared uniaxial order parameters via

$$S_l^2 = \frac{(2l-1)!!}{l!}\, \boldsymbol{a}_{[l]} \odot^l \boldsymbol{a}_{[l]}\,. \tag{10.41}$$

The director is obtained as the eigenvector corresponding to the largeset eigenvalue of $\boldsymbol{a}_{[2]}$. In case of uniaxial order, two of the eigenvalues of $\boldsymbol{a}_{[2]}$ must be equal. For the general biaxial case, corresponding order parameters involving the remaining eigenvectors of $\boldsymbol{a}_{[2]}$ have been iontroduced. Further, when combining (10.37) and (10.39), one has

$$S_l = \frac{(2l-1)!!}{l!}\, \boldsymbol{a}_{[l]} \odot^l \boldsymbol{n}_{[l]}\,. \tag{10.42}$$

10.6.2 Biaxial Phase

In the biaxial phase, the three principal values of the 2nd order alignment tensor $\boldsymbol{a}_{[2]}$ are distinct, cf. representation (7.23), and we have two 'directors' denoted as $\boldsymbol{n}$ and $\boldsymbol{m}$[4]

$$\overline{\boldsymbol{a}_{[2]}} = (S_2 + B_2/2)\,\overline{\boldsymbol{nn}} + B_2\,\overline{\boldsymbol{mm}}\,, \tag{10.43}$$

[4] The third one, say $\boldsymbol{l}$, can be always eliminated using the identity $\mathbf{1} = \boldsymbol{nn} + \boldsymbol{mm} + \boldsymbol{ll}$.

with order parameters S_2 and B_2, where B_2 is the second-order scalar biaxial order parameter

$$B_2 = \frac{2}{3}\left(\langle P_2(\boldsymbol{u}\cdot\boldsymbol{n})\rangle + 2\,\langle P_2(\boldsymbol{u}\cdot\boldsymbol{m})\rangle\right). \tag{10.44}$$

It ranges in value by $|\,B_2\,| \leq \frac{2}{3}(1-S_2) \leq 1$. For perfect uniaxial alignment in the $\boldsymbol{n}$ direction, $S_2 = 1$ and $B_2 = 0$. For perfect uniaxial alignment in the $\boldsymbol{m}$ direction, $B_2 = -2S_2 = 1$. For random alignment (hence, isotropic) $S_2 = B_2 = 0$.

Similarly, we obtain for the fourth-order alignment tensors

$$\boldsymbol{a}_{[4]} = \left(S_4 - \frac{3}{8}B_4 + \frac{1}{2}M_4\right)\boldsymbol{n}_{[4]} + B_4\boldsymbol{m}_{(4)} + M_4\,\overline{\boldsymbol{n}_{(2)}\boldsymbol{m}_{(2)}} \tag{10.45}$$

where[5]

$$B_4 = \frac{8}{35}\left[4\big(\langle P_4(\boldsymbol{m}\cdot\boldsymbol{u})\rangle + \langle P_4(\boldsymbol{l}\cdot\boldsymbol{u})\rangle\big) - 3\,\langle P_4(\boldsymbol{n}\cdot\boldsymbol{u})\rangle\right], \tag{10.50}$$

$$M_4 = \frac{8}{35}\left[11\,\langle P_4(\boldsymbol{l}\cdot\boldsymbol{u})\rangle - 3\big(\langle P_4(\boldsymbol{n}\cdot\boldsymbol{u})\rangle + \langle P_4(\boldsymbol{m}\cdot\boldsymbol{u})\rangle\big)\right]. \tag{10.51}$$

Note that there are 3 distinct fourth-order scalar measures of alignment: S_4, B_4, and M_4. In the uniaxial case with director $\boldsymbol{n}$, we have $B_4 = M_4 = 0$, so that these two can be interpreted as fourth-order measures of the deviation from uniaxiality. The 4th order Legendre polynomial $P_4(x)$ is bound to $-3/8 \leq P_4 \leq 1$. The lowest order Legendre polynomials are plotted in Fig. 10.1.

10.7 Tensor Invariants

The theorem of Caley and Hamilton states the following. Let $\boldsymbol{A}$ be a $d \times d$ matrix, and

$$\phi(\lambda) = \det(\boldsymbol{A} - \lambda\boldsymbol{1}) = \lambda^d + a_{d-1}\lambda^{d-1} + \ldots + a_0 \tag{10.52}$$

be the characteristic polynomial of $\boldsymbol{A}$ where λ_i are roots of $\phi(\lambda) = 0$, i.e., eigenvalues of $\boldsymbol{A}$, with multiplicity ν_i. For arbitrary matrices $\boldsymbol{A}$, not only traceless ones, which is however not of relevance in the context of this book, one has

[5] Equation (10.45) can be also rewritten more explicitely as

$$\boldsymbol{a}_{(4)} = \left(S_4 - \frac{3}{8}B_4 + \frac{1}{2}M_4\right)\boldsymbol{n}_{(4)} + B_4\boldsymbol{m}_{(4)} + M_4\{\boldsymbol{n}_{(2)}\boldsymbol{m}_{(2)}\}^{\mathrm{sym}}$$
$$+\alpha_1\{\boldsymbol{n}_{(2)}\boldsymbol{1}\}^{\mathrm{sym}} + \alpha_2\{\boldsymbol{m}_{(2)}\boldsymbol{1}\}^{\mathrm{sym}} + \alpha_3\{\boldsymbol{1}\,\boldsymbol{1}\}^{\mathrm{sym}}, \tag{10.46}$$

$$\alpha_1 \equiv \frac{1}{28}[24(S_2 - S_4) + 12B_2 + 9B_4 - 16M_4], \tag{10.47}$$

$$\alpha_2 \equiv \frac{1}{7}[6(B_2 - B_4) - M_4], \tag{10.48}$$

$$\alpha_3 \equiv \frac{1}{280}[4(14 + 5M_4 - 20S_2 + 6S_4) - 120B_2 + 15B_4]. \tag{10.49}$$

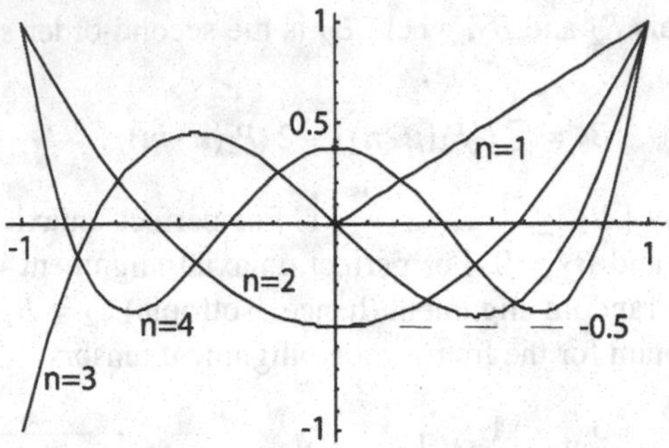

Fig. 10.1. Legendre polynomials $P_l(x)$ for $l = 1,2,3,4$. Uniaxial order parameters are defined as $S_l \equiv P_l(\boldsymbol{u}\cdot\boldsymbol{n})$ with director n. For nonpolar fluids (liquid crystal), only the even order parameters do not vanish, for polar fluids (ferrofluid) $S_1 \geq 0$ by convention. This figure shows, in particular, that order parameters can have either sign. Except for the uniaxial phase, we prefer order parameter tensors (alignment tensors) rather than scalar order parameters for the description of anisotropic fluids

$$\phi(\boldsymbol{A}) = \boldsymbol{A}^d - I_1\boldsymbol{A}^{d-1} - \ldots - I_d\mathbf{1} = 0\,. \tag{10.53}$$

In this manuscript we are concerned with three dimensional problems, $d = 3$, thus (10.53) becomes[6]

$$\boldsymbol{A}^3 = I_1\boldsymbol{A}^2 + I_2\boldsymbol{A} + I_3\,\mathbf{1}\,, \tag{10.54}$$

and alternatively, by multiplication with $\boldsymbol{A}^{-1}$,

$$\boldsymbol{A}^2 = I_1\boldsymbol{A}^1 + I_2\,\mathbf{1} + I_3\boldsymbol{A}^{-1}\,, \tag{10.55}$$

where the coefficients define the tensor invariants

$$I_1(\boldsymbol{A}) \equiv \mathrm{Tr}\boldsymbol{A}\,, \tag{10.56}$$

$$I_2(\boldsymbol{A}) \equiv \frac{1}{2}\{\mathrm{Tr}(\boldsymbol{A}^2) - I_1^2\}\,, \tag{10.57}$$

$$I_3(\boldsymbol{A}) \equiv \det\boldsymbol{A}\,. \tag{10.58}$$

For anisotropic tensors $\boldsymbol{A}$, such as the alignment tensor of rank 2, $\boldsymbol{a}_{[2]}$, for which I_1 vanishes, we have, by recursively multiplying (10.54) by $\boldsymbol{a}_{[2]}$,

$$\boldsymbol{a}_{[2]}{}^{i} = I_2\,\boldsymbol{a}_{[2]}{}^{i-2} + I_3\,\boldsymbol{a}_{[2]}{}^{i-3}\,, \tag{10.59}$$

for $i \geq 3$ and with $\boldsymbol{a}_{[2]}{}^0 = \mathbf{1}$. With the help of (10.59), supplemented with a $\overline{\cdots}$ on both sides, recursively calculating the anisotropic powers of $\boldsymbol{a}_{[2]}$ also poses no problem, as illustrated in Table 10.1.

---

[6] The implication (10.54) of the Caley-Hamilton theorem has been used to simplify the appearance of the analytic extension of the linear stress-optic rule in Sect. 4.7.

Table 10.1. The anisotropic part of any power of a dimensionless anisotropic (yet arbitrary 2nd rank) tensor $\boldsymbol{a}_{[2]}$ can be, for example, expressed as a linear combination of first and second order terms, as demonstrated by the lowest powers in this table. Further entries are immediately generated using the recursion (10.59)

$\overline{\boldsymbol{a}_{[2]}^{\;0}}$	$\overline{\boldsymbol{a}_{[2]}^{\;1}}$	$\overline{\boldsymbol{a}_{[2]}^{\;2}}$	$\overline{\boldsymbol{a}_{[2]}^{\;3}}$	$\overline{\boldsymbol{a}_{[2]}^{\;4}}$	$\overline{\boldsymbol{a}_{[2]}^{\;5}}$	$\overline{\boldsymbol{a}_{[2]}^{\;6}}$	$\ldots$	$\overline{\boldsymbol{a}_{[2]}^{\;-1}}$	$=$
0	1	0	I_2	I_3	I_2^2	$2I_2I_3$	$\ldots$	0	$\times\boldsymbol{a}_{[2]}+$
0	0	1	0	I_2	I_3	I_2^2	$\ldots$	I_3^{-1}	$\times\,\overline{\boldsymbol{a}_{[2]}^{\;2}}$

In (7.36), and often in the literature, a third invariant is defined differently as

$$I^{(3)} = \sqrt{6}\,\mathrm{Tr}\,\overline{\boldsymbol{a}_{[2]}^{\;3}}\,. \tag{10.60}$$

Using (10.59), we immediatly obtain the relationship between the two representations

$$I^{(3)} = \sqrt{6}(I_2\,\boldsymbol{a}_{[2]} + I_3)\,. \tag{10.61}$$

10.8 Solutions of the Laplace Equation

Of particular relevance is the property $\mathcal{L}^2\boldsymbol{u}_{[l]} = -l(l+1)\boldsymbol{u}_{[l]}$, thus $\boldsymbol{u}_{[l]}$ (or equally $\boldsymbol{Q}_{[l]}$) are the eigentensors of $\mathcal{L}^2$ with eigenvalues $-l(l+1)$ (as we are familiar with from quantum mechanics). The irreducible tensors $\boldsymbol{Q}_{[l]}$ are solutions of the Laplace equation,

$$\Delta\boldsymbol{Q}_{[l]} = 0\,, \tag{10.62}$$

since, using (10.9) and (10.2), we have

$$Q^2\Delta_Q\boldsymbol{Q}_{[l]} = Q^2\left(\frac{\partial^2}{\partial Q^2} + \frac{2}{Q}\frac{\partial}{\partial Q}\right)Q^l\boldsymbol{u}_{[l]} = l(l+1)\,\boldsymbol{u}_{[l]}\,. \tag{10.63}$$

Equation (10.63) combined with (10.36) and (10.9) proves (10.62). Further tensorial solutions of $\Delta\psi = 0$ with different radial dependence are the (descending) multipole potentials $\boldsymbol{X}_{[l]}$,

$$\boldsymbol{X}_{[l]}(\boldsymbol{Q}) \equiv \left(-\frac{\partial}{\partial\boldsymbol{Q}}\right)^l Q^{-1} = \frac{(2l-1)!!}{Q^{2l+1}}\,\boldsymbol{Q}_{[l]}\,. \tag{10.64}$$

This equation does not only offer the definition of the descending multipole potentials $\boldsymbol{X}_{[l]}(\boldsymbol{Q})$, but also the relationship with the anisotropic tensors $\boldsymbol{Q}_{[l]}$. The multipole potentials are symmetric, and traceless, since $\mathrm{Tr}(\boldsymbol{X}_{[l]}(\boldsymbol{Q})) = \Delta Q^{-1} = 0$.

Further, spherical surface tensors $\boldsymbol{Y}_{[l]}(\boldsymbol{u})$, which carry the orientational dependence of the multipole potentials, the cartesian counterpart of the spherical harmonics $Y_l^m(\boldsymbol{u})$, are related to the anisotropic tensors $\boldsymbol{u}_{[l]}$ as follows

$$\boldsymbol{Y}_{[l]}(\boldsymbol{u}) \equiv Q^{l+1}\boldsymbol{X}_{[l]}(\boldsymbol{Q}) = (2l-1)!!\,\boldsymbol{u}_{[l]}\,. \tag{10.65}$$

The factor Q^{l+1} is immediatly understood for dimensional reasons. We refer to [82] for the connection between cartesian and spherical tensors.

10.9 The Reverse $\boldsymbol{\Delta}^{(l)}$ Operation

The $\boldsymbol{\Delta}^{(l)}$-operator, for which we have the explicit expression (10.14) at hand generates the irreducible part for a given tensor,

$$\boldsymbol{\Delta}^{(l)} \odot^l \boldsymbol{u}_{(l)} = \boldsymbol{u}_{[l]}(\{\boldsymbol{u}_{(k)}\}, k \le l) = \boldsymbol{u}_{(l)} + (\boldsymbol{u}_{(l)} - \boldsymbol{u}_{[l]})(\{\boldsymbol{u}_{(k)}\}, k \le l-1)\,, \quad (10.66)$$

with the l-fold dyadic product $\boldsymbol{u}_{(l)} \equiv \boldsymbol{uu}..\boldsymbol{u}$, and $(\{\boldsymbol{u}_{(k)}\}, k \le l)$ helps to remind us to the dependence of an expression on the set of tensors $\boldsymbol{u}_{(k)}$ with $k \le l$. The last term in (10.66) has the important, with its bracket attached, property. One realizes, that applying many of the rules presented in this chapter requires performing the reverse, apparently more difficult task: rewrite the expression $\boldsymbol{u}_{(k)}$ solely in terms of anisotropic tensors $\{\boldsymbol{u}_{[l]}\}$ with $0 \le l \le k$ (including constants, '$l = 0$). While for scalars ($k = 0$), vectors ($k = 1$), tensors of rank 2 ($k = 2$) this is still trivial, it is generally achieved using the trivial identity once again from another perspective

$$\boldsymbol{u}_{(l)} = (\boldsymbol{u}_{(l)} - \boldsymbol{u}_{[l]}) + \boldsymbol{u}_{[l]} = (\mathbf{1}^{(l)} - \boldsymbol{\Delta}^{(l)}) \odot^l \boldsymbol{u}_{(l)} + \boldsymbol{u}_{[l]}\,, \quad (10.67)$$

with the important feature, that the term $(\mathbf{1}^{(l)} - \boldsymbol{\Delta}^{(l)}) \odot^l \boldsymbol{u}_{(l)}$, according to (10.66), does depend only (and explicitly!) on $\{\boldsymbol{u}_{(k)}\}$ with $k \le l-1$. We thus can iteratively express $\boldsymbol{u}_{(k)}$ for any k by applying (10.67) k times for $l = k, k-1, .., 1$, in terms of anisotropic tensors $\{\boldsymbol{u}_{[l]}\}$ with $l \le k$. With the definitions (10.1) and expression (10.14) we have an explicit expression to evaluate $(\mathbf{1}^{(l)} - \boldsymbol{\Delta}^{(l)}) \odot^l \boldsymbol{u}_{(l)} = \boldsymbol{u}_{(l)} - \boldsymbol{u}_{[l]}$, which we do not need to write down (again)[7].

10.10 Integrating Irreducible Tensors

Integrating dyadic products of irreducible tensors over the unit sphere evaluates as follows (and is nonzero only if ranks k and l are equal):[8]

$$\frac{1}{4\pi} \int \boldsymbol{u}_{[k]} \boldsymbol{u}_{[l]} \, \mathrm{d}^2 u = \frac{l!}{(2l+1)!!} \, \delta_{kl} \boldsymbol{\Delta}^{(l)}\,, \quad (10.68)$$

where $\boldsymbol{\Delta}^{(l)}$ is the isotropic tensor of Sect. 10.4.

---

[7] Examples: ($l = 1$) $\mathbf{1}^{(1)} - \boldsymbol{\Delta}^{(1)} = \mathbf{1} - \mathbf{1} = 0$. ($l = 2$) $(\mathbf{1}^{(2)} - \boldsymbol{\Delta}^{(2)}) : \boldsymbol{u}_{(2)} = \boldsymbol{u}_{(2)} - (\boldsymbol{u}_{(2)} - \mathbf{1}/3) = \mathbf{1}/3$, and thus $\boldsymbol{u}_{(2)} = \boldsymbol{u}_{[0]}\mathbf{1}/3 + \boldsymbol{u}_{[2]}$ is $\boldsymbol{u}_{(2)}$ expressed solely in terms of $\{\boldsymbol{u}_{[l]}\}$ with $0 \le l \le k$.

[8] Components: $(1/4\pi) \int \overline{u_{\mu..\nu}}\; \overline{u_{\lambda..\gamma}}\, \mathrm{d}^2 u \propto \Delta^{(l)}_{\mu..\nu,\lambda..\gamma}$.

11

Nonequilibrium Dynamics of Anisotropic Fluids

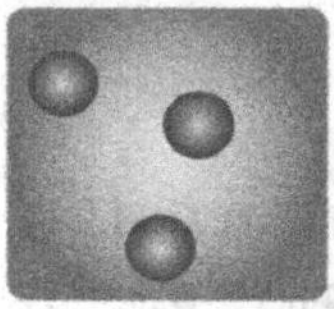

In Chap. 9 we summarized methods for analyzing the equilibrium statistics of physical system, using Monte Carlo methods. While at equilibrium complex fluids may be isotropic or anisotropic, under the influence of external fields, all systems tend to become anisotropic (isotropic external fields are seldom). The anisotropy is induced by the external fields (flow field, magnetic field etc.) and must reflect the tensorial symmetry of the applied field. We have to deal with the coupling of tensors of different ranks, with important implications on Onsager reciprocal relationships, a topic discussed elsewhere. This chapter is concerned with the dynamics, in particular, the orientational dynamics of structured fluids subjected to orienting fields. The dynamics and anisotropy is properly modeled by using orientational distribution functions, their equations of change, and the corresponding equations of change for the moments (here, alignment tensors) of the distribution function. We restrict ourself to discuss the case of one-particle (single-link) orientational distribution functions.

11.1 Orientational Distribution Function

Any orientational (part of a eventually space and time-dependent) distribution function $f(\boldsymbol{u})$ with $u^2 = 1$ can be expanded in terms of anisotropic tensors $\boldsymbol{u}_{[l]}$, and the tensorial coefficients in front of the $\boldsymbol{u}_{[l]}$'s are determined by multiplying f with $\boldsymbol{u}_{[l]}$ and subsequent integration over the tensor unit sphere, to yield

$$f(\boldsymbol{u}) = \frac{1}{4\pi}\left(1+\sum_{l=1}^{\infty}\langle\zeta_l \boldsymbol{u}_{[l]}\rangle \odot^l (\zeta_l \boldsymbol{u}_{[l]})\right) = \frac{1}{4\pi}\left(\sum_{l=0}^{\infty}\zeta_l^2\, \boldsymbol{a}_{[l]} \odot^l \boldsymbol{u}_{[l]}\right). \quad (11.1)$$

The constant $(4\pi)^{-1}$ ensures proper normalization $\langle 1\rangle = 1$, and the average $\langle ...\rangle$ is defined through $\langle ...\rangle \equiv \int ... f(\boldsymbol{u})\mathrm{d}^2u$. The prefactor

$$\zeta_l = \sqrt{\frac{(2l+1)!!}{l!}} \tag{11.2}$$

is immediatly[1] derived using (10.68).

With Sect. 10.5 we obtain

$$\mathcal{L}f = \sum_{l=1}^{\infty} \zeta_l^2 (\mathcal{L}\boldsymbol{u}_{[l]}) \odot^l \boldsymbol{a}_{[l]} = -\sum_{l=1}^{\infty} (2l+1)\zeta_l^2 (\boldsymbol{u} \times \boldsymbol{u}_{[l+1]}) \odot^l \boldsymbol{a}_{[l]} \,,$$

$$\mathcal{L}^2 f = -\sum_{l=1}^{\infty} l(l+1)\zeta_l^2 \boldsymbol{u}_{[l]} \odot^l \boldsymbol{a}_{[l]} \,, \tag{11.3}$$

where $\boldsymbol{a}_{[l]}$ stands for the lth rank alignment tensor.

11.1.1 Alignment Tensors

The moments $\boldsymbol{a}_{[l]} \equiv \langle \boldsymbol{u}_{[l]} \rangle = \langle \boldsymbol{u}_{[l]} \rangle (\boldsymbol{r},t)$ of the distribution function $f(\boldsymbol{r},\boldsymbol{u},t)$ (11.1) we refer to as 'alignment tensors' $\boldsymbol{a}_{[l]}$ of rank l', with the convention $\boldsymbol{a}_{[0]} \odot^0 \boldsymbol{u}_{[0]} = 1$. In the early literature the scaled moments $\zeta_l \langle \boldsymbol{u}_{[l]} \rangle$, for the reason shown in (11.1), were called alignment tensors. Notice the difference, relevant for the layout during analytic evaluations: $\zeta_1 = \sqrt{3}$, $\zeta_2 = \sqrt{15/2}$, $\zeta_3 = \sqrt{35/2}$.

11.1.2 Uniaxial Distribution Function

For the special case of uniaxial order, $f(\boldsymbol{u}) = f(\boldsymbol{u}\cdot\boldsymbol{n})$ with director $\boldsymbol{n}$ of Sect. 10.6, and order parameters $\boldsymbol{a}_{[l]} = S_l \boldsymbol{n}_{[l]}$, (11.1) simplifies, using (10.39), to

$$f_{\mathrm{uni}}(\boldsymbol{u}) = \frac{1}{4\pi}\left(1 + \sum_{l=1}^{\infty} (2l+1) S_l P_l(\boldsymbol{u}\cdot\boldsymbol{n})\right). \tag{11.4}$$

Biaxial distributions, and alignment tensors for biaxial distributions are outside the scope of this book. Viscosities for biaxial fluids made of uniaxial particles, along the lines indicated in Sect. 7.3 on uniaxial fluids can be found in [315], and references cited herein.

11.2 Fokker–Planck Equation, Smoluchowski Equation

An equation of change for distribution functions describing diffusion processes, the Fokker–Planck equation, was derived and presented in Sect. 8.5. Accordingly, the equation of change for the orientational distribution function $f(\boldsymbol{r},\boldsymbol{u},t)$ can be also written for the present purpose, cf. (11.44), as

[1] Proof: $\zeta_k \langle \boldsymbol{u}_{[k]} \rangle = \zeta_k \int \boldsymbol{u}_{[k]} f(\boldsymbol{u}) \mathrm{d}^2u = \zeta_k \sum_l \zeta_l^2 \langle \boldsymbol{u}_{[l]} \rangle (4\pi)^{-1} \int \boldsymbol{u}_{[k]} \boldsymbol{u}_{[l]} \mathrm{d}^2u = \zeta_k \sum_l \zeta_l^2 \langle \boldsymbol{u}_{[l]} \rangle \delta_{kl} \zeta_l^{-2} = \zeta_k \langle \boldsymbol{u}_{[k]} \rangle$.

$$\frac{\partial f}{\partial t} = \mathcal{L}_{\text{FP}} f = -\nabla \cdot (\dot{\boldsymbol{r}} f) - \frac{\partial}{\partial \boldsymbol{u}} \cdot (\dot{\boldsymbol{u}} f) , \tag{11.5}$$

or, equivalently, with angular momentum $\boldsymbol{\omega} = \boldsymbol{u} \times \dot{\boldsymbol{u}}$,

$$\frac{\partial f}{\partial t} = \mathcal{L}_{\text{FP}} f = -\nabla \cdot (\dot{\boldsymbol{r}} f) - \boldsymbol{\mathcal{L}} \cdot (\boldsymbol{\omega} f) . \tag{11.6}$$

Differential operators were defined in (10.6), and the equivalence between the two representations follows using the tables in Sect. 10.5, and $u^2 = 1$ which implies $\frac{1}{2} du^2/dt = \boldsymbol{u} \cdot \dot{\boldsymbol{u}} = 0$. In this chapter we always consider volume-conserving flows. Generally, the derivatives $\dot{\boldsymbol{r}}$, $\dot{\boldsymbol{u}}$, $\boldsymbol{\omega} = \boldsymbol{u} \times \dot{\boldsymbol{u}}$ can be split into two parts from which one is derived from a free energy (the 'F' part), plus an 'external', $\boldsymbol{\mathcal{L}}, \nabla$-divergence-free, remaining 'f' part, i.e., $\dot{\boldsymbol{x}} = \dot{\boldsymbol{x}}_F(\boldsymbol{u}) + \dot{\boldsymbol{x}}_f$ (compare with the GENERIC framework, cf. Sect. 8.3, and [54] for some more general considerations). A form including quite relevant cases in the theory of complex fluids – where $\boldsymbol{u}$ represents the symmetry axis of am uniaxial particle – is obtained using velocity $\dot{\boldsymbol{r}}$ (or friction force $\boldsymbol{\zeta} \cdot \dot{\boldsymbol{r}}$, with translational diffusion matrix $\boldsymbol{D}$ and friction matrix $\boldsymbol{\zeta} \equiv k_{\text{B}} T \boldsymbol{D}^{-1}$) [74,94,216,313,314]

$$\dot{\boldsymbol{r}} = -\frac{\boldsymbol{D}}{k_{\text{B}} T} \cdot \nabla \left(\frac{\delta F[f]}{\delta f(\boldsymbol{r},\boldsymbol{u})} \right) + \dot{\boldsymbol{r}}_f \Leftrightarrow \quad \boldsymbol{\zeta} \cdot \dot{\boldsymbol{r}} = -\nabla \left(\frac{\delta F[f]}{\delta f(\boldsymbol{r},\boldsymbol{u})} \right) + \boldsymbol{\zeta} \cdot \dot{\boldsymbol{r}}_f , \tag{11.7}$$

and angular velocity (featuring the angular operator $\boldsymbol{\mathcal{L}}$ instead of ∇, with rotational diffusion matrix $\boldsymbol{\mathcal{D}}$ and friction matrix $\boldsymbol{\xi} \equiv k_{\text{B}} T \boldsymbol{\mathcal{D}}^{-1}$)

$$\boldsymbol{\omega} = -\frac{\boldsymbol{\mathcal{D}}}{k_{\text{B}} T} \cdot \boldsymbol{\mathcal{L}} \left(\frac{\delta F[f]}{\delta f(\boldsymbol{r},\boldsymbol{u})} \right) + \boldsymbol{\omega}_f , \quad \dot{\boldsymbol{u}} = -\frac{\boldsymbol{\mathcal{D}}}{k_{\text{B}} T} \cdot \frac{\partial}{\partial \boldsymbol{u}} \left(\frac{\delta F[f]}{\delta f(\boldsymbol{r},\boldsymbol{u})} \right) + \dot{\boldsymbol{u}}_f \tag{11.8}$$

and free energy functional F, entropy S, potential U

$$\begin{aligned} F[f] &= V[f] - T\,S[f] , \\ S[f] &= -k_{\text{B}} \int f(\boldsymbol{u}) \ln f(\boldsymbol{u}) \mathrm{d}^2 u , \\ \frac{\delta F[f]}{\delta f(\boldsymbol{r},\boldsymbol{u})} &= \delta_f F[f] = \delta_f V[f] + k_{\text{B}} T \ln f . \end{aligned} \tag{11.9}$$

Here, $V[f]$ is an arbitrary potential functional, and $\boldsymbol{D}$ and $\boldsymbol{\mathcal{D}}$ are model dependent diffusion tensors, which may depend on space $\boldsymbol{r}$ and orientation $\boldsymbol{u}$. The entropic $\ln f$-terms ensure, with positive definite diffusion tensor, that the distribution function relaxes to its equilibrium value, once the potential (V) is removed. Inserting, (11.9) into (11.8) gives

$$\boldsymbol{\omega} = -\frac{\boldsymbol{\mathcal{D}}}{k_{\text{B}} T} \cdot \boldsymbol{\mathcal{L}}(\delta_f V[f]) - \frac{1}{f} \boldsymbol{\mathcal{D}} \cdot \boldsymbol{\mathcal{L}} f + \boldsymbol{\omega}_f , \tag{11.10}$$

and further inserting (11.7) and (11.9) into (11.6) yields

$$\begin{aligned} \frac{\partial f}{\partial t} &= \nabla \cdot (f \boldsymbol{\zeta}^{-1} \cdot \nabla(\delta_f V[f])) + (\nabla \cdot \boldsymbol{D} - \dot{\boldsymbol{r}}_f) \cdot \nabla f \\ &\quad + \boldsymbol{\mathcal{L}} \cdot (f \boldsymbol{\xi}^{-1} \cdot \boldsymbol{\mathcal{L}}(\delta_f V[f])) + (\boldsymbol{\mathcal{L}} \cdot \boldsymbol{\mathcal{D}} - \boldsymbol{\omega}_f) \cdot \boldsymbol{\mathcal{L}} f , \end{aligned} \tag{11.11}$$

with $\boldsymbol{u}$-independent angular momentum $\boldsymbol{\omega}_f$, rotary diffusion tensor $\mathcal{D}$, friction tensor $\boldsymbol{\xi} = k_B T \mathcal{D}^{-1}$, and functional $V[f]$.

In the absence of external field, $\dot{\boldsymbol{v}}_f = 0$, $\boldsymbol{\omega}_f = 0$, the stationary solution of (11.11) is

$$f_{\text{stat}} = \frac{1}{Z} \mathrm{e}^{-\delta_f V[f]/k_B T}, \tag{11.12}$$

since $\mathcal{L} f_{\text{stat}} = -f_{\text{stat}} \mathcal{L}(\delta_f V[f])/k_B T$. The denominator Z ensures proper normalizarion $\langle 1 \rangle = 1$.

11.2.1 Spatial Inhomogeneous Distribution

A common choice for a spatial homogeneous, anisotropic translational diffusion matrix is

$$\boldsymbol{D} = \left(\frac{D_\parallel}{k_B T} \boldsymbol{u}_{(2)} + \frac{D_\perp}{k_B T} (\boldsymbol{1} - \boldsymbol{u}_{(2)}) \right). \tag{11.13}$$

For rodlike particles with large axis ratio $r = a/b$, where a and b are the semiaxes, $D_\parallel \approx 2D_\perp \approx k_B T \ln(a/b)/(2\pi \eta_s a)$ with solvent viscosity η_s. For moderate axes ratios, the affine transformation model [421] predicts $D_\parallel = r^{-2/3} D_{\text{sp}}$ and $D_\perp = r^{4/3} D_{\text{sp}}$ with D_{sp} being the corresponding diffusion coefficient of spherical particles. This paricular choice for $\boldsymbol{D}$ has been used, e.g. in [422] to derive anisotropic diffusion coefficients in nematic liquid crystals and in ferrofluids, and in [311] to express Frank elasticity coefficients for nematic liquid crystals in terms of order parameters.

11.2.2 Flow Field

For the case of macroscopic, homogeneous flow, $\boldsymbol{v}(\boldsymbol{r}) = \boldsymbol{\kappa} \cdot \boldsymbol{r}$, and $\boldsymbol{\kappa} = (\nabla \boldsymbol{v})^T$ ($\boldsymbol{\kappa}$ may depend on time but not on position), we decompose $\boldsymbol{\kappa} = \boldsymbol{\gamma} + \boldsymbol{\Omega}$ into symmetric and antisymmetric parts, with $\boldsymbol{\Omega} = (\boldsymbol{\kappa} - \boldsymbol{\kappa}^T)/2$ and $\boldsymbol{\gamma} = (\boldsymbol{\kappa} + \boldsymbol{\kappa}^T)/2$. Hence the rotational part of velocity motion $\dot{\boldsymbol{r}}_f \equiv \boldsymbol{\Omega} \cdot \boldsymbol{r} = \boldsymbol{\omega}_f \times \boldsymbol{r}$ with vorticity

$$\boldsymbol{\omega}_f \equiv (\nabla \times \boldsymbol{v})/2, \quad \boldsymbol{v}(\boldsymbol{r}) = \boldsymbol{\kappa} \cdot \boldsymbol{r}, \quad \boldsymbol{\kappa} = (\nabla \boldsymbol{v})^T \tag{11.14}$$

cannot be adsorbed by a potential $V[f]$. This angular velocity corresponds with the equation of deterministic (flow-induced) motion of the axis of an ellipsoid of revolution

$$\dot{\boldsymbol{u}} = B(\boldsymbol{1} - \boldsymbol{u}\boldsymbol{u}) \cdot \boldsymbol{\gamma} \cdot \boldsymbol{u} + \boldsymbol{\omega}_f \times \boldsymbol{u}, \tag{11.15}$$

where the first part[2] can be adsorbed by a potential of the form (11.17) below, and B is the shape factor of Sect. 7.2. This is obvious with the following decomposition into irreducible tensors

[2] The structure of the determinsitic part of motion, (11.15) is obvious from the following argument: A rod $\boldsymbol{Q}$ ($B = 1$) affinely following the flow field obeys $\dot{\boldsymbol{Q}} = \boldsymbol{\kappa} \cdot \boldsymbol{Q}$ (so called upper-convected motion), the axis $\boldsymbol{Q}$ of a disk ($B = -1$) obeys $\dot{\boldsymbol{Q}} = -\boldsymbol{\kappa}^T \cdot \boldsymbol{Q}$ (lower convected motion, thus both $\boldsymbol{Q}$'s remain perpendicular to each other, if they were once perpendicular), and we further correct the corresponding equation for the pseudoaffine $\boldsymbol{u}$, using $u^2 = 1$ for the unit vecctor $\boldsymbol{u}$, i.e., $\dot{\boldsymbol{u}} \cdot \boldsymbol{u} = 0$ leading to the term $(\boldsymbol{\gamma} : \boldsymbol{u}\boldsymbol{u})\boldsymbol{u}$ in (11.15).

$$(\mathbf{1}-\boldsymbol{u}_{(2)})\cdot\boldsymbol{\gamma}\cdot\boldsymbol{u} = \frac{3}{5}\boldsymbol{\gamma}\cdot\boldsymbol{u} - \boldsymbol{\gamma} : \boldsymbol{u}_{[3]} . \qquad (11.16)$$

Section 10.5 has been used.

11.2.3 Spatial Homogeneous Distribution, *N*th Order Potential

For anisotropic bulk fluids we are often concerned with the homogeneous orientational (single-link, or single particle) distribution $f(\boldsymbol{u},t)$.

A general *N*th order (orientation-dependent) potential functional, with yet unspecified (irreducible, without any restriction) tensors $\boldsymbol{T}_{[l]}$ of rank l is

$$\begin{aligned} V[f] &= k_{\mathrm{B}}T \sum_{l=1}^{N} \boldsymbol{T}_{[l]} \odot^{l} \int \boldsymbol{u}_{[l]} f(\boldsymbol{u}) \mathrm{d}^2 u \\ &= k_{\mathrm{B}}T \sum_{l=1}^{N} \boldsymbol{T}_{[l]} \odot^{l} \boldsymbol{a}_{[l]} , \end{aligned} \qquad (11.17)$$

such that

$$\frac{\delta V[f]}{\delta f(\boldsymbol{u})} = k_{\mathrm{B}}T \sum_{l=1}^{N} \boldsymbol{T}_{[l]} \odot^{l} \boldsymbol{u}_{[l]} . \qquad (11.18)$$

Inserting (11.18) into the homogeneous part of (11.11) yields the Fokker–Planck equation for the orientational distribution function $f(\boldsymbol{u},t)$ with coefficients $\boldsymbol{T}_{[l]}$ characterizing the *N*th order potential,

$$\frac{\partial f}{\partial t} = \sum_{l=1}^{N} \mathcal{L}\cdot(f\mathcal{D}\cdot\mathcal{L}(\boldsymbol{T}_{[l]}\odot^{l}\boldsymbol{u}_{[l]})) + (\mathcal{L}\cdot\mathcal{D} - \boldsymbol{\omega}_f)\cdot\mathcal{L}f . \qquad (11.19)$$

Since $\boldsymbol{T}_{[l]}$ does not depend on orientation $\boldsymbol{u}$ of the particles, we use (10.35), to obtain

$$\mathcal{L}(\boldsymbol{T}_{[l]}\odot^{l}\boldsymbol{u}_{[l]}) = -(2l+1)\boldsymbol{u}\times\boldsymbol{u}_{[l+1]}\odot^{l}\boldsymbol{T}_{[l]} = -l\,\boldsymbol{u}_{[l]}\odot^{l}\boldsymbol{\Delta}^{(l,1,l)}\odot^{l}\boldsymbol{T}_{[l]} . \qquad (11.20)$$

Now assuming isotropic and orientation-independent diffusiion, $\mathcal{D} = \mathcal{D}\mathbf{1}$, we simplify (11.19) as follows

$$\frac{\partial f}{\partial t} = -\mathcal{D}\sum_{l=1}^{N} l\mathcal{L}\cdot(f\boldsymbol{u}_{[l]}\odot^{l}\boldsymbol{\Delta}^{(l,1,l)})\odot^{l}\boldsymbol{T}_{[l]} + \mathcal{D}\mathcal{L}^2 f - \boldsymbol{\omega}_f\cdot\mathcal{L}f , \qquad (11.21)$$

where the properties $\mathcal{L}(\boldsymbol{T}_{[l]}) = \mathbf{0}$, $\mathcal{L}(\boldsymbol{\Delta}^{(l,1,l)}) = \mathbf{0}$, cf. Sect. 10.5 have been used to simplify the expression.

Multiplication of the Fokker–Planck equation by $\boldsymbol{u}_{[l]}$ and subsequent integration over the unit sphere (averaging) yields coupled equations of change for the infinitely many moments of f; alignent tensors $\boldsymbol{a}_{[l]} = \langle \boldsymbol{u}_{[l]} \rangle$. Many of them have been used in this monograph. General analytic considerations are offered in Sect. 11.3. Chapter 10 provides all necessary equations to derive the coupled equations easily with the help of a symbolic programming language.

11.2.4 Examples for Potentials and Applications

- Mean-field potentials:

$$\boldsymbol{T}_{[l]} \propto \boldsymbol{a}_{[l]} \tag{11.22}$$

used to describe the physics and phase transitions in liquid crystals, polymeric liquid crystals [74,94,179,216,313,422], FENE-PM dumbbels (Sdc. 2.1), anisotropic fluids (Landau-de-Gennes potentials), cf. Chap. 7 for applications. For the case of strict uniaxial symmetry, even the most general mean-field potential of the form (11.22) is equivalent with a mean-field potential of the order $l = 3$, i.e., Landau-de-Gennes type, in view of the Caley-Hamilton theorem, cf. Sect. 10.7 and the equation of motion for moments, cf. (11.26).
- Magnetic (tensor) field, such as for liquid crystals (due to head-tail symmetry, the liquid crystal does not couple to the vector field $\boldsymbol{H}$),

$$\boldsymbol{T}_{[2]} = -\frac{1}{2}\chi_a \overline{\boldsymbol{H}\boldsymbol{H}}\,, \tag{11.23}$$

- Magnetic (vector) field, such as for ferrofluid and magnetorheological fluids (the magnetic moment is a vector and thus couples also to the magnetic field vector):

$$\boldsymbol{T}_{[1]} = -\boldsymbol{h} = -\mu \boldsymbol{H}/k_{\mathrm{B}}T\,, \tag{11.24}$$

with $\boldsymbol{h} = h\boldsymbol{H}/H$, $H = |\boldsymbol{H}|$, Langevin parameter $h = \mu H/k_{\mathrm{B}}T$, magnetic moment μ, external magnetic field $\boldsymbol{H}$, cf. Sect. 7.5 for applications.
- Flow (2nd rank tensor) field

$$\boldsymbol{T}_{[2]} = -\frac{1}{2\mathcal{D}} B\, \overline{\nabla \boldsymbol{v}} = -\tau B\, \overline{\nabla \boldsymbol{v}} \tag{11.25}$$

with shape coefficient $B = (1 - r_p^2)/(1 + r_p^2)$ for ellipsoids of revolution, cf. Sect. 7.2, isotropic orientational diffusion coefficient $\mathcal{D}$, and orientational relaxation time $\tau = 1/(2\mathcal{D})$, cf. Chaps. 6 and 7 for applications.

11.3 Coupled Equations of Change for Alignment Tensors

Let us consider as an application the implications of the quite general Fokker–Planck equation (11.19) for the orientational distribution function in case of isotropic diffusion matrix $\mathcal{D} = \mathcal{D}\mathbf{1}$ to simplify notation. Multiplication of (11.19) – with $\boldsymbol{T}_{[l]}$ characterizing the potential – by $\boldsymbol{u}_{[l]}$ and subsequent integration over the unit sphere (averaging) yields coupled equations of change for the infinitely many moments of f; alignent tensors $\boldsymbol{a}_{[l]} = \langle \boldsymbol{u}_{[l]} \rangle$. By applying rules of Chapt. 10 and some further manipulation we see, that the equation of change for the nth moment $\boldsymbol{a}_{[n]}$ in the presence of a potential (11.18) (defining $\boldsymbol{T}_{[l]}$) and vorticity $\boldsymbol{\omega}_f$ (11.14) can be cast into the following form (which is one of the main results of this section),

$$\frac{\partial}{\partial t}\boldsymbol{a}_{[n]} = n\, \overline{\boldsymbol{\omega}_f \times \boldsymbol{a}_{[n]}} - \mathcal{D}n(n+1)\boldsymbol{a}_{[n]} + \mathcal{D}\sum_{l=1}^{N} l \sum_{j=0}^{l} \alpha_{lj}^{n}\, \overline{\boldsymbol{T}_{[l]} \odot^{j} \boldsymbol{a}_{[n+2j-l]}}\,, \tag{11.26}$$

with coefficients α^n_{lj} nonzero only for $n > 0, l = 1,..,N$ and $j = 0,1,..l$, and the convention $\odot^j \boldsymbol{a}_{[0]} = 1$. We derived the result (11.26) by performing a partial integration for all terms on the right hand side of the integrated (11.19). Further, (10.36) and (10.24) have been used to derive the first two terms on the right hand side of (11.26). The term exhibiting the still unspecified coefficients α^n_{lj} in (11.26) covers all possible combinations to couple the tensor $\boldsymbol{T}_{[l]}$ with alignment tensors while producing an irreducible tensor of rank n. This requirement is obvious, since the left hand side of (11.26) is an irreducible tensor of rank n.[3] Expressions for the coefficients are obtained using the more explicit representation for the equation of change, derived using methods of Chapt. 10, which leads to the following linear system of equations for coefficients α^n_{lj}:

$$
\begin{aligned}
\forall_{n,l} \quad & \sum_{j=0}^{l} \alpha^n_{lj} \overline{\boldsymbol{T}_{[l]} \odot^j \boldsymbol{a}_{[n+2j-l]}} \\
&= -\left\langle \mathcal{L}(\boldsymbol{T}_{[l]} \odot^l \boldsymbol{u}_{[l]}) \cdot \mathcal{L}\boldsymbol{u}_{[n]} \right\rangle && (11.28) \\
&= -\frac{1}{k_B T} \left\langle \mathcal{L}\left(\frac{\delta V[f]}{\delta f(\boldsymbol{u})}\right) \cdot \mathcal{L}\boldsymbol{u}_{[n]} \right\rangle && (11.29) \\
&= -\frac{1}{k_B T} \left\langle \left(\frac{\partial}{\partial \boldsymbol{u}} \frac{\delta V[f]}{f(\boldsymbol{u})}\right) \cdot \frac{\partial}{\partial \boldsymbol{u}} \boldsymbol{u}_{[n]} \right\rangle && (11.30) \\
&= -(2n+1)(2l+1) \left\langle \left((\boldsymbol{u} \times \boldsymbol{u}_{[l+1]}) \odot^l \boldsymbol{T}_{[l]}\right) \cdot (\boldsymbol{u} \times \boldsymbol{u}_{[n+1]}) \right\rangle , && (11.31)
\end{aligned}
$$

which can be rewritten in components using the $\boldsymbol{\Delta}^{(l,1,l)}$-operator (10.18). The three interior lines in (11.31) are just identities which help to identify identical structures from equations in the literature.[4] For the cases discussed in this monography, which include potentials of the form (11.18) with $N = 1$ (magnetic vector field $\boldsymbol{T}_{[1]}$) and

[3] In (7.34) we obtained a structure similar to the one in (11.26). By comparing these equations we see, that the derivative of the Landau-de Gennes potential with respect to the alignment tensor of rank 2, $\boldsymbol{\Phi}$, has to be identified as follows

$$
\begin{aligned}
\boldsymbol{\Phi} &= \vartheta \boldsymbol{a}_{[2]} - 3\sqrt{6}\, \overline{\boldsymbol{a}_{[2]}{}^2} + 2a^2 \boldsymbol{a}_{[2]} \\
&\leftrightarrow D\tau_a (6\boldsymbol{a}_{[2]} - 2\alpha^2_{20} \boldsymbol{T}_{[2]} - 2\alpha^2_{21} \overline{\boldsymbol{T}_{[2]} \cdot \boldsymbol{a}_{[2]}} - 2\alpha^2_{22} \overline{\boldsymbol{T}_{[2]} : \boldsymbol{a}_{[4]}}) . \qquad (11.27)
\end{aligned}
$$

Therefore, $\boldsymbol{T}_{[2]} \propto \boldsymbol{a}_{[2]}$ and the fourth rank alignment tensor is virtually replaced by a closure relationship to arrive at the Landau-de Gennes potential (which prevents a coupling between 2nd rank and 3th rank alignment tensors in the moment equation (7.34).

[4] For example, for $n = 2$ (equation of change for the 2nd moment) and $l = 2$ (anisotropic flow gradient), we can use the identity

$$
\left(\frac{\partial}{\partial \boldsymbol{u}} U\right) \cdot \frac{\partial}{\partial \boldsymbol{u}} \boldsymbol{a}_{[2]} = \left(\frac{\partial}{\partial \boldsymbol{u}} U\right) \cdot (-2\boldsymbol{uuu} + \mathbf{1}\boldsymbol{u} + \boldsymbol{i}\boldsymbol{u}\boldsymbol{i}) = \left(\frac{\partial}{\partial \boldsymbol{u}} U\right) \boldsymbol{u} + \boldsymbol{u} \frac{\partial}{\partial \boldsymbol{u}} U , \qquad (11.32)
$$

valid for arbitrary scalar function U to arrive at the representation used in (7.9).

$N = 2$ (deformation rate tensor, alignment tensor $\boldsymbol{T}_{[2]}$), cf. Sect. 11.2.4, the solutions of (11.31) are, after some calculation,

$$
\begin{aligned}
\alpha_{10}^n &= \frac{n+1}{2n+1}\,, \\
\alpha_{11}^n &= -1\,, \\
\alpha_{20}^n &= \frac{n^2-1}{4n^2-1}\,, \\
\alpha_{21}^n &= \frac{6}{2n+3}\,, \\
\alpha_{22}^n &= -2\,.
\end{aligned}
\tag{11.33}
$$

At this point the reader may wish to convince her/himself, that the special case $n = 1$ and $l = 1,2$, by inserting (11.33) into (11.26) returns the equation of change for the magnetization of ferrofluids (7.32). The special cases $n = 2,4,6$ and $l = 2$ are the most relevant for polymer melts subjected to flow.

11.3.1 Dynamical Closures

With the form (11.26) at hand, it is easy to implement closure relationships which are simultaneously correct close to equilibrium, and in the fully oriented state. We need just to set $\boldsymbol{a}_{[k]} = \boldsymbol{0}$ for a certain k to obtain a systematic and 'admissible' set of closures of order k. An example for $k = 6$ is given in Sect. 6.3.

While the Fokker–Planck equation is a partial differential equation, which may be solved in different ways, including discrete methods, expansion methods (see Sect. 6.3 for a worked out example), finite element methods etc. it is important to know about the alternative way to look at these equations as being equivalent with stochastic differential, so called Langevin equations to be presented in the next section. The current section offered coupled equations of change for alignment tensors, and strategies to close the system.

11.3.2 Equations of Change for Order Parameters

Assuming uniaxial symmetry of the alignment tensor, the simplest way to write down equations of change for the order parameters is to use the representation (10.41), which implies, for all $n \geq 0$,

$$
S_n \dot{S}_n = \frac{1}{2}\frac{d}{dt}(S_n^2) = \frac{(2n-1)!!}{n!}\, \boldsymbol{a}_{[n]} \odot^n \frac{\partial}{\partial t}\boldsymbol{a}_{[n]}\,. \tag{11.34}
$$

We then replace $(\partial/\partial t)\boldsymbol{a}_{[n]}$ by the expression given in (11.26) and perform the n-fold contraction. To this end, we make use the trivial identity

$$
\forall_{n,k}\; \overline{\boldsymbol{n}_{[n]}\boldsymbol{n}_{[k]}} = \boldsymbol{n}_{[n+k]} \tag{11.35}
$$

and of our formula (10.29) upon replacing $\boldsymbol{u}_{(l)}$ by the director $\boldsymbol{n}_{(l)}$,[5] which becomes

$$\boldsymbol{n}_{[l]} \odot^l \boldsymbol{n}_{[l+k]} = \frac{(l+k)!(2k-1)!!}{k!(2l+2k-1)!!}\,\boldsymbol{n}_{[k]}\,. \tag{11.36}$$

With the help of (11.35) and (11.36), (11.26) we evaluate (11.34) as

$$\begin{aligned}\dot{S}_n &= \frac{(2n-1)!!}{n!}\Bigg(nS_n\,\overline{\boldsymbol{\omega}_f \times \boldsymbol{n}_{[n]}} \odot^n \boldsymbol{n}_{[n]} - \mathcal{D}n(n+1)S_n\boldsymbol{n}_{[n]} \odot^n \boldsymbol{n}_{[n]}\\ &\quad + \mathcal{D}\sum_{l=1}^{N} l \sum_{j=0}^{l} \alpha^n_{lj} S_{n+2j-l}\,\overline{\boldsymbol{T}_{[l]} \odot^j \boldsymbol{n}_{[n+2j-l]}} \odot \boldsymbol{n}_{[n]}\Bigg)\\ &= -\mathcal{D}n(n+1)S_n +\\ &\quad \frac{1}{n^2_{[n]}}\left(\mathcal{D}\sum_{l=1}^{N} l \sum_{j=0}^{l} \alpha^n_{lj} S_{n+2j-l}\,\overline{\boldsymbol{T}_{[l]} \odot^j \boldsymbol{n}_{[n+2j-l]}} \odot \boldsymbol{n}_{[n]}\right),\end{aligned} \tag{11.37}$$

with the coefficients α^n_{lj} given by (11.33), generally defined through (11.31), and we have used the identity[6]

$$\overline{\boldsymbol{\omega}_f \times \boldsymbol{n}_{[n]}} \odot^n \boldsymbol{n}_{[n]} = 0 \tag{11.39}$$

and the abbreviation

$$n^2_{[n]} \equiv \boldsymbol{n}_{[n]} \odot^n \boldsymbol{n}_{[n]} = \frac{n!}{(2n-1)!!}\,, \tag{11.40}$$

a relationship derived earlier in this monograph. Equation (11.37) is the main result of this section, the equation of change for nematic order parameters $S_1, S_2, \ldots$ for the Fokker–Planck equation (11.19) with a potential of the form (11.18).

For the case $l = 1$ ('ferrofluid'), we use the identities

$$\overline{\boldsymbol{T}_{[1]}\boldsymbol{n}_{[n-1]}} \odot^n \boldsymbol{n}_{[n]} = n^2_{[n]}(\boldsymbol{T}\cdot\boldsymbol{n})\,, \tag{11.41}$$

$$\overline{\boldsymbol{T}_{[1]}\cdot\boldsymbol{n}_{[n+1]}} \odot^n \boldsymbol{n}_{[n]} = n^2_{[n+1]}(\boldsymbol{T}\cdot\boldsymbol{n})\,, \tag{11.42}$$

[5] Notice, that we can replace the unit vector $\boldsymbol{u}$ by $\boldsymbol{n}$ in this formula, and $\boldsymbol{u}_{(l)}$ by $\boldsymbol{n}_{(l)}$, but we cannot replace $\boldsymbol{u}_{(l)}$ by $\boldsymbol{a}_{(l)}$ since the alignment tensor is an averaged quantity. For the case of uniaxial symmetry, this different transformation behavior between the average $\boldsymbol{a}_{[l]}$ and the dyadics $\boldsymbol{n}_{[l]}$ is captured by the order parameters.

[6] Some further identities to facilitate the comparison with equations presented in Part I of this monograph:

$$\overline{\boldsymbol{\gamma}\cdot\boldsymbol{n}_{[2]}} : \boldsymbol{n}_{[2]} = \frac{1}{3}\boldsymbol{n}_{[2]}\,,$$

$$\overline{\boldsymbol{\gamma}:\boldsymbol{n}_{[4]}} : \boldsymbol{n}_{[2]} = \frac{12}{35}\boldsymbol{n}_{[2]}\,, \tag{11.38}$$

with traceless, symmetric $\boldsymbol{\gamma}$.

to specialize the general result (11.37), and explicitly write down the equations of motion for all order parameters. Here, for $l = 1$, they couple to the director field via vorticity $\boldsymbol{\omega}_f$ and vector field $\boldsymbol{T}_{[1]}$. The general case $l > 1$ can be handled in straightforward manner.

11.4 Langevin Equation

Langevin equations can be regarded as alternative, equivalent approach to determine the distribution function governing the Fokker–Planck equation from the previous section. Both equations produce the same moments, and thus, the same distribution function. We won't go into detail here but refer the reader to [54, 368] for details on stochastic variables and Langevin equations, in general.

A Langevin equation for the stochastic variable $\boldsymbol{X}_t$ is often introduced as

$$\frac{d\boldsymbol{X}}{dt} = \boldsymbol{A}(\boldsymbol{X}) + \boldsymbol{B}\cdot\tilde{\boldsymbol{\eta}}, \quad \boldsymbol{B}\cdot\boldsymbol{B}^T \equiv \tilde{\boldsymbol{D}}, \tag{11.43}$$

with a deterministic part $\boldsymbol{A}$, plus some kind of stochastic (usually white) noise $\boldsymbol{\eta}$. Using the Itó interpretation for mixed moments of the type $\langle\tilde{\boldsymbol{\eta}}\boldsymbol{X}\rangle$ it has been shown that (11.43) is equivalent to the Fokker–Planck equation for the distribution $f(\boldsymbol{x},t)$ of the form

$$\frac{\partial}{\partial t} f(\boldsymbol{x},t) = -\nabla_{\boldsymbol{x}}\cdot(\boldsymbol{A}(\boldsymbol{x})f) + \frac{1}{2}\nabla_{\boldsymbol{x}}\cdot(\tilde{\boldsymbol{D}}\cdot\nabla_{\boldsymbol{x}} f(\boldsymbol{x},t)), \tag{11.44}$$

which immediately compares with (11.11) by identifying $\boldsymbol{A} = \dot{\boldsymbol{x}}_f - \boldsymbol{\xi}^{-1}\cdot\delta_f V[f]$ and $\tilde{\boldsymbol{D}} = 2\boldsymbol{D}$, with $\boldsymbol{x} = \boldsymbol{u}$ and $\boldsymbol{x} = \boldsymbol{r}$.

While macroscopic quantities $\langle Q\rangle$ (in particular, moments) are obtained from the solution $f(\boldsymbol{x},t)$ of (11.44) as

$$\langle Q(\boldsymbol{x})\rangle(t) = \int f(\boldsymbol{x},t)\,Q(\boldsymbol{x})\,d\boldsymbol{x}, \tag{11.45}$$

the Langevin equation can be solved using random numbers, cf. Sect. 9, and macroscopic quantities after obtained from time series. After N time steps, let $t = N\Delta t$, one estimates

$$\langle Q\rangle(t) = \frac{1}{N}\sum_{i=1}^{N} Q(\boldsymbol{X}(i\Delta t)), \tag{11.46}$$

with a statistical error

$$\sigma^2(t) = \frac{1}{N}[\langle Q^2\rangle - \langle Q\rangle^2] = \frac{1}{N^2}\sum_{i=1}^{N}[Q(\boldsymbol{X}(i\Delta t)) - \langle Q\rangle]^2. \tag{11.47}$$

11.4.1 Brownian Dynamics Simulation

Brownian dynamics simulation solves the Langevin equation (11.43) by discrete integration in time (time step $\triangle t$)

$$\begin{aligned} \triangle \boldsymbol{X}_t &= \boldsymbol{A}(\boldsymbol{X}_t, t)\triangle t + \triangle \boldsymbol{\Omega}\,, \\ \triangle \boldsymbol{\Omega} &= \boldsymbol{B}\cdot\boldsymbol{\eta}\,\sqrt{\triangle t}, \quad \text{white noise } \boldsymbol{\eta}\,, \end{aligned} \tag{11.48}$$

which ensures the required properties

$$\begin{aligned} \langle \triangle \mathbf{W} \rangle &= 0\,, \\ \langle (\triangle \boldsymbol{\Omega})^2 \rangle &= \tilde{\boldsymbol{D}}\triangle t\,. \end{aligned} \tag{11.49}$$

A sample brownian dynamics code (including nonequilibrium brownian dynamics) is given in Sect. 12.6.1.

If $\boldsymbol{B}$ is constant, and $\boldsymbol{A}$ linear in $\boldsymbol{x}$, there is no need to perform a simulation, the solution is analytically known [368].

12

Simple Simulation Algorithms and Sample Applications

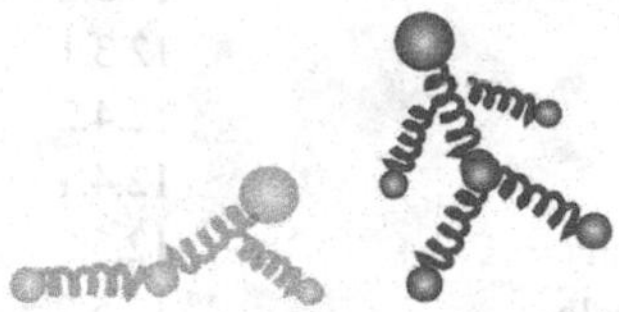

This section offers basic recipes and sample applications which allow the reader to immediately start his/her own simulation project on topics we dealt with in this book. Concerning molecular dynamics and Monte Carlo simulation there are, of course, several useful books already available which describe the 'art of simulation' [141, 156, 256] in an exhaustive way. The reason we print some simple codes is that we skipped algorithmic details in the foregoing chapters. Simulations are always performed using dimensionless numbers, and all dimensional quantities can be expressed in terms of reduced units, cf. Sect. 4.3 for conventional Lennard–Jones units. In this chapter, we concentrate on the necessary, and skip anything more sophisticated. Codes have been used in classrooms, they are obviously open for modifications and extensions, and offer not only an executable, but all necessary formulas for doing simulations in the correct (which is often essential) order. The overall spirit is as follows: codes are short, run without changes, demonstrate the main principle in a modular fashion, and are thus in particular open regarding efficiency issues and extensions. Algorithms are presented in the Matlab^{TM} language, which is mostly directly portable to programming languages like fortran, c, or Mathematica^{TM}. For an introduction we refer to [423]. Additional commands needed to visualize the results are given in the figure title for each application. Simulation codes, in a less modular fashion, are also available online at www.complexfluids.ethz.ch. Functions are shared over sections, for that reason we begin with an alphabetic list of all (non-builtin) functions in this chapter.

12.1 Index of Programs

Table 12.1 provides a list of functions contained in this chapter.

Table 12.1. These and further codes are also available at www.complexfluids.ethz.ch. Codes werewritten by the author of this book, and if distributed, should be supplemented by the following line which we skip here to save space:
% jan 2005 written by martin kroger, mk@mat.ethz.ch

Function	Section	Function	Section
all_interactions	12.4.1	coeff	12.6.2
all_interactions_shear	12.5.1	energy_flip	12.3.2
boundary_periodic	12.2.2	f	12.3.1
boundary_periodic_centered	12.2.2	force_FENE	12.4.2
boundary_periodic_pore	12.5.2	force_LJ	12.4.1
boundary_periodic_shear	12.5.1	force_LJ_wall	12.5.2
boundary_reflection	12.2.2	force_basiscell	12.5.2
brownian_propagate	12.6.1	forces	12.4.1
coarse_grain	12.7.1	forces_shear	12.5.1
code_MC_standard_1D	12.3.1	init_basiscell	12.5.2
code_MC_standard_1D_howto	12.3.1	random_vector_2D	12.2.1
code_NEMD	12.5.1	random_vector_3D	12.2.1
code_NEMD_howto	12.5.1	random_vector_howto	12.2.1
code_brownian_dynamics_howto	12.6.1	random_walk_2D	12.2.1
code_chebyshev	12.6.2	random_walk_3D	12.2.1
code_chebyshev_howto	12.6.2	temperature_control	12.4.1
code_coarse_grain_howto	12.7.1	temperature_control_pore	12.5.2
code_equilibrium_FENE	12.4.2	temperature_control_shear	12.5.1
code_equilibrium_FENE_howto	12.4.2	useful_initial_configuration	12.2.3
code_flow_through_pore	12.5.2	useful_initial_configuration_pore	12.5.2
code_flow_through_pore_howto	12.5.2	velocity_verlet	12.4.1
code_ising_2D	12.3.2	velocity_verlet_FENE	12.4.2
code_ising_2D_howto	12.3.2	velocity_verlet_pore	12.5.2
code_molecular_dynamics_howto	12.4.1	velocity_verlet_shear	12.5.1
coeff	12.6.2	visualize_particles	12.2.4

12.2 Recipes

12.2.1 Random Vectors, Random Paths (2D, 3D)

Create isotropically distributed vectors of given norm (length) in 2 and 3 dimensions. Such vectors are needed to generate initial, polymeric (random walk) configurations.

main routine

```
function random_vector_3D= random_vector_3D (veclength)    3D random vector of length veclength
    znorm=1;
    while znorm >= 1, z=1-2*rand(1,2); znorm=sum(z.*z); end;    accept suitable
    random_vector_3D=veclength*[2*z*sqrt(1-znorm) 1-2*znorm];    scale unit vector
function random_vector_2D= random_vector_2D (veclength)    2D random vector
    phi=rand*2*pi;    random angle
    random_vector_2D=veclength*[cos(phi) sin(phi)];    scale unit vector
function x= random_walk_3D (N,xstart)    3D random walk with N nodes starting at xstart
    x=zeros(N,3); x(1,:)=xstart; for i=2:N, x(i,:)=x(i-1,:)+random_vector_3D(1); end;
function x= random_walk_2D (N,xstart)    2D random walk with N nodes starting at xstart
    x=zeros(N,2); x(1,:)=xstart; for i=2:N, x(i,:)=x(i-1,:)+random_vector_2D(1); end;
```

sample application, embedding & testing

```
function random_vector_howto (N)
    uu=zeros(3,3); for i=1:N, U=random_vector_3D(1); UU=U'*U; uu=uu+UU; end;    3D case
    3*uu/N,    average over N realizations (3x3 unity matrix)
    uu=zeros(2,2); for i=1:N, U=random_vector_2D(1); UU=U'*U; uu=uu+UU; end;    2D case
    2*uu/N,    aberage over N realizations (2x2 unity matrix)
```

12.2.2 Periodic and Reflecting Boundary Conditions (*n*D)

Transformation rules for particles reaching or crossing boundaries of the simulation cell. The same rules are used to employ the nearest image convention.

main routine

```
function x= boundary_periodic (x,xmin,xmax)    peridic bounds:
    L = xmax-xmin; x = x-round((x-xmin)./L-0.5).*L;    box size (L), xmax+eps -> xmin+eps etc.

function x= boundary_periodic_centered (x,L)    periodic bounds:
    x = x-round(x./L).*L;    if box centered at origin, L/2+eps -> -L/2+eps etc.
function x= boundary_reflection (x,xmin,xmax)    reflecting bounds:
    dummy = min(xmax,max(xmin,x)); x = 2*dummy-x;    xmax+eps -> xmax-eps etc.
```

sample application, embedding & testing

see code_ising_2D

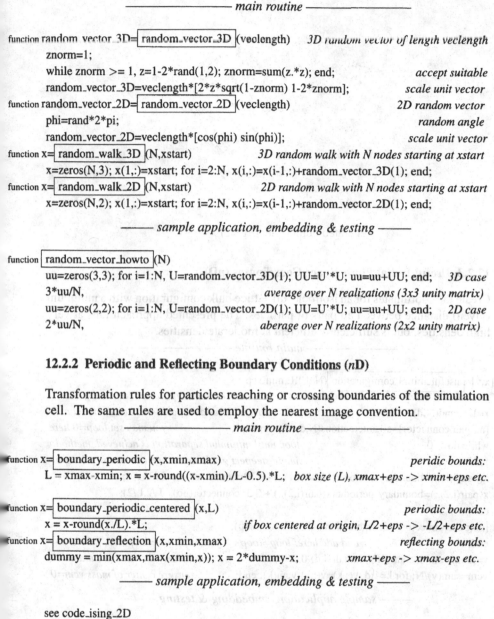

```
for i=1:1000; U=random_vector_3D(1); plot3(U(1),U(2),U(3)); hold on; end;
```

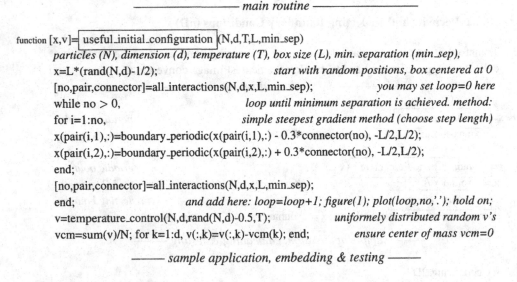

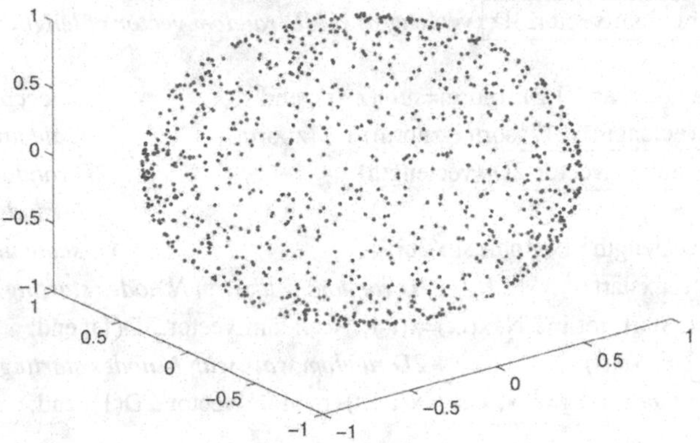

generated output (Sect. 12.2)

12.2.3 Useful Initial Phase Space Coordinates (*n*D)

Create the coordinates for an isotropic, off-lattice bulk configuration with a minimum (given) distance between all pairs of particles. The presented approach fails at very high densities, but is sufficiently efficient at moderate densities.

——————— *main routine* ———————

```
function [x,v]= useful_initial_configuration (N,d,T,L,min_sep)
particles (N), dimension (d), temperature (T), box size (L), min. separation (min_sep),
x=L*(rand(N,d)-1/2);                        start with random positions, box centered at 0
[no,pair,connector]=all_interactions(N,d,x,L,min_sep);        you may set loop=0 here
while no > 0,                          loop until minimum separation is achieved. method:
for i=1:no,                         simple steepest gradient method (choose step length)
x(pair(i,1),:)=boundary_periodic(x(pair(i,1),:) - 0.3*connector(no), -L/2,L/2);
x(pair(i,2),:)=boundary_periodic(x(pair(i,2),:) + 0.3*connector(no), -L/2,L/2);
end;
[no,pair,connector]=all_interactions(N,d,x,L,min_sep);
end;                  and add here: loop=loop+1; figure(1); plot(loop,no,'.'); hold on;
v=temperature_control(N,d,rand(N,d)-0.5,T);          uniformely distributed random v's
vcm=sum(v)/N; for k=1:d, v(:,k)=v(:,k)-vcm(k); end;      ensure center of mass vcm=0
```

—— *sample application, embedding & testing* ——

see molecular dynamics

12.2.4 Visualization, Animation & Movies (*n*D)

Visualize N points (1D), circles (2D), spheres (3D) with given radius at given positions. Visualize a single unfolded chain in 2D and 3D as thick path. Avi movies can be generated in Matlab™ via M=avifile('file.avi'), .., figure(1), F=getframe; M=addframe(M,F), ... figure(1), M=addframe(M,F), .. and M=close(M). See www.complexfluids.ethz.ch for worked out examples.

——— *main routine* ———

```
function [visualize_particles](N,d,x,L,resolution,r,mytitle)  visualize N spheres, radius r, at positions x
[X,Y,Z]=sphere(resolution); figure(1); hold off;     in 1,2,3D. parameters: radius (r), box size (L)
if d==3, for i=1:N, surf(X*r+x(i,1),Y*r+x(i,2),Z*r+x(i,3)); hold on; end; end;
if d==2, for i=1:N, surf(X*r+x(i,1),Y*r+x(i,2),0*X); view(2); hold on; end; end;
if d==1, for i=1:N, surf(X*r+x(i,1),Y*r ,0*X); view(2); hold on; end; end;
axis([-L L -L L -L L]/2); title(mytitle); pause(0.001);     customize axes and title, pause
```

——— *sample application, embedding & testing* ———

The visualization routine can be added within most of the codes of this chapter.
Create animations/movies by using getframe and frame2avi matlab commands.

12.3 Monte Carlo

12.3.1 Standard Monte Carlo Integration (nD)

Evaluate the integral $\int f(x)\,dx$ by using (pseudo) random numbers, cf. 9.2.1. The code is immediately adopted to compute high-dimensional integrals for which Monte Carlo methods (or quasi Monte Carlo methods using quasi random numbers) are the method of choice. The error scales with the number of N realizations for random numbers as $N^{-1/2}$ independent of the dimensionality of the integral.

——— *main routine* ———

```
function [I,E]=[code_MC_standard_1D](N,a,b)          integral I, error estimate E
X =a + (b-a)*rand(1,N);                             N random X values in [a,b]
F =f(X);                                            corresponding f values
I =(b-a) /N*sum(F);                                 Integral estimate
I2=(b-a)^2/N*sum(F.*F); E =sqrt((I2-I^2)/N);        Error estimate
This 1D example is immediately adopted for high-dimensional integration
```

——— *sample application, embedding & testing* ———

```
function [code_MC_standard_1D_howto](N,a,b)       N random shots, integral bounds [a,b]
N=10, a=0, b=1, [Integral,Error]=code_MC_standard_1D(N,a,b),      (π)
function f=[f](x)                                  specifies integrand
f=4*sqrt(1-x.^2);                sample integrand -> ∫ f(x) dx = π (a=0 & b=1)
```

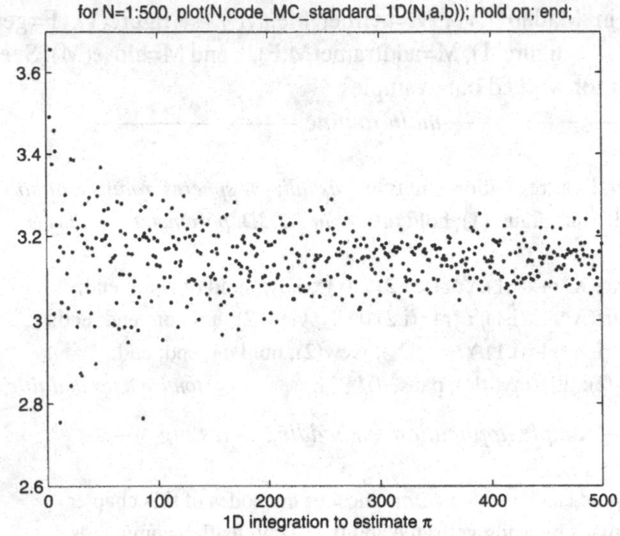

generated output (Sect. 12.3)

12.3.2 Ising Model via Metropolis Monte Carlo (2D)

Numerically solve the Ising model with Hamiltonian $H = -B\sum_i x_i - \frac{J}{2}\sum_{\langle i,j\rangle} x_i x_j$, spins $x_i \in \{-1,+1\}$ on a two-dimensional ($L \times L$) lattice, for which the exact analytic (Onsager) solution is known for $L \to \infty$. In particular, the phase transition isotropic - magnetic occurs at the critical, inverse dimensionless temperature $\beta J = 0.4406868$. Implement the Metropolis method of Chap. 9.1

——— *main routine* ———

```
function code_ising_2D (beta,N);                                    Metropolis scheme
    global L x;                                                     shared, 'global' variables
    RANDOMSITE = randint(N,2,[1 L]);                                cpu efficient, but
    RANDOMNUMBER = rand(N,1);                                       memory consuming
    for i=1:N, site=RANDOMSITE(i,:);                                choose site randomly
    if RANDOMNUMBER(i) < exp(-beta*energy_flip(site(1),site(2))),
    x(site(1),site(2))=-x(site(1),site(2)); end;                    flip site
    end;
function energy_flip= energy_flip (sx,sy);
    global J H x; sum_n = xs(sx+1,sy)+xs(sx-1,sy)+xs(sx,sy+1)+xs(sx,sy-1);  sum over neighbors
    energy_flip = 2*x(sx,sy)*(H+J*sum_n);                           Ising Hamiltonian
function xs= xs (sx,sy); global L x;                                periodic boundary conditions
    site = boundary_periodic([sx sy],1,L);                          a routine of this section
    xs = x(site(1),site(2));                                        spin at corrected coordinate (site)
```

——— *sample application, embedding & testing* ———

```
function code_ising_2D_howto (L,J,H,beta);        lattice size L (30), coupling coefficient J (0.43)
```

```
global L J H x;                                    magnetic field H (0), 1/kT (1)
x=2*round(rand(L,L))-1;                  initial random LxL lattice x=-1 or x=+1
N=3000; code_ising_2D(beta,N);                      perform N Monte Carlo steps
M=mean(x(:)),                                         calculate magnetization <x>
```

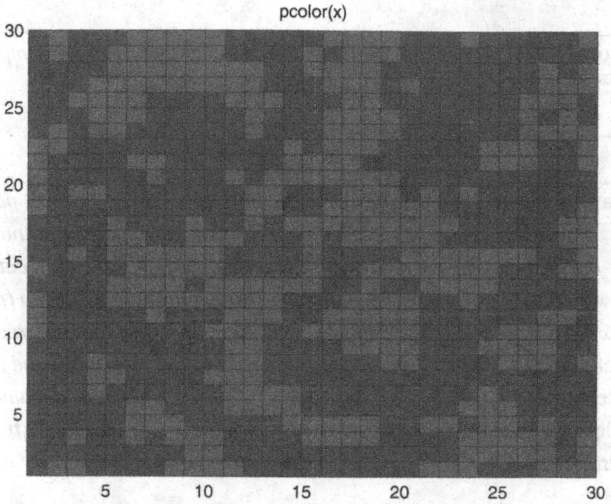

generated output (Sect. 12.3)

12.4 Molecular Dynamics

12.4.1 Molecular Dynamics of a Lennard–Jones System (*n*D)

Calculate phase space trajectories for a system of mass points interacting via given two-body interaction forces (Lennard–Jones, for example). Employ the velocity Verlet algorithm to integrate the equations of motion and a introduce a thermostat, cf. Sect. 4.1.

main routine

```
function [x,v,F]= velocity_verlet (N,d,x,v,F,dt,L,cutoff)      velocity Verlet integrator
x = boundary_periodic(x + dt*v + dt^2/2*F,-L/2,L/2);    new x with boundary conditions
v = v + dt/2*F; F = forces(N,d,x,L,cutoff); v = v + dt/2*F   new v, new forces, new v
function [ip,pair,connector]= all_interactions (N,d,x,L,cutoff)  obtain interacting pairs
ip=0; connector=zeros(1,d); pair=zeros(1,2);
for i=1:N-1, for j=i+1:N,
distance = boundary_periodic(x(j,:)-x(i,:),-L/2,L/2);    (true) connecting vector x_j-x_i
if norm(distance) < cutoff,     only interacting pairs (cutoff+shell for neighbor lists)
ip = ip + 1;                                               interaction pair counter
pair(ip,:) = [i j];                   particle numbers (i,j) belonging to pair (ip)
connector(ip,:) = distance; end;           connecting vector x_j - x_i for pair (i,j)
```

```
        end; end;                                                         end both 'for' loops
function F= forces (N,d,x,L,cutoff)                  clear forces F, then calculate them using ..
        F=zeros(N,d); [no,pair,connector]=all_interactions(N,d,x,L,cutoff);      interacting pairs
        for i=1:no, FORCE=force_LJ(connector(i,:));
        F(pair(i,1),:)=F(pair(i,1),:)-FORCE; F(pair(i,2),:)=F(pair(i,2),:)+FORCE;    actio=reactio
        end;
function v= temperature_control (N,d,v,T)      rescaling velocities according 'wanted' temperature
        T_measured=sum(v(:).^2)/(d*N); v=v*sqrt(T/T_measured);
```

——— *sample application, embedding & testing* ———

```
function code_molecular_dynamics_howto (N,d,n,T,dt,MDsteps,cutoff,min_sep)     N (10) particles,
                dimension d (1-3), particle number density n (0.5), temperature T (1),
              integration time step dt (0.005), time steps MDsteps (2000), cutoff (2.5)
        L=(N/n)^(1/d); min_sep =0.85;   box size (L), minimum pair separation at startup (min_sep)
        [x,v] =useful_initial_configuration(N,d,T,L,min_sep);                 init. coordinates x,
        F =forces(N,d,x,L,cutoff);                                       velocities v and forces F
        for MDstep=1:MDsteps,                                              molecular dynamics loop
        [x,v,F]=velocity_verlet(N,d,x,v,F,dt,L,cutoff);                         propagate trajectory
        v =temperature_control(N,d,v,T);                                         velocity rescaling
        end;
function force_LJ= force_LJ (r_vector); r=norm(r_vector);                           two-body force
        force_LJ = 24*(2*r.^(-14)-r^(-8)) * r_vector;                          here: Lennard–Jones
```

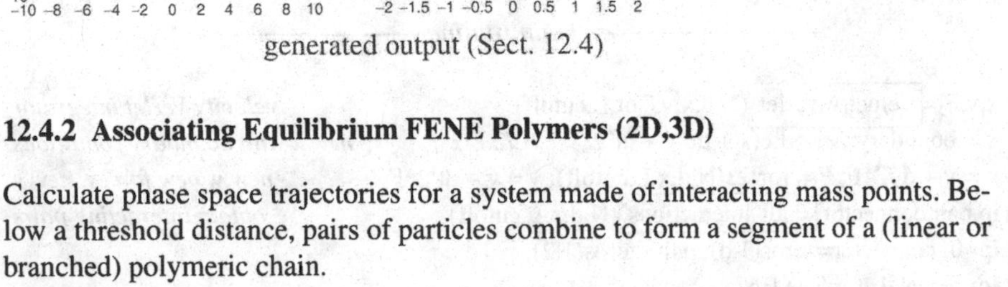

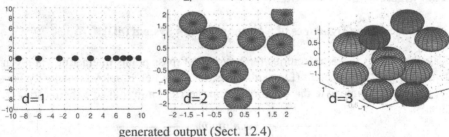

generated output (Sect. 12.4)

12.4.2 Associating Equilibrium FENE Polymers (2D,3D)

Calculate phase space trajectories for a system made of interacting mass points. Below a threshold distance, pairs of particles combine to form a segment of a (linear or branched) polymeric chain.

——— *main routine* ———

```
function code_equilibrium_FENE (N,d,x,v,L,cutoff,MDsteps,dt,T,Qmax_FENE,k_FENE)
        total number of beads (N), space dimension (d), phase space coordinates (x,v),
        box size (L), potential cutoff (cutoff), MD steps (steps), time step (dt), temperature
        (T), FENE spring coefficient (k_FENE), FENE max. extension (Qmax_FENE)
```

```
global QmaxFENE kFENE bonds
QmaxFENE=Qmax_FENE; kFENE=k_FENE; time=0; F=forces(N,d,x,L,cutoff);
for MDstep=1:MDsteps, time=time+dt;                                     molecular dynamics loop
[x,v,F]=velocity_verlet_FENE(N,d,x,v,F,dt,L,cutoff);                    propagate trajectory
v =temperature_control(N,d,v,T);                                        velocity rescaling
end;

function force_FENE= force_FENE (r_vector)                              FENE force
global QmaxFENE kFENE
r2=sum(r_vector.^2); force_FENE = kFENE*r_vector/(1-r2/QmaxFENE^2);

function [x,v,F]= velocity_verlet_FENE (N,d,x,v,F,dt,L,cutoff)          velocity Verlet integrator
global pol QmaxFENE kFENE
x = boundary_periodic(x + dt*v + dt^2/2*F,-L/2,L/2);                    new x
v = v + dt/2*F;                                                         new v
F = forces(N,d,x,L,cutoff);                                             LJ forces
for i=1:N,                                  FENE forces between neighbors along pol backbone
j=pol(i,1); if j>0, FF=force_FENE(pol(i,3:5)); F(i,:)=F(i,:)+FF; F(j,:)=F(j,:)-FF; end;
j=pol(i,2); if j>0, FF=force_FENE(pol(i,6:8)); F(i,:)=F(i,:)+FF; F(j,:)=F(j,:)-FF; end;
end;
v = v + dt/2*F;                                                         new v

function [ip,pair,connector]= all_interactions (N,d,x,L,cutoff)         obtain interacting pairs
global pol QmaxFENE bonds
ip=0; connector=zeros(1,d); pair=zeros(1,2); C1=3:(2+d); C2=2+d+(1:d);
for i=1:N-1, for j=i+1:N,
distance = boundary_periodic(x(j,:)-x(i,:),-L/2,L/2);                   (true) connecting vector
if norm(distance) < cutoff,                                             collect interacting pairs
ip = ip + 1;                                                            interaction pair counter
pair(ip,:) = [i j];                                  particle numbers (i,j) belonging to pair (ip)
connector(ip,:) = distance;                          connecting vector x_j - x_i for pair (i,j)
end;
if norm(distance) < 1,                                                  polymerization distance (USER)
if pol(i,2)+pol(j,2)==0 & pol(i,1) ~= j,                                check for single free, new neighbor
if pol(i,1) ~= pol(j,1) — pol(i,1) == 0,                                prevent stable trimers
pol(i,[2 C2])=pol(i,[1 C1]);                                            keep first neighbor
pol(i,1)=j;                                                             add new neighbor
end; end;
end
if pol(i,1) == j, pol(i,C1)=distance; end;                              update connectors for
if pol(i,2) == j, pol(i,C2)=distance; end;                              all existent neigbhors
end; end;
bonds=(sum(sign(pol(:,1)))+sum(sign(pol(:,2)))),        # of FENE bonds (increasing with time)
```

—— sample application, embedding & testing ——

```
function code_equilibrium_FENE_howto
global pol;
N=20; d=3; L=10; pol=zeros(N,8);          bead no., max. 2 neighbor sites, connector vectors
[x,v]=useful_initial_configuration(N,d,1,10,0.9);
code_equilibrium_FENE(N,d,x,v,L,2^(1/6),1000,0.005,1,1.5,30);
```

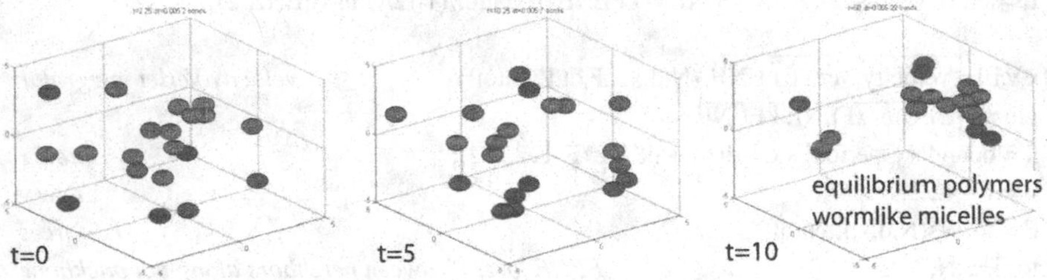

generated output (Sect. 12.4)

12.5 NonEquilibrium Molecular Dynamics

12.5.1 NonEquilibrium Molecular Dynamics (*n*D)

Calculate phase space trajectories for a system of interacting mass points subjected to shear flow at given rate, cf. Sect. 4.1.

——————— main routine ———————

```
function [x,v]= code_NEMD (N,d,L,x,v,n,T,dt,MDsteps,cutoff,shearrate),
        number of particles (N), space dimension (d>1), bos xize (L), number density (n),
        temperature (T), integration time step (dt), time steps (MDsteps),
        interaction potential cutoff distance (cutoff), shear rate (shearrate)
F=forces_shear(N,d,x,L,cutoff,0,0);                                    initialize forces
for time=0:dt:dt*MDsteps,                                        molecular dynamics loop
[x,v,F]=velocity_verlet_shear(N,d,x,v,F,dt,L,cutoff,shearrate,time);          trajectory
v =temperature_control_shear(N,d,x,v,T,shearrate);                NEMD velocity rescaling
visualize_particles(N,d,x,L,20,0.5,['t=' num2str(time)]);                       optionally
end;
function [x,v,F]= velocity_verlet_shear (N,d,x,v,F,dt,L,cutoff,shearrate,time)  integrator
x = boundary_periodic_shear(x + dt*v + dt^2/2*F,L,shearrate,time);                  new x
v = v + dt/2*F; F = forces_shear(N,d,x,L,cutoff,shearrate,time); v = v + dt/2*F;   id MD
v(:,1) = v(:,1)-round(x(:,2)./L).*shearrate*L;                  shift v and x upon leaving
function F= forces_shear (N,d,x,L,cutoff,shearrate,time)     forces using interacting pairs
                                                  identical with routine forces, except
        all_interactions replaced by all_interactions_shear(N,d,x,L,cutoff,shearrate,time)
```

```
function [ip,pair,connector]= all_interactions_shear (N,d,x,L,cutoff,shearrate,time)
        identical with routine all_interactions, except
        boundary_periodic() replaced boundary_periodic_shear(x(j,:)-x(i,:),L,shearrate,time);
function v= temperature_control_shear (N,d,x,v,T,shearrate)
    v(:,1)=v(:,1)-shearrate*x(:,2);                     peculiar velocities
    T_measured=sum(v(:).^2)/(N*d);                      peculiar temperature
    v=v*sqrt(T/T_measured);                             rescale peculiar velocities
    v(:,1)=v(:,1)+shearrate*x(:,2);                     recover true velocities
function x= boundary_periodic_shear (x,L,shearrate,time)  assumes simulation box centered at origin,
    box size L
    x(:,1) = x(:,1)-round(x(:,2)./L).*shearrate*time*L;  box in flow gradient direction
    x = x-round(x./L).*L;                               fold back to box
```

——— sample application, embedding & testing ———

```
function code_NEMD_howto (N,d,n,T,dt,MDsteps,cutoff,shearrate);
    min_sep =0.85;              minimum particle separation kept by initial configuration
    L=(N/n)^(1/d);                                      box size
    [x,v] = useful_initial_configuration(N,d,T,L,min_sep);  initialize x and v
    [x,v] = code_NEMD(N,d,L,x,v,n,T,dt,MDsteps,cutoff,shearrate);  NEMD
```

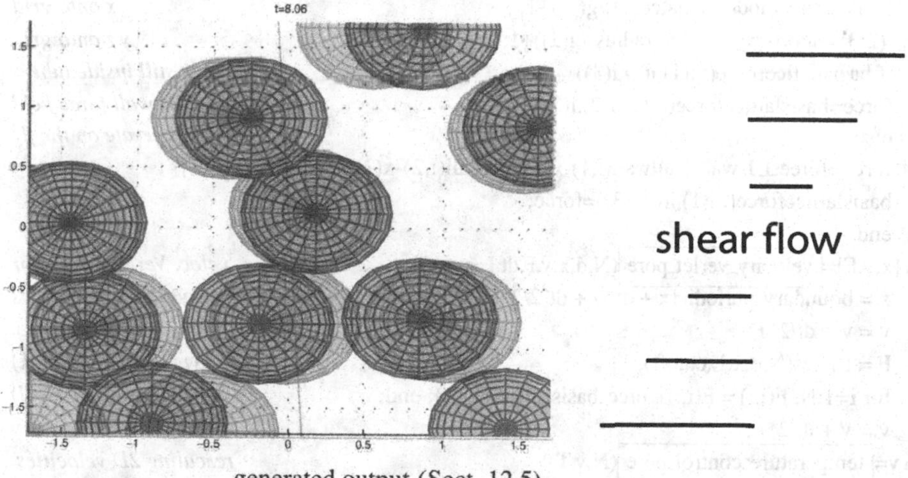

generated output (Sect. 12.5)

12.5.2 Flow through Nanopore (3D)

Calculate phase space trajectories for a system of interacting mass points subjected to flow through a circular pore with structured surface.

——— main routine ———

```
function code_flow_through_pore (N,d,L,radius,cutoff,resolution,v_wanted,...
```

```
MDsteps,min_sep,latconst,dt,T,wallcutoff)
number particles (N), box size (L), pore radius (radius), pot. cutoff (cutoff),
wall force (resolution), wanted mean flow velocity (v_wanted), MD steps (steps),
minimum particle separation at startup (min_sep), wall lattice constant
(latconst), time step (dt), temperature (T), pot. cutoff wall particles (wallcutoff)

fill wall force lattice on the fly (time/step thus reducing with time)
(c) 9 feb 2005 mk@mat.ethz.ch
global wallcutoff radius resolution v_wanted latconst;
init_basiscell(radius,L,latconst,resolution);                  prepare for wall force field
[x,v] =useful_initial_configuration_pore(N,d,T,L,min_sep);                    initialize x, v
v(:,1) =2*v_wanted*(1-(x(:,2).^2+x(:,3).^2)/radius^2);                 initial macro v profile
F=forces(N,d,x,L,cutoff);                                                        initial force
for MDstep=1:MDsteps,                                                molecular dynamics loop
[x,v,F]=velocity_verlet_pore(N,d,x,v,F,dt,L,cutoff);                   propagate trajectory
v(:,1) =v(:,1)-mean(v(:,1))+v_wanted;                                          control <v_x>
v =temperature_control_pore(N,v,T);                                   2D velocity rescaling
if mod(MDstep,10)==0, visualize_particles(N,d,x,L,20,0.5,'mytitle'); end;
end;
function force= force_basiscell (x,L);                       lookup or generate force field
global bin basislatticeforce g fullwall latconst radius fullwall;
i(1) =floor( mod(x(1),latconst)/g(1))+1;                                         x onto grid
i(2:3)=floor(([x(2) x(3)]+radius)/g(2))+1;                                     y,z onto grid
if basislatticeforce(:,i(1),i(2),i(3)),                           particle still inside tube?
force=basislatticeforce(:,i(1),i(2),i(3))';                        lookup (3D) wall force field
else                                                                       generate on the fly
force=force_LJ_wall([fullwall(:,1)-x(1) fullwall(:,2)-x(2) fullwall(:,3)-x(3)])';
basislatticeforce(:,i(1),i(2),i(3))=force;
end;
function [x,v,F]= velocity_verlet_pore (N,d,x,v,F,dt,L,cutoff)       veloc. Verlet integrator
x = boundary_periodic(x + dt*v + dt^2/2*F,-L/2,L/2);          new x with boundary conditions
v = v + dt/2*F;                                                                        new v
F = forces(N,d,x,L,cutoff);                                             new forces (bulk-bulk)
for i=1:N, F(i,:) = F(i,:)+force_basiscell(x(i,:),L); end;              new forces (bulk-wall)
v = v + dt/2*F;                                                                        new v
function v= temperature_control_pore (N,v,T)                          rescaling 2D velocities
T_measured =sum(v(:,2).^2+v(:,3).^2)/(2*N); v(:,2:3) =v(:,2:3)*sqrt(T/T_measured);
function x= boundary_periodic_pore (x,L)
global radius
x(:,1) = x(:,1)-round(x(:,1)./L).*L;                            fold back to simulation cell
r = sqrt(x(:,2).^2+x(:,3).^2);                                            all radial distances
init_dist_wall=1;                                             initial min. distance from wall
W = find(r > radius-init_dist_wall);                                find all interacting sites
if W,
rW = (radius-init_dist_wall)*rand(length(W),1)./r(W);
```

```
x(W,2) = rW.*x(W,2); x(W,3) = rW.*x(W,3);                    random fold back ..
end;
function [x,v]= useful_initial_configuration_pore (N,d,T,L,min_sep)
as useful_initial_configuration, but
replace boundary_periodic(..,-L/2,L/2) by boundary_periodic_pore(..,L)
```

——— *sample application, embedding & testing* ———

```
function code_flow_through_pore_howto
code_flow_through_pore(10,3,10,4,2^(1/6),20,0.5,1000,0.8,1,0.005,1,Inf)       sample call
function force_LJ_wall= force_LJ_wall (R)             specify bulk - fluid interaction force
global wallcutoff                                       R is a N x d list of wall particle coords
r2 = sum(R'.^2);                                        N squared norms if R is N x d matrix
indices = find(r2 < wallcutoff^2);                      interacting indices
if indices, r2=r2(indices);                             obtain wall force from all interacting sites
fac = -24*(2-r2.^4)./(r2.^7);                           Lennard–Jones force (USER)
force_LJ_wall=[sum(fac.*R(indices,1)');sum(fac.*R(indices,2)');sum(fac.*R(indices,3)')];
else force_LJ_wall = [0;0;0]; end;
function init_basiscell (radius,L,latconst,resolution);  specify wall particle sites
global bin basislatticeforce g fullwall latconst wallcutoff fullwall;
bin(1)=resolution; bin(2)=floor(bin(1)*radius/latconst)+1;     grid sizes
basislatticeforce=zeros(3,bin(1),bin(2),bin(2));              tube x direction
g(1)=latconst/bin(1); g(2)=2*radius/bin(2);                   lattice constants
phi=latconst:latconst:2*pi*radius; phi=phi/radius; wall=[];   tube structure (USER)
for x=-L/2:latconst:L/2-latconst,                             tube structure (USER)
fullwall=[fullwall; x*ones(length(phi),1),radius*cos(phi'),radius*sin(phi')];
end;
```

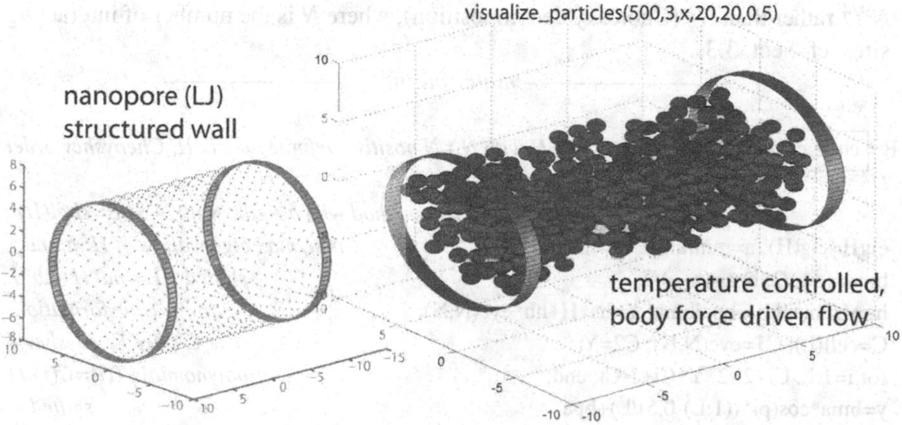

generated output (Sect. 12.5)

12.6 Brownian Dynamics

12.6.1 Brownian Dynamics of a Lennard–Jones System (*n*D)

Calculate phase space trajectories for a given Langevin equation, cf. Sect. 8.5.

——— main routine ———

```
function [x,F]= brownian_propagate (N,d,x,dt,friction_coeff,T,L,cutoff)
  F = forces(N,d,x,L,cutoff);                                        deterministic forces
  x = x + F/friction_coeff * dt + sqrt(2*T*dt/friction_coeff) * randn(N,d)   F + stochastic
  x = boundary_periodic(x,-L/2,L/2);                                 convenient, but not necessary
```

——— sample application, embedding & testing ———

```
function code_brownian_dynamics_howto (N,d,n,T,dt,BDsteps,cutoff,friction_coeff)
  number of particles (N), dimension d (1-3), particle number density n (0.5), temperature T (1),
      integration time step dt (0.005), time steps BDsteps (2000), cutoff (2.5),
                                            friction coefficient friction_coeff (1)
  L=(N/n)^(1/d); min_sep =0.85;   box size (L), minimum pair separation at startup (min_sep)
  [x,v] =useful_initial_configuration(N,d,T,L,min_sep);              init. coordinates x,
  F =forces(N,d,x,L,cutoff);                                         velocities v and forces F
  for BDstep=1:BDsteps, [x,F]=brownian_propagate(N,d,x,dt,friction_coeff,T,L,cutoff); end;
```

12.6.2 Hydrodynamic Interaction via Chebyshev Polynomials (3D)

Calculate Chebyshev approximation of a given positive definite matrix, at given expansion order. The recursion property of the Chebyshev polynomials is used to devise simulation algorithms for the study of hydrodynamic interactions which scale with $N^{9/4}$ rather than N^3 (Cholesky decomposition), where N is the number of interacting sites, cf. Sect. 3.3.

——— main routine ———

```
function B= code_chebyshev (H,N,L)  B*B'=H with N x N positive definite matrix H, Chebyshev order L
                              alternate direct method which scales with N^3: B=chol(H)'
  eigH=eig(H); a=min(eigH); z=max(eigH);          min, max eigenvalues of H: a and z
  bma=(z-a)/2; bpa=(z+a)/2;                        shifted HI = ha*H+hb*1
  ha=1/bma; hb=-bpa/bma; Y=ha*H+hb*eye(N,N);       transformation
  C=cell(L); C1=eye(N,N); C2=Y;                    The C's are Chebyshev's
  for i=1:L, Ci+2=2*Y*Ci+1-Ci; end;                polynomials T(i)=C(i+1)
  y=bma*cos(pi*((1:L)-0.5)/L)+bpa;                 shifted x
  fH=sqrt(y);                                      since B=sqrt(H)
  B=-1/2*coeff(1,L,fH)*eye(N,N);                   Chebyshev reconstruction
  for i=1:L, B=B+coeff(i,L,fH)*Ci; end;
function coeff= coeff (j,L,fH)                     Chebyshev coefficients
```

```
coeff=sum(fH(1:L).*cos(pi*(j-1)*((1:L)-1/2)/L))*2/L;
```

——— *sample application, embedding & testing* ———

```
function code_chebyshev_howto (R,L)          rank (R>1), Chebyshev expansion order (L)
X=rand(1,R);                   R uniformely distributed random numbers in [0,1]
H=X'*X,                              generate sample pos. semidef. RxR (H) matrix
B=code_chebyshev(H,R,L),                                     approximation for B
H_recalculated=B*B'                                      recalculate H using B
```

Note: Effcient implementation of these ideas in a brownian dynamics code does not require explicitely calculating B!

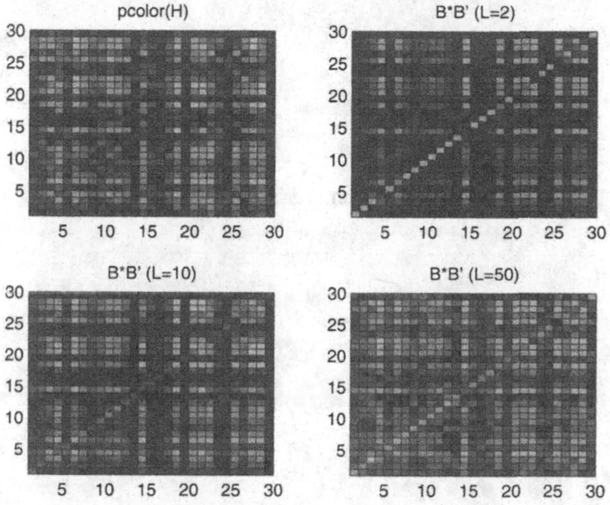

generated output (Sect. 12.6)

12.7 Coarse-Graining

12.7.1 Coarse-Graining Polymer Chains (nD)

Calculate the coarse-grained representation (parameterized by ξ) of a given (atomistic) configuration of a linear polymer, cf. Sect. 8.10.1

——— *main routine* ———

```
function x= coarse_grain (x,xi)              returns coarse-grained x from atomistic x
if xi==0, x=x; return; end;                                      parameterized by xi
N=length(x); y=zeros(N,3);                           number of nodes N, initialize y
m(1)=1+xi; for i=2:N-1, m(i)=(1+2*xi)-xi^2/m(i-1); end;
m(N)=1+xi-xi^2/m(N-1);                          calculate inverse band-diagonal matrix
y(1,:)=x(1,:); for i=2:N, y(i,:)=x(i,:)+xi*y(i-1,:)/m(i-1); end;
x(N,:)=y(N,:)/m(N); for i=N-1:-1:1, x(i,:)=(y(i,:)+xi*x(i+1,:))/m(i); end;
```

——— *sample application, embedding & testing* ———

```
function code_coarse_grain_howto
N=200; x=random_walk_3D(N,[0 0 0]);        create random path with N=200 nodes
xi=20; x=coarse_grain(x,xi);               coarse-grained path (parameter xi)
```

plot3(x(:,1),x(:,2),x(:,3))

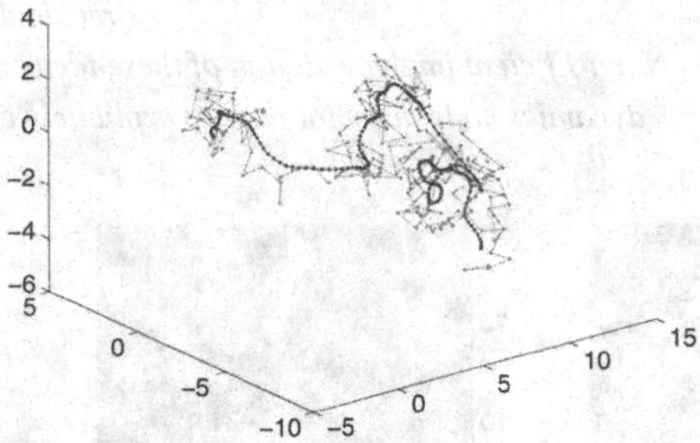

generated output (Sect. 12.7)

13 Concluding Remarks

The development of constitutive relationships which connect strain or strain rate with stress and material behavior is at the heart of a successful macroscopic modeling of complex fluids, and soft matter in general. We reviewed simple physical models which allow to find such relationships for the case of neutral bulk polymeric fluids, nematic fluids, ferrofluids, colloidal suspensions. We have shown that the simplest approximate treatments (Chap. 2) such as the Peterlin approximation turn out to be insufficiently precise. More detailed chain models which allow to capture molecular architecture, finite extensibility, bending stiffness and interchain interactions without approximation, on the other hand, are computationally expensive while remaining conceptually simple (Chaps. 3–5). In order to predict rheooptic behaviors on time and length scales relevant for applications chemical details are shown to be not essential. These models serve to make progress towards appropriate decoupling approximations for stochastic differential equations, and a reduced description using relevant (slow) variables (Chaps. 6–8). Most interestingly, they provide deep insight into the microscopic origins of viscoelastic behavior.

This monograph contains an introduction to the microscopic modeling of anisotropic, in particular, polymeric fluids involving FENE chain models, tube models, and and elongated particle models and may serve as a starting point to devise appropriate models and to understand soft matter and nonequilibrium complex fluids as encountered in applications and current experiments. We discussed several efficient strategies to solve microscopic models such as the Cholesky decomposition or variance reduction methods for FENE solutions with HI. We provided examples which demonstrated how to attack the non-analytical solvable models in approximate, and less approximate fashion. Coarseing procedures have been applied to microscopic trajectories onto objects which can be treated within the framework of primitive path models. Using the coarsening procedure of Sect. 8.10.1 one should be able to extract the parameters of tube models directly from atomistic simulation on the nanosecond scale, i.e., small compared to the reptation time scale. Insight from the microscopic FENE chain level – such as anisotropic tube renewal, stress-optic failures – have been used to refine these theories and to work out consequences in Chap. 6. The rheological crossover observed for FENE chain melts allowed to discuss and interpret characteristic lengths scales in polymer melts. These scales can be expressed in terms of density, molecular weight, and flexibilty, i.e. based on geometric or 'topological' quantities and independent of chemical details. The soft ellipsoid model [3, 424] is another representative of a coarseing strategy from many monomers to many polymers. Elongated (rigid) particle models have been characterized in detail and connection was made to macroscopic description such as the EL theory for nematics. Inhomogeneous extensions of Fokker–Planck discussed in this monograph have been studied, e.g., for liquid crystals in order to calculate elastic coefficients [311].

The formulation of new models for nonequilibrium fluids remains a difficult task but should be guided through frameworks ensuring their thermodynamically admissible, intrinsically consistent, description. The corresponding GENERIC approach reviewed in Sect. 8.3 has not yet been extended to describe nonholonomic constraints or boundary conditions. It may be interesting to show, e.g., how the simple model for polymer melts considering anisotropic tube renewal (Sect. 6.2) may be cast into a suitable generalized framework.

This monograph did certainly not provide sufficiently detailed information on how to implement efficient and advanced simulations, but original articles for each application have been cited, where missing details can be found. Chapter 12 further reduced the gap by providing full simulation codes demonstrating various simulation, visualization and animation techniques. Standard textbooks such as [4, 58, 141, 156, 179, 216] contain background and supplementary information on the modeling of nonequilibrium fluids. An overview of some of the more popular computational models and methods used today in the field of molecular and mesoscale simulation of polymeric materials, ranging from molecular models and methods that treat electronic degrees of freedom to mesoscopic field theoretic methods can be also found in [3, 12, 425].

13.1 Acknowledgement

It is my great pleasure to thank Avinoam Ben-Shaul, Wilfried Carl, Peter J. Daivies, Masao Doi, Pep Espanol, Marco Ellero, Denis J. Evans, Jiannong Fang, Peter Fischer, William Gelbart, Siegfried Hess, Heinz Hoffmann, William G. Hoover, Markus Hütter, Patrick Ilg, Asja Jelic, Iliya V. Karlin, Nikos Karayiannis, Kurt Kremer, Klaus Kroy, Ronald G. Larson, Manolo Laso, Werner Loose, Clarisse Luap, Rachid Makhloufi, Vlasis Mavrantzas, Réne Muller, Hans Christian Öttinger, Juanjo de Pablo, Jorge Ramirez, Götz Rienäcker, Erich Sackmann, Christoph Schneggenburger, Howard Shaun Sellers, Nic Spencer, Igor Stankovic, Theo Tervoort, Doros N. Theodorou, Billy D. Todd, Harald Voigt, Jan Wilhelm, Hiromichi Yanagihara for collaborations and anonymous referees for very inspiring discussions on the physics and simulation of soft matter. Supercomputer facilities at ETH Zürich and ZIB Berlin have been extensively used.

14

Notation

14.1 Special Symbols

!	Factorial ($N! = 1 \times 2 \times 3 \times \cdots \times N$ and $N!! = 1 \times 3 \times \cdots \times N$)
:	Two subsequent tensorial products (: is equivalent with $\cdot\cdot$)
$=$	Equal to (as a result of basic operations)
$\approx$	Approximately equal to
$\langle\rangle$	Ensemble or time average
$\cdot$	Scalar or tensorial product
$\equiv$	Left hand side defined as expression on right hand side
$\propto$	Proportional to
∇	Gradient, sometimes labeled with the variable to avoid confusion
$\boldsymbol{A}$	(boldface) vector or tensor $\boldsymbol{A}$, norm denoted as A
1	Unity matrix
$\boldsymbol{AB}$	Dyadic product between $\boldsymbol{A}$ and $\boldsymbol{B}$
$\overline{\boldsymbol{A}}$	Anisotropic (symmetric traceless) part of tensor $\boldsymbol{A}$
$\boldsymbol{A}^{\mathrm{sym}}$	Symmetric, normalized, part of tensor $\boldsymbol{A}$
$\boldsymbol{\nabla}$	Nabla-Operator (gradient with respect to spatial coordinates)
$\dot{A}$	Total derivative with respect to time, $\dot{A} = dA/dt$
$\odot^{l}$	l-fold contratcion
A_{μ}	μ-component of vector $\boldsymbol{A}$
$A_{\mu\nu}$	(μ,ν)-component of 2nd rank tensor $\boldsymbol{A}$

14.2 Tensor Symbols

$\boldsymbol{u}_{(l)} \equiv \boldsymbol{uu}..\boldsymbol{u}$ $[\boldsymbol{n}_{(l)} \equiv \boldsymbol{nn}..\boldsymbol{n}]$ (l-fold dyadics made of vector $\boldsymbol{u}$, [director $\boldsymbol{n}$], tensor of rank l)

$\boldsymbol{u}_{[l]} \equiv \overline{\boldsymbol{u}_{(l)}} = \overline{\boldsymbol{uu}\ldots\boldsymbol{u}}$, $[\boldsymbol{n}_{[l]} = \overline{\boldsymbol{n}_{(l)}}]$ (anisotropic, symmetric traceless tensor)

$\boldsymbol{a}_{(l)} \equiv \langle \boldsymbol{u}_{(l)} \rangle = \langle \boldsymbol{uu}\ldots\boldsymbol{u} \rangle$ (average value)

$\boldsymbol{a}_{[l]} \equiv \langle \boldsymbol{u}_{[l]} \rangle = \left\langle \overline{\boldsymbol{uu}\ldots\boldsymbol{u}} \right\rangle$ (so called alignment tensor of rank l)

14.3 Upper Case Roman Symbols

B	Shape factor for uniaxial, elongated particles (1: rod, 0: sphere, -1: disk)
BD	brownian dynamics
BEMD	Beyond-equilibrium molecular dynamics
C_∞	characteristic ratio, cf. Sect. 4.5
$D_{..}$	Diffusion coefficient (usually labeled by a model)
DE	Doi-Edwards
DSMC	Density of states Monte Carlo
E	Energy
EL	Ericksen-Leslie
FENE	Finitely extendable nonlinear elastic
G' (G'')	Storage (loss) modulus ($G^* = G' + iG'' = i\omega\eta^*$)
GENERIC	General equation for the nonequilibrium reversible-irreversible coupling
H	Hookean spring coefficient
$\boldsymbol{H}$	External magnetic field
$\boldsymbol{H}^{ij}$	Hydrodynamic interaction matrix
HL	Hinch-Leal
$\mathbf{1}$	Unit tensor of rank 3
$I_n(\boldsymbol{A})$	nth order scalar invariant of $mbfA$ defined in (10.58)
L	Chain contour length
$\boldsymbol{L}$	GENERIC building block characterizing reversible dynamics
MC	Monte Carlo
MD	Molecular dynamics
$\boldsymbol{M}$	Magnetization
$\boldsymbol{M}$	GENERIC building block characterizing irreversible dynamics
N	Number of beads within a single chain
NEMD	Nonequilibrium molecular dynamics
N_e	Entanglement number of beads ($\propto$ entanglement molecular weight)
N_c	Critical number of beads ($\propto$ critical molecular weight)
N_s	Number of solvent particles
N_t	Total number of beads (polymer plus solvent)
Q	Tube segment length
Q_0	Maximum extension for FENE spring
$\boldsymbol{Q}, \boldsymbol{Q}_j$	Connector(s) between adjacent beads within primitive chain ($j = 1 \ldots N-1$)
$\boldsymbol{Q}_{[l]}$	$= \overline{\boldsymbol{QQ}..\boldsymbol{Q}}$ (irreducible tensor of rank l)
S	Entropy
$S_{1,2,..}$	Order parameters
$\boldsymbol{u}_{(l)}$	$= \boldsymbol{uu}..\boldsymbol{u}$ (dyadics of rank l)
T	Temperature
Tr	Trace with respect to an arbitrary pair of indices
$\boldsymbol{T}^{(k)}$	Symmetry adapted basis tensors, 2.9, 2.112.12. $k \in \{0, 1, 2, 3, 4, \mathrm{Tr}\}$
$\boldsymbol{T}_{..}$	Orienting torque entering the Fokker–Planck equation
WCA	Weeks-Chandler-Anderson

14.4 Lower Case Roman Symbols

$a_{+,-,0}$	Shear flow adapted components of the alignment tensor, (6.2)
a_k	Components of the alignment tensor with respect to $\boldsymbol{T}^{(k)}$
$\boldsymbol{a}_{(l)}$	$= \langle \boldsymbol{u}_{(l)} \rangle = \langle \boldsymbol{uu..u} \rangle$
$\boldsymbol{a}_{[l]}$	Anisotropic alignment tensor (of rank l)
$\boldsymbol{u}_{[l]}$	$= \overline{\boldsymbol{u}_{(l)}} = \overline{\boldsymbol{uu..u}}$ (irreducible tensor of rank l)
b	FENE parameter ($b = HQ_0^2/k_\mathrm{B}T$)
d	Space dimension
f	Orientational distribution function $f(\boldsymbol{u},t)$
g_k	Components of the gyration tensor with respect to $\boldsymbol{T}^{(k)}$
$\boldsymbol{g}$	Dimensionless anisotropic 2nd moment of f for FENE dumbbells
h	Dimensionless Langevin parameter characterizing magnetic field
$\boldsymbol{i}$	'Symbolic half' unity matrix, $\boldsymbol{1} = \boldsymbol{ii}$
k_B	Boltzmann constant $k_\mathrm{B} = 1.38 \times 10^{-23}$ J K^{-1}
m	Bead mass
n	Bead number density: number of beads per volume
n_max	Maximum birefringence (chemistry dependent)
n_p	Polymer number density ($n_p = n/N$ for monodisperse systems)
$\boldsymbol{n}$	Director in the EL theory
$p_{..}$	Orientational distribution, statistical weighting factor
r	Axis ratio r for uniaxial, elongated particles
r_cut	Cut-off distance for interaction potential
$\boldsymbol{R}, R$	End-to-end vector, end-to-end distance
$\boldsymbol{R}_g, R_g$	Gyration tensor, radius of gyration
s	Dimensionless chain contour position $0 < s < 1$
t	Time
$\boldsymbol{u}, \boldsymbol{u}_j$	Unit vector(s) tangential to the primitive path (normalized $\boldsymbol{Q}$)
$\boldsymbol{v}$	Macroscopic flow field (shear flow $v_x = \dot{\gamma}y$, $\boldsymbol{v} = \dot{\gamma}\boldsymbol{e}^{(2)}$ for convenience)
$\boldsymbol{x}_i$	Position vector of bead i ($i = 1 \ldots N$)
ζ_l	Defined in (11.2) (alignment tensor definition)

14.5 Greek Symbols

$\alpha_{..}$	Ericksen-Leslie (EL) viscosity coefficients
α^n_{lj}	Coefficients (11.33) of the coupled moment equations (11.26)
β	$\beta = (k_B T)^{-1}$
$\gamma_{..}$	EL rotational viscosity coefficients
$\boldsymbol{\gamma}$	Symmetric part of the velocity gradient ∇v
$\dot{\gamma}$	Shear rate
Γ	Dimensionless shear rate
Δ	Laplace-Operator ($\Delta = \boldsymbol{\nabla} \cdot \boldsymbol{\nabla}$)
$\boldsymbol{\Delta}^{(l,k,l)}$	Isotropic tensor, cf. (10.14). $\boldsymbol{\Delta}^{(l)} \equiv \boldsymbol{\Delta}^{(l,0,l)}$
δ_{ij}	Kronecker symbol $\delta_{i,j} \equiv 1$ $(i = j)$ and 0 otherwise
ε	Characteristic energy of the Lennard–Jones and WCA potentials
$\boldsymbol{\varepsilon}$	Total antisymmetric tensor of rank three
ζ	Friction coefficient
η	Shear viscosity
η' (η'')	Real (negative imaginary) part of the complex viscosity $\eta^* = \eta' - i\eta''$
η^*	Complex (shear) viscosity
$\eta_{1,2,3}$	Miesowicz viscosities (1: flow, 2: gradient, 3: vorticity direction)
$\boldsymbol{\kappa}$	Transposed macroscopic velocity gradient $(\nabla \boldsymbol{v})^T$
λ	Tumbling parameter or relaxation time (depends on context)
Λ	Eigenvalue
$\lambda_{\max}$	Defined in (4.16)
ξ	Parameter for the coarse-graining from atomistic to tapeworm (Sect. 8.10.1)
ρ	Volume density
σ	Characteristic length of the Lennard–Jones and WCA potentials
σ_{xy}	Shear stress
$\boldsymbol{\sigma}$	Stress tensor
τ	Relaxation time (sometimes labeled by a model)
ϕ	Concentration
χ	Flow alignment angle
Ψ_1 (Ψ_2)	1st (2nd) viscometric function
f	Configurational distribution function
$\boldsymbol{\omega}, \boldsymbol{\omega}_f$	Macroscopic flow vorticity $(\nabla \times \boldsymbol{v})/2$
ω	Oscillation frequency
$\boldsymbol{\Omega}$	Antisymmetric part of the velocity gradient ∇v
ζ_l	Coefficient defined in (11.2)

14.6 Caligraphic Symbols

$\mathcal{L}_{\mathrm{FP}}$	Differential operator characterizing the Fokker–Planck equation
$\mathcal{D}_{..}$	Damping term entering the Fokker–Planck equation
$\boldsymbol{\mathcal{L}}$	Angular operator $\boldsymbol{\mathcal{L}} = \mathbf{u} \times \partial/\partial \boldsymbol{u}$

14.7 FENE Models

Table 14.1. Recommended nomenclature for finitely extendable nonlinear elastic (FENE) models (for dilute/concentrated solutions, melts, etc., cf. Fig. 1.3). Models based on the Peterlin approximation should carry a 'P', models for branched macromolecules should be suffixed by the maximum functionality considered (for linear chains, $n = 2$, it is understood that the suffix 2 is skipped). Many of the proposed simulation models have not been extensively studied, and analytic approaches such as FENE-Pn (Peterlin approximation for branched FENE chains) are missing

Model	For Finitely Extendable ..	Reference
	simulation (linear or branched chains) NEMD/NEBD	
FENE	linear flexible classical polymers incl. dumbbells ($N = 2$)	[59] or (2.1)
FENE-n	branched flexible classical polymers, incl. H-shaped (maximum functionality $n = 3$), star polymers ($n > 3$)	[59] or (2.1)
FENE-B	linear semiflexible (B for 'bend') classical polymers, actin filaments	(5.33)
FENE-Bn	branched FENE-B, maximum functionality n, semiflexible classicial networks	(5.33)
FENE-C	FENE which allows for unimolecular scission and recombination (C for 'cut'), wormlike micelles, equilibrium polymers	[29] or (5.18)
FENE-Cn	FENE-C, maximum functionality n, living flexible and saturated networks	[29] or (5.18)
FENE-CB	semiflexible FENE-C, associative polymer networks	(5.32)
FENE-CBn	semiflexible FENE-Cn, living semiflexible non-saturated networks	(5.32)
	theory (linear chains), approximate constitutive equation	
FENE-P	(P for 'Peterlin) appoximation for FENE dumbbells, second moment as single state variable	[60,78,79]
FENE-P^2	Second-order Peterlin model	[35]
FENE-PM	Small set of equations approximating FENE-P chains	[69,426]
FENE-PCR	also known as FENE-CR, Peterlin approximation plus a non-constant diffusion coefficient	[427]
FENE-PCD	also known as FENE-CD, Peterlin approximation plus a configuration dependent diffusion coefficient	[428]
FENE-L	Second-order L-shaped closure model for FENE chains	[35]
FENE-LS	Simplified version of FENE-L	[429]
FENE-PMF	FENE-P supplemented by a mean-field (MF) interaction term modeling concentration effects	[80], Sect. 2.1

14.8 Gaussian Integrals

Gaussian integrals over the infinite three-dimensional space arise at several places in the text. Let α be a real positive number, $\boldsymbol{\alpha}$ a positive-definite, 2nd order tensor, and both $\boldsymbol{\alpha}$ and $\boldsymbol{\beta}$ symmetric.

$$\int e^{-\alpha r^2}\, d^3r = \left(\frac{\pi}{\alpha}\right)^{3/2}, \tag{14.1}$$

$$\int \frac{1}{r} e^{-\alpha r^2}\, d^3r = \frac{2\pi}{\alpha}, \tag{14.2}$$

$$\int e^{-(\boldsymbol{\alpha}:\boldsymbol{rr})}\, d^3r = \frac{\pi^{3/2}}{\sqrt{\det \boldsymbol{\alpha}}}, \tag{14.3}$$

$$\int \frac{1}{r^3} \boldsymbol{rr}\, e^{-\alpha r^2}\, d^3r = \frac{2\pi}{3\alpha}\boldsymbol{1}, \tag{14.4}$$

$$\int e^{-\alpha r^2 - \beta(\boldsymbol{r}\cdot\boldsymbol{s})}\, d^3r = \left(\frac{\pi}{\alpha}\right)^{3/2} e^{\frac{(\beta s)^2}{4\alpha}}, \tag{14.5}$$

$$\int e^{-(\boldsymbol{\alpha}:\boldsymbol{rr})-(\boldsymbol{\beta}:\boldsymbol{rs})}\, d^3r = \frac{\pi^{3/2}}{\sqrt{\det \boldsymbol{\alpha}}}\, e^{\frac{1}{4}(\boldsymbol{\alpha}^{-1}\cdot\boldsymbol{\beta}^2):\boldsymbol{ss}}. \tag{14.6}$$

By making use of the identity

$$\frac{d}{d\boldsymbol{\alpha}} \det \boldsymbol{\alpha} = \boldsymbol{\alpha}^{-1} \det \boldsymbol{\alpha}, \tag{14.7}$$

we obtain, by differentiation of (14.3) with respect to $\boldsymbol{\alpha}$,

$$\int \boldsymbol{rr}\, e^{-(\boldsymbol{\alpha}:\boldsymbol{rr})}\, d^3r = -\frac{d}{d\boldsymbol{\alpha}} \int e^{-(\boldsymbol{\alpha}:\boldsymbol{rr})}\, d^3r = \frac{\pi^{3/2}\boldsymbol{\alpha}^{-1}}{2\sqrt{\det \boldsymbol{\alpha}}}, \tag{14.8}$$

and so on for tensors of arbitrary rank. Further, for $\boldsymbol{r} \in \Re^n$, in (14.3) we have to replace $\pi^{3/2}$ by $\pi^{n/2}$. Notice, that Gaussian integrals over the unit sphere (d^2u), cf. (10.68), are more difficult to compute (involve Bernoulli integrals), the above formulas hold for integration over $\Re^3$.

References

1. N. Goldenfeld and L.P. Kadanoff, Science **284** (1999) 87.
2. S.S. Saarman, Phys. Rep. **305** (1998) 1.
3. K. Kremer and F. Muller-Plathe, Molec. Simul. **28** (2002) 729.
4. R.B. Bird, O. Hassager, R.C. Armstrong and C.F. Curtiss, *Dynamics of Polymeric Liquids*, Volumes 1+2 (John Wiley & Sons, NY, 1987).
5. T.C.B. McLeish, Adv. Phys. **51** (2002) 1379.
6. R.G. Larson, T.T. Perkins, D.E. Smith and S. Chu, Phys. Rev. E **56** (1997) 1794 & refs. cited herein.
7. R.B. Bird and J.M. Wiest, Annu. Rev. Fluid Mech. **27** (1995) 169.
8. P.S. Doyle, E.S.G. Shaqfeh and A.P. Gast, J. Fluid. Mech. **334** (1997) 251.
9. B.H.A.A. Van den Brule, J. Non-Newtonian Fluid Mech. 47 (1993) 357.
10. B. Dünweg, D. Reith, M. Steinhauser, J. Chem. Phys. **117** (2002) 914.
11. K. Kremer, Macromol. Chem. Physic. **204** (2003) 257.
12. K. Binder and G. Ciccotti (Eds), *Monte Carlo and Molecular Dynamics of Condensed Matter Systems* (IPS Conf. Proc., Bologna, 1996) pp. 669–723 and 825–841.
13. J.W. Rudisill and P.T. Cummings, J. Non-Newtonian Fluid Mech. **41** (1992) 275.
14. H.X. Guo, K. Kremer and T. Soddemann, Phys. Rev. E **66** (2002) 061503.
15. M. Murat, G.S. Grest and K. Kremer, Europhys. Lett. **42** (1998) 401.
16. G.S. Grest, Curr. Opin. Coll. Interf. Sci. **2** (1997) 271.
17. M. Kröger, W. Loose and S. Hess, J. Rheol. **37** (1993) 1057.
18. P.J. Daivis, M.L. Matin and B.D. Todd, J. Non-Newtonian Fluid. Mech. **111** (2003) 1.
19. J. Koplik and J.R. Banavar, Phys. Rev. Lett. **84** (2000) 4401.
20. G.S. Grest, Adv. Polym. Sci. **138** (1999) 149.
21. P.G. Khalatur, A.R. Khokhlov and D.A. Mologin, J. Chem. Phys. **109** (1998) 9602; 9614
22. M.J. Stevens, M. Mondello, G.S. Grest, S.T. Cui, H.D. Cochran and P.T. Cummings, J. Chem. Phys. **106** (1997) 7303.
23. Z.F. Xu, R. Khare, J.J. de Pablo and S. Kim, J. Chem. Phys. **106** (1997) 8285.
24. P. Padilla and S. Toxvaerd, J. Chem. Phys. **104** (1996) 5956.
25. J. Koplik and J.R. Banavar, Phys. Rev. Lett. **78** (1997) 2116.
26. A. Uhlherr and D.N. Theodorou, Curr. Opin. Solid State Mech. **3** (1998) 544.
27. W.M. Gelbart, A. Ben-Shaul and D. Roux, *Micelles, membranes, microemulsions and monolayers* (Springer, NY, 1994).
28. R.G. Larson, Curr. Opin. Coll. Interf. Sci. **2** (1997) 361.
29. M. Kröger and R. Makhloufi, Phys. Rev. E **53** (1996) 2531.

30. W. Carl, R. Makhloufi and M. Kröger, J. Phys. France II (Paris) **7** (1997) 931.
31. S.R. Zhao, C.P. Sun and W.X. Zhang, J. Chem. Phys. **106** (1997) 2520; 2530.
32. M. Kröger, Comput. Phys. Commun. **118** (1999) 278.
33. F. Affouard, M. Kröger and S. Hess, Phys. Rev. E **54** (1996) 5178.
34. T. Odijk, Curr. Opin. Coll. Interf. Sci. **1** (1996) 337.
35. G. Lielens, P. Halin, I. Jaumain, R. Keunings, V. Legat, J. Non-Newton. Fluid Mech. **76** (1998) 249.
36. M. Dressler, B.J. Edwards, H.C. Öttinger, Rheol. Acta **38** (1999) 117.
37. P. Wapperom and M.A. Hulsen, J. Rheol. **42** (1998) 999.
38. H.C. Öttinger and M. Grmela, Phys. Rev. E **56** (1997) 6633.
39. T. Soddemann, B. Dünweg, K. Kremer, Eur. Phys. J. E **6** (2001) 409.
40. R.R. Netz and D. Andelman, Phys. Rep. **380** (2003) 1.
41. G.G. Fuller, *Optical rheometry of complex fluids* (Oxford Univ. Press, U.K., 1995).
42. B. Dünweg and K. Kremer, Phys. Rev. Lett. **61** (1991) 2996.
43. B. Dünweg and K. Kremer, J. Chem. Phys. **99** (1993) 6983.
44. S. Hess, C. Aust, L. Bennett, M. Kröger, C. Pereira Borgmeyer, T. Weider, Physica A **240** (1997) 126.
45. C. Aust, M. Kröger and S. Hess, Macromolecules **32** (1999) 5660.
46. J.K.C. Suen, Y.L.J. Joo and R.C. Armstrong, Ann. Rev. Fluid Mech. **34** (2002) 417.
47. M. Melchior and H.C. Öttinger, J. Chem. Phys. **103** (1995) 9506.
48. B.Z. Dlugogorski, M. Grmela and P.J. Carreau, J. Non-Newtonian Fluid Mech. **49** (1993) 23.
49. G.B. Jeffrey, Proc. R. Soc. London, Ser. A **102** (1922) 161.
50. E.J. Hinch and L.G. Leal, J. Fluid Mech. **71** (1975) 481.
51. M.A. Peterson, Am. J. Phys. **47** (1979) 488.
52. A.N. Kaufman, Phys. Lett. A **100** (1984) 419.
53. A. Beris and B.J. Edwards, *Thermodynamics of Flowing Systems with Internal Microstructure*, vol. **36** of *Engineering and Science series* (Oxford Univ. Press, New York, 1994).
54. H.C. Öttinger, *Beyond equilibrium thermodynamics* (Wiley, Hoboken, New Jersey, 2005).
55. R.J.J. Jongschaap, J. Non-Newtonian Fluid. Mech. **96** (2001) 63.
56. B.J. Edwards, H.C. Öttinger and R.J.J. Jongschaap, J. Non-Equil. Thermodyn. **22** (1997) 356.
57. B.J. Edwards and H.C. Öttinger, Phys. Rev. E **56** (1997) 4097.
58. H.C. Öttinger, *Stochastic processes in polymeric fluids, Tools and Examples for Developing Simulation Algorithms* (Springer, Berlin, 1996).
59. H.R. Warner, Ind. Eng. Chem. Fundam. **11** (1972) 379.
60. X. Fan, J. Non-Newtonian Fluid Mech. **17** (1985) 125.
61. H.C. Öttinger, J. Non-Newtonian Fluid Mech. **26** (1987) 207.
62. L.E. Wedgewood and H.C. Öttinger, J. Non-Newtonian Fluid Mech. **27** (1988) 245.
63. J. Bossart and H.C. Öttinger, Macromolecules **28** (1995) 5852; **30** (1997) 5527.
64. C. Pierleoni and J.-P. Ryckaert, Phys. Rev. Lett. **71** (1993) 1724.
65. C. Pierleoni and J.-P. Ryckaert, Macromolecules **28** (1995) 5097.
66. M. Herrchen and H.C. Öttinger, J. Non-Newtonian Fluid Mech. **68** (1997) 17.
67. C.D. Dimitropoulos, R. Sureshkumar and A.N. Beris, J. Non-Newtonian Fluid Mech. **79** (1998) 433.
68. Q. Zhou and R. Akhavan, J. Non-Newtonian Fluid. Mech. **109** (2003) 115.
69. L.E. Wedgewood, D.N. Ostrov and R.B. Bird, J. Non-Newtonian Fluid Mech. **40** (1991) 119.

70. R. Keunings, J. Non-Newtonian Fluid Mech. 68 (1997) 85.
71. P. Ilg, I.V. Karlin, Phys. Rev. E **62** (2000) 1441.
72. A. Link and J. Springer, Macromolecules **26** (1993) 464.
73. N. Kuzuu and M. Doi, J. Phys. Soc. Jpn. **52** (1983) 3486.
74. S. Hess, Z. Naturforsch. **31**a (1976) 1034.
75. W. Maier and A.Z. Saupe, Z. Naturforsch. **14**a (1959) 882.
76. W. Hess, J. Polym. Sci. Polym. Symp. **73** (1985) 201.
77. S. Hess, J. Non-Newtonian Fluid Mech. **23** (1987) 305.
78. A. Peterlin, Makromol. Chem. **44** (1961) 338.
79. A. Peterlin, J. Polym. Sci. Polym. Lett. **48** (1966) 287.
80. C. Schneggenburger, M. Kröger and S. Hess, J. Non-Newtonian Fluid Mech. **62** (1996) 235. Addenda see [81].
81. M. Kröger and E. de Angelis, J. Non-Newtonian Fluid Mech. **125** (2005) 87.
82. S. Hess and W. Köhler, *Formeln zur Tensorrechnung* (Palm & Enke, Erlangen, 1980)
83. P. Kaiser, W. Wiese and S. Hess, J. Non-Equilib. Thermodyn. **17** (1992) 153.
84. J. Remmelgas and L.G. Leal, J. Non-Newtonian Fluid Mech. **89** (2000) 231.
85. E.C. Lee, M.J. Solomon and S.J. Muller, Macromolecules **30** (1997) 7313.
86. J. Honerkamp and R. Seitz, J. Chem. Phys. **87** (1987) 3120.
87. P. Van der Schoot, Macromolecules **25** (1992) 2923 & refs. cited herein.
88. F. Boue and P. Lindner, Europhys. Lett. **25** (1994) 421.
89. T. Kume, K. Asakawa, E. Moses, K. Matsuzaka and T. Hashimoto, Acta Polym. **46** (1995) 79.
90. S.A. Patlazhan and P. Navard, J. Phys. France II (Paris) **5** (1995) 1017.
91. M. Zisenis and J. Springer, Polymer **36** (1995) 3459.
92. Y. Tsunashima, J. Chem. Phys. **102** (1995) 4673.
93. R. Balian, *From microphysics to macrophysics*. Vol. 2 (Springer, Berlin, 2nd Ed., 1992).
94. P. Ilg, I.V. Karlin, M. Kröger and H.C. Öttinger, Physica A **319** (2003) 134.
95. A.N. Gorban, I.V. Karlin, P. Ilg and H.C. Öttinger, J. Non-Newtonian Fluid Mech. **96** (2001) 203.
96. P. Ilg, I.V. Karlin, H.C. Öttinger, Physica A **315** (2002) 367.
97. P. Ilg, M. Kröger and S. Hess, J. Chem. Phys. **116** (2002) 9078.
98. A.N. Gorban, I.V. Karlin and A.Y. Zinovyev, *Constructive methods of invariant manifolds for kinetic problems*, Phys. Rep. **396** (2004) 197.
99. G. Strobl, *The Physics of Polymers* (Springer-Verlag, Heidelberg, 1996)
100. J.G. Kirkwood and J. Riseman, J. Chem. Phys. **16** (1948) 565.
101. S.W. Fetsko and P.T. Cummings, Int. J. Thermophys. **15** (1994) 1085.
102. M. Kröger, A. Alba, M. Laso and H.C. Öttinger, J. Chem. Phys. **113** (2000) 4767.
103. H.C. Öttinger, J. Chem. Phys. **86** (1987) 3731; **87** (1987) 1460.
104. J. Rotne, S. Prager, J. Chem. Phys. **50** (1969) 4831.
105. G.B. Thurston and A. Peterlin, J. Chem. Phys. **46** (1967) 4881.
106. H.C. Öttinger and Y. Rabin, J. Non-Newtonian Fluid Mech. **33** (1989) 53.
107. R.G. Larson, *Constitutive Equations for Polymer Melts and Solutions* (Butterworths, Boston, 1988).
108. M. Fixman, Macromolecules **19** (1986) 1195; 1204.
109. W. Press, S.A. Teukolsky, W.T. Vetterling and B.P. Flannery, *Numerical Recipes in Fortran*, 2nd Ed. (Cambridge Univ. Press, U.K., 1992) pp. 184.
110. D. Dahlquist and A. Bork, *Numerical Methods* (Prentice-Hall, Englewood Cliffs, NJ, 1974).
111. R.M. Jendrejack, M.D. Graham and J.J. de Pablo, J. Chem. Phys. **113** (2000) 2894.

112. R. Rzehak, D. Kienle, T. Kawakatsu and W. Zimmermann, Europhys. Lett. **46** (1999) 821.
113. R. Rzehak, W. Kromen, T. Kawakatsu and W. Zimmermann, Eur. Phys. J. E **2** (2000) 3.
114. M. Melchior and H.C. Öttinger, J. Chem. Phys. **105** (1996) 3316.
115. S. Navarro, M.C.L. Martinez and J. Garcia de la Torre, J. Chem. Phys. **103** (1995) 7631.
116. P.S. Grassia, E.J. Hinch and L.C. Nitsche, J. Fluid Mech. **282** (1995) 373.
117. J.D. Schieber, J. Non-Newtonian Fluid Mech. **45** (1992) 47.
118. C. Elvingston, Biophys. Chem. **43** (1992) 9.
119. W. Zylka, J. Chem. Phys. **94** (1991) 4628.
120. C. Elvingston, J. Comput. Chem. **12** (1991) 71.
121. M. Fixman, J. Chem. Phys. **89** (1988) 2442.
122. M. Fixman, Faraday Discuss. **83** (1987) 199.
123. D.-J. Yang and Y.-H. Lin, Polymer **44** (2003) 2807.
124. A.J. Banchio and J.F. Brady, J. Chem. Phys. **118** (2003) 10323.
125. J. Huang and T. Schlick, J. Chem. Phys. **117** (2002) 8573.
126. M. Fixman, J. Chem. Phys. **78** (1983) 1594.
127. J. Garcia de la Torre, M.C. Lopez, M.M. Terado and J. Freire, Macromolecules **17** (1984) 2715.
128. J.M. Garcia Bernal, M.M. Lopez, M.M. Tirado and J. Freire, Macromolecules **24** (1991) 693.
129. J.R. Prakash and H.C. Öttinger, J. Non-Newtonian Fluid Mech. **71** (1997) 245.
130. Y. Oono and M. Kohmoto, J. Chem. Phys. **78** (1983) 1.
131. H.C. Öttinger, Phys. Rev. A **40** (1989) 2664.
132. M. Schmidt and W. Burchard, Macromolecules **14** (1981) 210.
133. A.Z. Akcasu and C.C. Han, Macromolecules **12** (1979) 276.
134. Y. Miyaki, Y. Einaga, H. Fujita and M. Fakuda, Macromolecules **13** (1980) 588.
135. A.M. Rubio, J.J. Freire, J.H.R. Clarke, C.W. Yong and M. Bishop, J. Chem. Phys. **102** (1995) 2277.
136. B.H. Zimm, G.M. Roe and L.F. Epstein, J. Chem. Phys. **24** (1956) 279.
137. M. Daoud and G. Jannink, J. Phys. Lett. **39** (1976) 1045.
138. P.G. de Gennes, Macromolecules **9** (1976) 587; 594.
139. K. Kremer, G.S. Grest and I. Carmesin, Phys. Rev. Lett. **61** (1988) 566.
140. K. Kremer and G.S. Grest, J. Chem. Phys. **92** (1990) 5057.
141. K. Binder (Ed) *Monte Carlo and molecular dynamics simulations in polymer science* (Oxford Univ. Press, U.S.A., 1996).
142. T. Aoyagi and M. Doi, Comput. Theor. Polym. Sci. **10** (2000) 317.
143. J. Gao and J.H. Weiner, J. Chem. Phys. **90** (1989) 6749.
144. M. Kröger, C. Luap and R. Muller, Macromolecules **30** (1997) 526.
145. B.D. Todd, Comput. Phys. Commun. **142** (2001) 14.
146. J.G. Hernández Cifre, S. Hess and M. Kröger, Macromol. Theory Simul. **13** (2004) 748.
147. M. Pütz, K. Kremer and G..S. Grest, Europhys. Lett. **49** (2000) 735; **52** (2000) 721.
148. Y. Masubuchi, J.I. Takimoto, K. Koyama, G. Ianniruberto, G. Marrucci and F. Greco, J. Chem. Phys. **115** (2001) 4387.
149. T.P. Lodge, N.A. Rotstein and S. Prager, Adv. Chem. Phys. **79** (1990) 1.
150. L.R.G. Treloar, Trans. Faraday. Soc. **36** (1940) 538.
151. F. Bueche, J. Chem. Phys. **20** (1952) 1959.
152. S.F. Edwards, Proc. Phys. Soc. **91** (1967) 513.
153. P.G. de Gennes, J. Chem. Phys. **55** (1971) 572.
154. A.L. Khodolenko and T.A. Vilgis, Phys. Rep. **298** (1998) 251.

155. J.D. Weeks, D. Chandler and H.C. Andersen, J. Chem. Phys. **54** (1971) 5237.
156. M. P. Allen and D.J. Tildesley, *Computer Simulations of Liquids* (Oxford Science, U.K., 1990).
157. D.J. Evans and G.P. Morris, *Statistical Mechanics of Nonequilibrium Liquids* (Academic Press, London, 1990).
158. W.G. Hoover, Physica A **240** (1997) 1.
159. S. Hess and M. Kröger, Phys. Rev. E **61** (2000) 4629.
160. S. Hess, in: Computational Physics, K.H. Hoffmann and M. Schreiber (ed) (Springer, Berlin, 1996) pp. 268–293.
161. G.S. Grest, B. Dünweg and K. Kremer, Comput. Phys. Commun. **55** (1989) 269.
162. M. Kröger and S. Hess, Phys. Rev. Lett. **85** (2000) 1128.
163. R. Muller and J.J. Pesce, Polymer **35** (1994) 734.
164. J.D. Ferry, *Viscoelastic Properties of Polymers,* (J. Wiley & Sons, NY, 3. ed., 1980)
165. L.J. Fetters, D.J. Lohse, S.T. Milner and W.W. Graessley, Macromolecules **32** (1999) 6847.
166. C. Luap, R. Muller and C. Picot, ILL exp. rep. **9-11-305**;
167. M. Kröger and H. Voigt, Macromol. Theory Simul. **3** (1994) 639.
168. J. Gao and J.H. Weiner, Macromolecules **27** (1994) 1201 & refs. cited herein.
169. M.N. Hounkonnou, C. Pierleoni and J.-P. Ryckaert, J. Chem. Phys. **97** (1992) 9335.
170. J.T. Padding and W.J. Briels, J. Chem. Phys. **117** (2002) 925.
171. D.J. Evans, W.G. Hoover, B.H. Failor, B. Moran and A.J.C. Ladd, Phys. Rev. A **28** (1983) 1016.
172. H.J. Janeschitz-Kriegl, *Polymer melt rheology and flow birefringence* (Springer, Berlin, 1983).
173. L.S. Priss, I.I. Vishnyakoov and I.P. Pavlova, Int. J. Polym. Mater **8** (1980) 85.
174. B.E. Read, Polym. Eng. Sci. **23** (1983) 835.
175. J. Fang, M. Kröger and H.C. Öttinger, J. Rheol. **44** (2000) 1293.
176. R.T. Bonnecaze and J.F. Brady, J. Rheol. **36** (1992) 73.
177. M. Warner and X.J. Wang, Macromolecules **24** (1991) 4932.
178. J.P. Rothstein and G.H. McKinley, J. Non-Newt. Fluid Mech. **108**, 275 (2002).
179. R.G. Larson, *The structure and rheology of complex fluids* (Oxford Univ. Press, U.K., 1999)
180. H.C. Öttinger, J. Rheol. **43** (1999) 1461.
181. W. Kuhn and F. Grün, Kolloid Z. **101** (1942) 248.
182. J.v. Meerveld, J. Non-Newtonian Fluid Mech. 123 (2004) 259.
183. L.R.G. Treloar, *The physics of rubber elasticity* (Clarendon Press, Oxford, 1975).
184. G. Marrucci and G. Ianniruberto, Phil. Trans. R. Soc. Lond. A **361** (2003) 677.
185. J. Wilhelm and E. Frey, Phys. Rev. Lett. **77** (1996) 2581.
186. T.B. Liverpool, R. Golestanian and K. Kremer, Phys. Rev. Lett. **80** (1998) 405.
187. R. Everaers, F. Jülicher, A. Ajdari and A.C. Maggs, Phys. Rev. Lett. **82** (1999) 3717.
188. Y. Kats, D.A. Kessler and Y. Rabin, Phys. Rev. E **65** (2002) 020801(R).
189. S.F. Edwards, Proc. Phys. Soc. London **85** (1965) 613.
190. N. Saito, K. Takahashi and Y. Yunoki, J. Phys. Soc. Jpn **22** (1967) 219.
191. K.F. Freed, Adv. Chem. Phys. **22** (1972) 1.
192. J. des Cloizeaux, Phys. Rev. A **10** (1974) 1665.
193. M. Muthukumar and B.G. Nickel, J. Chem. Phys. **80** (1984) 5839; **86** (1987) 460.
194. Y. Oono, Adv. Chem. Phys. **61** (1985) 301.
195. O. Kratky and G. Porod, Recueil Trav. Chim. **68** (1949) 1106.
196. R.P. Feynman and A.R. Hibbs, *Quantum Mechanics and Path Integrals* (McGraw-Hill, New York, 1975).

197. J.B. Lagowski and J. Noolandj, J. Chem. Phys. **95** (1991) 1266.
198. S.M. Bhattacharjee and M. Muthukumar, J. Chem. Phys. **86** (1987) 411.
199. A.D. Egorov, P.I. Sobolevsky and L.A. Yanovich, *Functional Integrals: Approximate Evaluation and Applications*, in: *Mathematics and Its Applications* **249** (Kluwer, Dordrecht, 1993).
200. D.J. Amit, *Field Theory, The Renormalization Group and Critical Phenomena* (McGraw-Hill, London, 1978).
201. S. Panyukov and Y. Rabin, Europhys. Lett. **57** (2002) 512.
202. P.G. de Gennes, *Scaling concepts in polymer physics*, (Cornell University Press, Ithaca, NY, 1979).
203. W.M. Gelbart and A. Ben-Shaul, J. Phys. Chem. **100** (1996) 13169.
204. J.P. Wittmer, A. Milchev and M.E. Cates, J. Chem. Phys. **109** (1998) 834.
205. A.T. Bernardes, V.B. Henriques and P.M. Bisch, J. Chem. Phys. **101** (1994) 645.
206. Y. Rouault and A. Milchev, Phys. Rev. E **51** 1995) 5905.
207. K. Binder, Rep. Progr. Phys. **60** (1997) 487.
208. P.J. Flory, *Principles of Polymer Chemistry* (Cornell Univ. Press, Ithaca, 1953).
209. T.L. Hill, *Statistical Mechanics* (McGraw-Hill, NY, 1956).
210. Y.A. Shchipunov and H. Hoffmann, Rheol. Acta **39** (2000) 542.
211. P. Boltenhagen, Y.T. Hu, E.F. Matthys and D.J. Pine, Europhys. Lett. **38** (1997) 389.
212. T. Koga and F. Tanaka, Macromol. Rapid Commun. **26** (2005) 701.
213. S.-Q. Wang, W.M. Gelbart and A. Ben-Shaul, J. Phys. Chem. **94** (1990) 2219.
214. J.N. Israelachvili, D.J. Mitchell and B.W. Ninham, J. Chem. Soc. Faraday Trans. II **72** (1976) 1525.
215. H. C. Booij, J. Chem. Phys. **80** (1984) 4571.
216. M. Doi and S.F. Edwards, *The Theory of Polymer Dynamics* (Clarendon, Oxford, 1986)
217. W. Carl, J. Chem. Soc. Faraday Trans. **91** (1995) 2525.
218. W. Carl and W. Bruns, Macromol. Theory Simul. **3** (1994) 295.
219. M.E. Cates, J. Phys. France (Paris) **49** (1988) 1593.
220. R. Makhloufi, J.P. Decruppe, A. Aitali and R. Cressely, Europhys. Lett. **32** (1995) 253.
221. M.E. Cates, Phys. Scripta **49** (1993) 107.
222. P. Van der Schoot and M.E. Cates, Europhys. Lett. **25** (1994) 515.
223. Y. Bohbot, A. Ben-Shaul, R. Granek and W.M. Gelbart, J. Chem. Phys. **103** (1995) 8764.
224. E. Sackmann, Macromol. Chem. Physic. **195** (1994) 7–28.
225. J. Käs, H. Strey and E. Sackmann, Nature **368** (1994) 226.
226. J. Käs, H. Strey, J.X. Tang, D. Finger, R. Ezzell, E. Sackmann and P.A. Janmey, Biophys. J. **70** (1996) 609–625.
227. P.A. Janmey, S. Hvidt, J. Käs, D. Lerche, A. Maggs, E. Sackmann, M. Schliwa and T.P. Stossel, J. Biol. Chem. **269** (1994) 32503.
228. H.J.C. Berendsen, Science **271** (1996) 954.
229. I. Carmesin and K. Kremer, Macromolecules **21** (1988) 2819.
230. H.P. Deutsch and K. Binder, Chem. Phys. **94** (1991) 2294.
231. G.S. Grest, K. Kremer and E.R. Duering, Physica A **194** (1993) 330.
232. P. Debnath and B.J. Cherayil, J. Chem. Phys. **118** (2003) 1970.
233. M. Kröger, Phys. Rep. **390** (2004) 453.
234. P.G. de Gennes, *The physics of liquid crystals* (Clarendon Press, Oxford, 1974).
235. G. Vertogen and W.H. de Jeu, *Thermotropic Liquid Crystals, Fundamentals* (Springer, Berlin, 1988).
236. S. Chandrasekhar, *Liquid Crystals* (Cambridge Univ. Press, U.K., 1971).
237. J.P. Bareman, G. Cardini and M.L. Klein, Phys. Rev. Lett. **60** (1988) 2152.

238. J. Baschnagel and K. Binder, Physica A **204** (1994) 47.
239. M.E. Mann, C.H. Marshall and A.D.J. Haymet, Mol. Phys. **66** (1989) 493.
240. C.W. Greeff and M.A. Lee, Phys. Rev. **49** (1994) 3225.
241. D. Frenkel, Mol. Phys. **60** (1987) 1.
242. L.F. Rull, Physica A **80** (1995) 113.
243. D.J. Adams, G.R. Luckhurst and R.W. Phippen, Mol. Phys. **61** (1987) 1575.
244. E. Egberts and H.J.C. Berendsen, J. Chem. Phys. **89** (1988) 3718.
245. R.R. Netz and A.N. Berker, Phys. Rev. Lett. **68** (1992) 333.
246. D. Levesque, M. Mazars and J.-J. Weis, J. Chem. Phys. **103** (1995) 3820.
247. A.L. Tsykalo, Mol. Cryst. Liq. Cryst. **129** (1985) 409.
248. S. Hess, D. Frenkel and M.P. Allen, Molec. Phys. **74** (1991) 765.
249. M.A. Glaser, R. Malzbender, N.A. Clark and D.M. Walba, Molec. Simul. **14** (1995) 343.
250. R.D. Kamien and G.S. Grest, Phys. Rev. E **55** (1997) 1197.
251. H. Heller, M. Schaefer and K. Schulten, J. Phys. Chem. **97** (1993) 8343.
252. M.P. Allen, in: *Observation, Prediction and Simulation of Phase Transitions on Complex Fluids*, M.Baus, L.F. Rull and J.P. Ryckaert, eds. (Kluwer Academic Publ., Dordrecht, 1995).
253. B.H. Zimm and W.H. Stockmayer, J. Chem. Phys. **17** (1949) 1301.
254. C.W. Cross and B.M. Fung, J. Chem. Phys. **101** (1994) 6839.
255. A. V. Komolkin and A. Maliniak, Mol. Phys. **84** (1995) 1227.
256. D.C. Rapaport, *The Art of Molecular Dynamics Simulation* (Cambridge Univ. Press, U.K., 1995).
257. J.A. Board, J.W. Causey, J.F. Leathrum, A. Windemuth and K. Schulten, J. Chem. Phys. Lett. **198** (1992) 89.
258. A. Windemuth and K. Schulten, Molec. Simul. **5** (1991) 353.
259. E. Paci and M. Marchi, J. Phys. Chem. **100** (1996) 4314.
260. G.V. Paolini, G. Ciccotti and M. Ferrario, Mol. Phys. **80** (1993) 297.
261. A. L. Tsykalo, *Thermophysical Properties of Liquid Crystals* (Golden and Breach Science, 1991).
262. M. Kröger, Makromol. Chem. Macromol. Symp. **133** (1998) 101;
263. R.H.W. Wientjes, R.J.J. Jongschaap, M.H.G. Duits and J. Mellema, J. Rheol. **43** (1999) 375.
264. S.H. Kim, H.G. Sim, K.H. Ahn and S.J.Lee, Korea-Australia Rheol. J. **14** (2002) 49.
265. K.R. Geurts and L.E. Wedgewood, J. Chem. Phys. **106** (1997) 339.
266. R.D. Groot, A. Bot and W.G.M. Agterof, J. Chem. Phys. **104** (1996) 9202.
267. M. Doi, J. Polym. Sci. Polym. Phys. Ed. **21** (1983) 667.
268. R. Ketzmerick and H.C. Öttinger, Continuum Mech. Thermodyn. **1** (1989) 113.
269. C. Tsenoglou, ACS Polym. Preprints **28** (1987) 185.
270. J. des Cloizeaux, Europhys. Lett. **5** (1988) 437.
271. N.P.T. O'Connor and R.C. Ball, Macromolecules **25** (1992) 5677.
272. D.C. Venerus and H. Kahvand, J. Rheol. **38** (1994) 1297 & refs. cited herein.
273. J.P. Oberhauser, L.G. Leal and D.W. Mead, J. Polym. Sci. Polym. Phys. **36** (1998) 265.
274. C.C. Hua and J.D. Schieber, J. Chem. Phys. **109** (1998) 10018.
275. C.C. Hua, J.D. Schieber and D.C. Venerus, J. Chem. Phys. **109** (1998) 10028.
276. C.C. Hua, J.D. Schieber and D.C. Venerus, J. Rheol. **43** (1999) 701.
277. J.D. Schieber, J. Neergard and S. Gupta, J. Rheol. **47** (2003) 213.
278. D.W. Mead, R.G. Larson and M. Doi, Macromolecules **31** (1998) 7895.
279. M. Doi and J. Takimoto, Phil. Trans. Roy. Soc. London Ser. A **361** (2003) 641.
280. M.H. Wagner, J. Rheol. **38** 1994) 655.

281. M.H. Wagner, V. Schulze and A. Gottfert, Polym. Eng. Sci. **36** (1996) 925.
282. A. Peterlin and H.A. Stuart, *Hand- und Jahrbuch d. Chem. Phys.* **8**, A. Eucken and K.I. Wolf, eds. (1943) 113.
283. M. Kröger and S. Hess, Physica A **195** (1993) 336.
284. S. Hess, Physica A **86** (1977) 383; **87** (1977) 273; **112** (1982) 287.
285. H. Giesekus, Rheol. Acta **2** (1962) 50.
286. T. Masuda, K. Kitagawa, I. Inoue and S. Onogi. Macromolecules **3** (1970) 109.
287. L.A. Holmes, S. Kusamizu, K. Osaki and J.D. Ferry. J. Polym. Sci. A **2** (1971) 2009.
288. M. Kröger, Makromol. Chem. Macromol. Symp. **81** (1994) 83.
289. A. Lozinski, C. Chauviere, J. Fang and R.G. Owens, J. Rheol. **47** (2003) 535.
290. M. Laso, M. Picasso and H.C. Öttinger, AIChE Journal **43** (1997) 877.
291. M. Abramowitz and I.A. Stegun, *NBS Handbook of mathematical functions* (Washington D.C., 1964) 804.
292. M. Kröger and H.S. Sellers, J. Chem. Phys. **103** (1995) 807.
293. F.M. Leslie, Arch. Rat. Mech. Anal. **28** (1968) 265.
294. F.M. Leslie, Cont. Mech. Thermodyn. **4** (1992) 167.
295. J.L. Ericksen, Arch. Rat. Mech. Anal. **113** (1991) 197.
296. D. Baalss and S. Hess, Z. Naturforsch. **43**a (1988) 662.
297. G.L. Hand, Arch. Rat. Mech. Anal. **7** (1961) 81.
298. S.-D. Lee, J. Chem. Phys. **88** (1988) 5196.
299. L.A. Archer and R.G. Larson, J. Chem. Phys. **103** (1995) 3108.
300. G.L. Hand, J. Fluid Mech. **13** (1962) 33.
301. S. Hess, Z. Naturforsch. **30**a (1975) 728.
302. C. Pereira Borgmeyer and S. Hess, J. Non-Equilib. Thermodyn. **20** (1995) 359.
303. I. Pardowitz and S. Hess, Physica A **100** (1980) 540.
304. S. Hess and I. Pardowitz, Z. Naturforsch. **36**a (1981) 554.
305. E.H. Macmillan, J. Rheol. **33** (1989) 1071.
306. J.L. Ericksen, Arch. Rat. Mech. Anal. **4** (1969) 231.
307. G. Marrucci, Mol. Cryst. Liq. Cryst. **72** (1982) 153.
308. A.N. Semonov, Zh. Eksp. Teor. Fiz. **85** (1983) 549.
309. S. Yamamoto and T. Matsuoka, J. Chem. Phys. **100** (1994) 3317.
310. H. Brenner and D.W. Condiff, J. Coll. Interf. Sci. **47** (1974) 199.
311. M. Kröger and H.S. Sellers, *A molecular theory for spatially inhomogeneous, concentrated solutions of rod-like liquid crystal polymers,* in: *Complex Fluids* (L. Garrido ed.) from the series *Lecture notes in physics* **415** (Springer, NY, 1992) 295.
312. G.K. Batchelor, J. Fluid Mech. **46** (1971) 813.
313. M. Doi, J. Polym. Sci. Polym. Phys. **19** (1981) 229.
314. S. Hess, Z. Naturforsch. **31**a (1976) 1507.
315. M. Kröger and H.S. Sellers, Phys. Rev. E **56** (1997) 1804.
316. T. Carlsson, Mol. Cryst. Liq. Cryst. **89** (1982) 57.
317. T. Carlsson, J. Phys. **44** (1983) 909.
318. D.J. Ternet, R.G. Larson and R.G. Leal, Rheol. Acta. **38** (1999) 183.
319. A.V. Zakharov, A.V. Komolkin and A. Maliniak, Phys. Rev. E **59** (1999) 6802.
320. M. Fialkowski, Phys. Rev. E **58** (1998) 1955.
321. T. Carlsson and K. Skarp, Mol. Cryst. Liq. Cryst. **78** (1981) 157.
322. H. Ehrentraut and S. Hess, Phys. Rev. E **51** (1995) 2203.
323. F.P. Bretherton, J. Fluid Mech. **14** (1962) 284.
324. E.J. Hinch and L.G. Leal, J. Fluid Mech. **76** (1976) 187.
325. B.J. Edwards and A.N. Beris, J. Rheol. **33** (1989) 537.

326. M. Kröger, P. Ilg and S. Hess, J. Phys.: Condens. Matter **15** (2003) S1403.
327. P. Ilg and M. Kröger, Phys. Rev. E **66** (2002) 021501; Phys. Rev. E **67** (2003) 049901.
328. M.A. Martsenyuk, Yu L. Raikher and M.I. Shliomis, Sov. Phys. JETP **38** (1974) 413.
329. P. Ilg, M. Kröger, S. Hess and A.Y. Zubarev, Phys. Rev. E **67** (2003) 061401.
330. S. Odenbach (Ed), *Ferrofluids* (Springer, Berlin, 2002).
331. G. Rienäcker, M. Kröger and S. Hess, Phys. Rev. E **66** (2002) 040702.
332. R.G. Larson and H.C. Öttinger, Macromolecules **24** (1991) 6270.
333. P.L. Maffettone and S. Crescitelli, J. Rheol. **38** (1994) 1559.
334. M. Grosso, R. Keunings, S. Crescitelli and P.L. Maffettone, Phys. Rev. Lett. **86** (2001) 3184.
335. A. L. Yarin, O. Gottlieb and I.V. Roisman, J. Fluid Mech. **340** (1997) 83.
336. G. Rienäcker, M. Kröger and S. Hess, Physica A **315** (2002) 537.
337. P.D. Olmsted and P. Goldbart, Phys. Rev. A **41** (1990) 4578; **46** (1992) 4966.
338. J. Mewis, M. Mortier, J. Vermant and P. Moldenaers, Macromolecules **30** (1997) 1323.
339. G. Rienäcker and S. Hess, Physica A **267** (1999) 294.
340. N.C. Andrews, A.J. McHugh and B.J. Edwards, J. Rheol. **40** (1996) 459.
341. G. Marrucci and F. Greco, Mol. Cryst. Liq. Cryst. **206** (1991) 17.
342. G. Sgalari, G.L. Leal and J.J. Feng, J. Non-Newtonian Fluid Mech. **102** (2002) 361.
343. R. Bandyopadhyay, G. Basappa and A.K. Sood, Phys. Rev. Lett. **84** (2002) 2022.
344. K.S. Kumar and T.R. Ramamohan, J. Rheol. **39** (1995) 1229.
345. M.E. Cates, D.A. Head and A. Ajdari, Phys. Rev. E **66** (2002) 025202(R).
346. A.S. Wunenburger, A. Colin, J. Leng, A. Arneodo and D. Roux, Phys. Rev. Lett. **86** (2001) 1374.
347. H.C. Öttinger, J. Non-Equilib. Thermodyn. **22** (1997) 386.
348. W.S. Koon and J.E. Marsden, Rep. Math. Phys. **40** (1997) 21.
349. V.O. Soloviev, J. Math. Phys. **34** (1993) 5747.
350. D. Lewis, J.Marsden and R. Montgomery, Physica D **18** (1986) 391.
351. J.A. De Azcarraga, J.M. Izquierdo, A.M. Perelomov and J.C. Perez-Bueno, J. Math. Phys. **38** (1997) 3735.
352. L. Takhtajan, Commun. Math. Phys. **160** (194) 295.
353. J. Grabowski and G. Marmo, J. Phys. A **32** (1999) 4239.
354. M. Kröger, M. Hütter and H.C. Öttinger, Comput. Phys. Commun. **137** (2001) 325.
355. G. Ianniruberto and G. Marrucci, J. Non-Newtonian Fluid Mech. **79** (1998) 225.
356. H.C. Öttinger, J. Non-Newtonian Fluid Mech. **89** (1999) 165.
357. H.C. Öttinger, J. Non-Equilib. Thermodyn. **22** (1997) 386.
358. H. Grad, *Principles of the Kinetic Theory of Gases*, in: *Thermodynamics of Gases* of the series *Encyclopedia of Physics*, Vol. XII, S. Flügge (Ed) (Springer, 1958, Berlin) pp. 205–294.
359. U. Geigenmüller, U.M. Titulaer and B.U. Felderhof, Physica A **119** (1983) 53.
360. M.P. Allen and D.J. Tildesley, *Computer Simulation of Liquids* (Clarendon Press, Oxford, 1987).
361. S. Hess and D.J. Evans, Phys. Rev. E **64** (2001) 011207.
362. S. Hess, M. Kröger and D.J. Evans, Phys. Rev. E **67** (2003) 042201.
363. E.G. Flekkoy and P.V. Coveney, Phys. Rev. Lett. **83** (1999) 1775.
364. M. Serrano and P. Espanol, Phys. Rev. E **64** (2001) 046115.
365. S. Succi, I.V. Karlin and H. Chen, Rev. Mod. Phys. **74** (2002) 1203.
366. I.V. Karlin, A. Ferrante and H.C. Öttinger, Europhys. Lett. **47** (1999) 182.
367. W. Dzwinel and D.A. Yuen, J. Colloid Interf. Sci. **225** (2000) 179.
368. J. Honerkamp, *Stochastische Dynamische Systeme* (VCH, Weinheim, 1990).

369. H. Risken, *The Fokker–Planck equation* (Springer, Berlin, 1984).
370. A. Schenzle and H. Brand, Phys. Rev. A **20** (1979) 1628.
371. S. Nakajima, Prog. Theor. Phys. **20** (1958) 948.
372. R.W. Zwanzig, J. Chem. Phys. **33** (1960) 1338.
373. R.W. Zwanzig, *Lectures in Theoretical physics*, Vol. 3 (Wiley-Interscience, New York, 1961).
374. H. Grabert, *Projection Operator Techniques in Nonequilibrium Statistical Mechanics* (Springer, NY, 1982).
375. H.C. Öttinger, Phys. Rev. E **57** (1998) 1416.
376. V.G. Mavrantzas and H.C. Öttinger, Macromolecules **35** (2002) 960.
377. V.A. Harmandaris, V.G. Mavrantzas, D.N. Theodorou, M. Kröger, J. Ramirez, H.C. Öttinger and D. Vlassopoulos, Macromolecules **36** (2003) 1376.
378. J.D. Schieber and H.C. Öttinger, J. Rheol. **38** (1994) 1909.
379. H.C. Öttinger, Phys. Rev. E **57** (1998) 1416.
380. D. Burnett, Proc. Math. Soc. **40** (1935) 382.
381. N. Herdegen and S. Hess, Physica A **115** (1982) 281.
382. S. Hess and H.J.M. Hanley, Phys. Lett. A **105** (1984) 238.
383. W. Loose, Phys. Lett. A **128** (1988) 39.
384. G. Ciccotti, G. Jacucci and I.R. McDonald, J. Stat. Phys. **21** (1979) 1.
385. G.P. Morriss and D.J. Evans, Phys. Rev. A **35** (1987) 792.
386. H. Grad, Commun. Pure Appl. Math. **2** (1949) 331.
387. H.J. Kreuzer, *Nonequilibrium Thermodynamics and its Statistical Foundations*, Clarendon Press, Oxford, 1981.
388. S. Hess, J. Non-Newtonian Fluid Mech. **23** (1987) 305.
389. M. Kröger and H.C. Öttinger, J. Non-Newtonian Fluid Mech. **120** (2004) 175.
390. V. Mavrantzas and H.C. Öttinger, Macromolecules **35** (2002) 960.
391. S. Nosé, J. Chem. Phys. **81** (1984) 511.
392. W.G. Hoover, Phys. Rev. A **31** (1985) 1695.
393. A.W. Lees and S.F. Edwards, J. Phys. C **5** (1972) 1921.
394. S. Melchionna, G. Ciccotti and B.L. Holian, Mol. Phys. **78** (1993) 533.
395. P.J. Daivis, M.L. Matin and B.D. Todd, J. Non-Newtonian Fluid Mech. **111** (2003) 1.
396. J. Feng, C.V. Chaubal and L.G. Leal, J. Rheol. **42** (1998) 1095.
397. B.J. Edwards and M. Dressler, J. Non-Newtonian Fluid Mech. **96** (2001) 163.
398. M.E. Tuckerman, C.J. Mundy, S. Balasubramanian and M.L. Klein, J. Chem. Phys. **106** (1997) 5615.
399. B.D. Todd and P.J. Daivis, Comput. Phys. Commun. **117** (1999) 191.
400. M. Kröger, J. Ramirez and H.C. Öttinger, Polymer **43** (2002) 477.
401. H.C. Öttinger, J. Non-Newtonian Fluid Mech. **120** (2004) 207.
402. M. Kröger, Comput. Phys. Commun. **168** (2005) 209.
403. R. Everaers, S. K. Sukumaran, G. S. Grest, C. Svaneborg, A. Sivasubramanian, K. Kremer, Science **303** 823 (2004).
404. S. K. Sukumaran, G. S. Grest, K. Kremer, R. Everaers, J. Pol. Sci. B: Pol. Phys. **43** (2005) 917.
405. M. Rubinstein, R. H. Colby, *Polymer Physics* (Oxford Univ. Press, Oxford, 2003).
406. A.E. Ismail, G. Stephanopoulos and G.C. Rutledge, J. Polym. Sci. B, Polym. Phys. **43** (2005) 897.
407. Y.G. Meng and W.H. Wong, Statistica Sinica **6** (1996) 831–860.
408. F. Wang and D.P. Landau, Phys. Rev. Lett. **86** (2001) 2050.
409. F. Wang and D.P. Landau, Phys. Rev. E **64** (2001) 056101.

410. Q. Yan and J.J. de Pablo, Phys. Rev. Lett. **90** (2003) 035701.
411. Q. Yan, T.S. Jain and J.J. de Pablo, Phys. Rev. Lett. **92** (2004) 235701.
412. B.D. Butler, O. Ayton, O.G. Jepps and D.J. Evans, J. Chem. Phys. **109** (1998) 6519.
413. J. Delhommelle and D.J. Evans, J. Chem. Phys. **13** (2002) 6015.
414. A. Papageorgiou, J. Complexity **19** (2003) 332.
415. X.Q. Wang and K.T. Fang, J. Complexity **19** (2003) 101.
416. J. Spanier and J.H. Pengilly (Eds), Math. Comput. Model. **23**:8–9 (1996) 1–179.
417. H. Jeffreys, *Cartesian tensors* (Cambridge Univ. Press, Cambridge, 1931)
418. H. Grad, Pure Appl. Math. **3** (1949) 325.
419. C. Zemach, Phys. Rev. B **140** (1965) 97.
420. J.R. Coope et al., J. Chem. Phys. **43** (1965) 2269; J. Math. Phys. **11** (1970) 1003; 1591.
421. D. Baalss and S. Hess, Phys. Rev. Lett. **57** (1986) 86.
422. P. Ilg, *Anisotropic diffusion in nematic liquid crystals and in ferrofluids*, Phys. Rev. E (2005) in press.
423. A. Quarteroni and F. Saleri, *Scientific computing with MATLAB* (Springer, Berlin, 2000)
424. M. Murat and K. Kremer, J. Chem. Phys. **108** (1998) 4340.
425. S.C. Glotzer, W. Paul, Ann. Rev. Mater. Res. **32** (2002) 401.
426. P.S. Doyle, E.S.G. Shaqfeh, G.H. McKinley, S.H. Spiegelberg, J. Non-Newton. Fluid Mech. 76 (1998) 79.
427. M.D. Chilcott and J.M. Rallison, J. Non-Newtonian Fluid Mech. **29** (1988) 381.
428. J. Remmelgas, P. Singh and L.G. Leal, J. Non-Newton. Fluid Mech. **88** (1999) 31.
429. G. Lielens, R. Keunings, V. Legat, J. Non-Newtonian Fluid Mech. **87** (1999) 179.

Author Index

—— Author [reference number] section number(s)

Index

Lecture Notes in Physics

For information about earlier volumes please contact your bookseller or Springer LNP Online archive: springerlink.com

Vol.631: D. J. W. Giulini, C. Kiefer, C. Lämmerzahl (Eds.), Quantum Gravity, From Theory to Experimental Search

Vol.632: A. M. Greco (Ed.), Direct and Inverse Methods in Nonlinear Evolution Equations

Vol.633: H.-T. Elze (Ed.), Decoherence and Entropy in Complex Systems, Based on Selected Lectures from DICE 2002

Vol.634: R. Haberlandt, D. Michel, A. Pöppl, R. Stannarius (Eds.), Molecules in Interaction with Surfaces and Interfaces

Vol.635: D. Alloin, W. Gieren (Eds.), Stellar Candles for the Extragalactic Distance Scale

Vol.636: R. Livi, A. Vulpiani (Eds.), The Kolmogorov Legacy in Physics, A Century of Turbulence and Complexity

Vol.637: I. Müller, P. Strehlow, Rubber and Rubber Balloons, Paradigms of Thermodynamics

Vol.638: Y. Kosmann-Schwarzbach, B. Grammaticos, K. M. Tamizhmani (Eds.), Integrability of Nonlinear Systems

Vol.639: G. Ripka, Dual Superconductor Models of Color Confinement

Vol.640: M. Karttunen, I. Vattulainen, A. Lukkarinen (Eds.), Novel Methods in Soft Matter Simulations

Vol.641: A. Lalazissis, P. Ring, D. Vretenar (Eds.), Extended Density Functionals in Nuclear Structure Physics

Vol.642: W. Hergert, A. Ernst, M. Däne (Eds.), Computational Materials Science

Vol.643: F. Strocchi, Symmetry Breaking

Vol.644: B. Grammaticos, Y. Kosmann-Schwarzbach, T. Tamizhmani (Eds.) Discrete Integrable Systems

Vol.645: U. Schollwöck, J. Richter, D. J. J. Farnell, R. F. Bishop (Eds.), Quantum Magnetism

Vol.646: N. Bretón, J. L. Cervantes-Cota, M. Salgado (Eds.), The Early Universe and Observational Cosmology

Vol.647: D. Blaschke, M. A. Ivanov, T. Mannel (Eds.), Heavy Quark Physics

Vol.648: S. G. Karshenboim, E. Peik (Eds.), Astrophysics, Clocks and Fundamental Constants

Vol.649: M. Paris, J. Rehacek (Eds.), Quantum State Estimation

Vol.650: E. Ben-Naim, H. Frauenfelder, Z. Toroczkai (Eds.), Complex Networks

Vol.651: J. S. Al-Khalili, E. Roeckl (Eds.), The Euroschool Lectures of Physics with Exotic Beams, Vol.I

Vol.652: J. Arias, M. Lozano (Eds.), Exotic Nuclear Physics

Vol.653: E. Papantonoupoulos (Ed.), The Physics of the Early Universe

Vol.654: G. Cassinelli, A. Levrero, E. de Vito, P. J. Lahti (Eds.), Theory and Appplication to the Galileo Group

Vol.655: M. Shillor, M. Sofonea, J. J. Telega, Models and Analysis of Quasistatic Contact

Vol.656: K. Scherer, H. Fichtner, B. Heber, U. Mall (Eds.), Space Weather

Vol.657: J. Gemmer, M. Michel, G. Mahler (Eds.), Quantum Thermodynamics

Vol.658: K. Busch, A. Powell, C. Röthig, G. Schön, J. Weissmüller (Eds.), Functional Nanostructures

Vol.659: E. Bick, F. D. Steffen (Eds.), Topology and Geometry in Physics

Vol.660: A. N. Gorban, I. V. Karlin, Invariant Manifolds for Physical and Chemical Kinetics

Vol.661: N. Akhmediev, A. Ankiewicz (Eds.) Dissipative Solitons

Vol.662: U. Carow-Watamura, Y. Maeda, S. Watamura (Eds.), Quantum Field Theory and Noncommutative Geometry

Vol.663: A. Kalloniatis, D. Leinweber, A. Williams (Eds.), Lattice Hadron Physics

Vol.664: R. Wielebinski, R. Beck (Eds.), Cosmic Magnetic Fields

Vol.665: V. Martinez (Ed.), Data Analysis in Cosmology

Vol.666: D. Britz, Digital Simulation in Electrochemistry

Vol.667: W. D. Heiss (Ed.), Quantum Dots: a Doorway to Nanoscale Physics

Vol.668: H. Ocampo, S. Paycha, A. Vargas (Eds.), Geometric and Topological Methods for Quantum Field Theory

Vol.669: G. Amelino-Camelia, J. Kowalski-Glikman (Eds.), Planck Scale Effects in Astrophysics and Cosmology

Vol.670: A. Dinklage, G. Marx, T. Klinger, L. Schweikhard (Eds.), Plasma Physics

Vol.671: J.-R. Chazottes, B. Fernandez (Eds.), Dynamics of Coupled Map Lattices and of Related Spatially Extended Systems

Vol.672: R. Kh. Zeytounian, Topics in Hyposonic Flow Theory

Vol.673: C. Bona, C. Palenzula-Luque, Elements of Numerical Relativity

Vol.674: A. G. Hunt, Percolation Theory for Flow in Porous Media

Vol.675: M. Kröger, Models for Polymeric and Anisotropic Liquids